Advanced Calculus

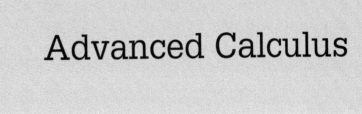

Advanced Calculus
A Transition to Analysis

Joseph B. Dence
St. Louis, MO

Thomas P. Dence
Ashland University
Ashland, OH

ELSEVIER

AMSTERDAM • BOSTON • HEIDELBERG • LONDON
NEW YORK • OXFORD • PARIS • SAN DIEGO
SAN FRANCISCO • SINGAPORE • SYDNEY • TOKYO
Academic Press is an imprint of Elsevier

AP

Academic Press is an imprint of Elsevier
30 Corporate Drive, Suite 400, Burlington, MA 01803, USA
525 B Street, Suite 1900, San Diego, California 92101-4495, USA
84 Theobald's Road, London WC1X 8RR, UK

Library of Congress Cataloging-in-Publication Data
Application submitted.

British Library Cataloguing-in-Publication Data
A catalogue record for this book is available from the British Library.

ISBN: 978-0-12-374955-0

For information on all Academic Press publications
visit our Web site at *www.elsevierdirect.com*

Typeset by: diacriTech, India

Printed and bound by CPI Group (UK) Ltd, Croydon, CR0 4YY

Transferred to digital print 2012

Contents

Preface

The virtues of the beauty and the power of calculus were not lost upon the creators of the subject. In particular, the physicist, **Sir Isaac Newton** was eminently cognizant of the great utility of calculus,[1] although it would be a historical mistake to suppose that he invented calculus so that he could do physics.[2] However, despite some efforts, neither Newton nor any of his contemporaries or immediate successors were able to provide satisfactory foundations for the new mathematics.[3] This would come much later.

In the rush to master the mechanics and to learn some of the applications of calculus, it is easy to gloss over the bases of the subject and the limitations of the results. A book on advanced calculus should attempt to correct these deficiencies. The title "Advanced Calculus" connotes a more penetrating study of sequences (Chapter 2), infinite series (Chapter 3), derivatives (Chapter 5), and integrals (Chapter 6). These are the principal topics of the present book.

In fact, the title of our book carries other layers of meaning besides that just stated and those that would be familiar to instructors but perhaps not expected by students. First, although we assume that readers are familiar with many manipulations of the standard functions, there is a decided deemphasis on numerical computations and a strong stress on conceptual relationships, for *theorems* are the real jewels of mathematics. For many (perhaps most) students using this book as a course text, such a course might well be the first in which they are asked to prove a great many statements. For this purpose, at the beginning of Section 1.4, we have presented some (but not too much)

---

[1] Florian Cajori remarks that "It is generally supposed that he [Newton] deduced many of his theorems in the Principia with the aid of his theory of fluxions and fluents [calculus], and afterwards translated his results into the synthetic form [geometry]." See Cajori, F., *Newton's Principia*, University of California Press, Berkeley, 1947, p. 654.

[2] See remarks in Katz, V. J., "Review of De Gandt, F., *Force and Geometry in Newton's Principia*, Princeton University Press, Princeton, 1995," *Amer. Math. Monthly*, **105**, 386–392 (1998).

[3] Newton provided four axioms to serve as foundations of his differential calculus, but these were not statements about the real number system. See Westfall, R. S., *Never at Rest: A Biography of Isaac Newton*, Cambridge University Press, Cambridge, 1980, pp. 226–231.

introductory material on methods of proof. We view this option as preferable to either including nothing at all on proof (because students "should already know it") or squirreling the material away in an appendix in the back of the book.

Second, this book really does focus on the relatively familiar topics of sequences, series, differentiation, and integration, and does not extend into the higher levels of abstraction that are proper for authentic courses in real analysis. To us, advanced calculus is a *bridge* between the standard three-semester introduction to calculus and real analysis. Thus, whereas we do bits and pieces of topology in this book, especially in Chapter 4 on continuity, the extent of this is kept small. Also, we do only the Riemann theory of the integral in any detail. There is plenty of time in later courses for measure theory and the Lebesgue integral, or for still other kinds of integrals.

Third, although due attention is paid to securing the foundations of calculus, it is apparent that any such exposition must always be a compromise between rigor and pedagogy in order to avoid producing an unusable tome. It would be nice, for example, to *construct* the system of real numbers. Or, in principle, complete rigor ought to take us all the way back to doing logic from scratch. We do not have the luxury of space and time to carry out these interesting projects. In this regard, the bibliographies may be useful.

The objectives of our book are threefold:

1. Reexamine many of the topics from standard calculus but at a level with *sufficient* rigor to make them intellectually stimulating.
2. Present an introduction to modern patterns of thought in mathematics that will be beneficial to those students who may pursue more advanced work.
3. Instill in the student the notion that the subject is not "contained" between the covers of any one textbook, and that outside reading (the more, the better) is essential for maturation in mathematical thinking.

One way to assist students in connection with the first objective is to assign *many* problems. We subscribe to the Ben Hogan philosophy that learning mathematics is a lot like learning to play golf—it requires tons of practice. We have written a generous (but not overabundant) supply of problems, most of them conceptually oriented. Additionally, Appendix C contains about five dozen supplementary problems on sequences, series, and integrals for challenge and stimulation.

We have tried to meet the second objective in the following two ways: (1) occasionally proving a theorem by a less traditional route in order to illustrate the flexibility of thinking that is essential in creative mathematics and (2) occasionally connecting advanced calculus to some other area of mathematics in order to illustrate the interpenetration of all of mathematical knowledge.[4] For example, we have proved the countability of the rationals by appeal to a standard multiplicative function from *number theory*, and we have proved the Mean-Value Theorem by a method that makes contact with *linear algebra*.

The third objective is very important. Each chapter concludes with numerous book and journal references pertinent to topics in the chapter. Many more such references could have been included. We strongly urge instructors to use ways to get their students to read more mathematics outside of class; this is a way to get them more "involved" in mathematics. If necessary, dangle some carrots. Even the historical aspects of developments in calculus are worth reading about, for although the masters were only human, there is no limit to what we can learn from them. The book is sprinkled with historical asides.

There is intentionally more material in the book than can be covered in one semester. This is because (1) the book reflects, in part, the predilections of the authors, and (2) there is no consensus as to what constitutes core material for a one-semester course. The following syllabus is a suggested one; for some classes less material may be advisable.

Chapter 1	Lecture on Sections 1.2–1.3, 1.7–1.8, 1.9; assign all other sections as reading.	3 lectures
Chapter 2	Cover all seven sections.	6 lectures
Chapter 3	Cover all six sections.	6 lectures
Chapter 4	Cover Sections 4.1, 4.3–4.9; assign Section 4.2 as reading and Section 4.10 only if you wish.	8 lectures
Chapter 5	Cover most of this chapter, but omit any sections that you wish.	7 lectures
Chapter 6	Cover most of this chapter also, picking just those threads that you want to emphasize.	8 lectures
Chapter 7	Some sections from this chapter should be done, but time will be your determining factor.	2 lectures
Total:		**40 lectures**

[4]In regard to the whole philosophy of the nature of mathematical knowledge, creativity, and proof, a terrific set of essays is contained in Ayoub, R. J. (ed.), *Musings of the Masters*, Mathematical Association of America, Washington, D.C. 2004. The review of this book in Kennedy, S., *Amer. Math. Monthly*, **113**, 575–580 (2006) is a gem.

If the class meets three times a week in a 15-week semester, then four to six class periods remain for examinations and review.

All authors have been influenced by certain prior writings in their field; we have felt most keenly the inspiration received from G. H. Hardy, *A Course of Pure Mathematics*, 10th ed., Cambridge University Press, Cambridge, 1967. We also acknowledge gratefully the numerous comments about the manuscript from several readers. They helped make this a better book. Deficiencies that remain (and there surely are some) are our responsibility alone. We welcome any and all comments from readers and users of the book.

Finally, we are indebted to Lauren Schultz Yuhasz, Senior Acquisitions Editor at Elsevier Science, and to the expert editorial staff there for their skill and assistance in bringing this project to fruition. We are honored to have been allowed to publish under the Elsevier umbrella.

One final note to instructors: A copy of the *Instructor's Solutions Manual* can be found at http://textbooks.elsevier.com/.

Joseph B. Dence
Thomas P. Dence

To The Student

Introductory calculus traditionally has been oriented toward mechanics, from which you learned how to compute limits, differentiate and integrate standard functions, and apply these techniques to various interesting problems. Now you are about to begin a deeper look into calculus that may entail some modification in your thinking about mathematics.

A holistic point of view that might be useful is to regard mathematics as a language (the nineteenth-century Yale chemist J. W. Gibbs said so). The following analogies seem to emerge.

A Spoken/Written Language	Mathematics
Letters of the alphabet, correctly written	Fundamental terms, carefully defined; depending upon scope, some terms may be left undefined
Words, correctly spelled and with diacritical or other marks such as accents, umlauts, apostrophes, as required	More complicated terms, concepts compounded of simpler ones and with symbols such as $+$, $\{\}$, f, as required
Sentences, including capitalizations and punctuation, as required, and constructed in conformity to standard rules of grammar	Lemmas, theorems, corollaries, stated unambiguously
Paragraphs, written coherently	Proofs of the above, written with correct reasoning and in conformity to standard rules of logic
Sets of paragraphs that develop a principal theme	Sets of theorems, plus supporting numerical work, if any, that develop a subject area of mathematics

Although in this book we assume that you are familiar with various manipulations of many standard functions, a more penetrating study of calculus will require mastering definitions and the careful wording of theorems. Especially important in the latter are the hypotheses, because these indicate the conditions under which the conclusions are valid. Assertions such as $\frac{\partial^2 f}{\partial x \partial y} = \frac{\partial^2 f}{\partial y \partial x}$ or $\frac{d}{dy} \int_a^b f(x,y)dx = \int_a^b \left(\frac{\partial f}{\partial y}\right)_x dx$ should be regarded not as identities, but rather as conditional statements that are true only under certain conditions.

The purpose of formal proof is to establish the logical connection between the hypotheses and the conclusions of theorems. Accordingly, this book is largely proof-oriented, and most of the exercises will ask you to show or prove certain statements. Commonly, proofs will take the form of direct proof, indirect proof, or proof by mathematical induction. We review these briefly in the first chapter.

One final, practical point is this: An intensive subject can never be mastered as fully as when you become maximally involved in it. You therefore should do more problems than are assigned by your wise instructor. You should also do more reading than is assigned. In this regard, you will discover that the chapters of the book have been generously referenced. References are collected at the ends of the chapters. You will find most of these references to be approachable; make a point of consulting some of them and working through them. Perhaps you can convince your instructor to give you extra credit for some of this work. Here's what the great Newton had to say:

> "A Vulgar Mechanick can practice what he has been taught or seen done, but if he is in error he knows not how to find it out and correct it, and if you put him out of his road, he is at a stand; Whereas he that is able to reason nimbly and judiciously about figure, force and motion, is never at rest till he gets over every rub." [1]
>
> **Newton, 1694**

Good luck in Advanced Calculus!

[1] Westfall, R.S., *Never at Rest: A Biography of Isaac Newton*, Cambridge University Press, Cambridge, 1980, p. 499.

PART 1

Fundamentals

CHAPTER 1

Sets, Numbers, and Functions

"The calculus was the first achievement of modern mathematics...."
John von Neumann

CONTENTS

Reviewed in this chapter	Real numbers; methods of proof; general nature of functions.
New in this chapter	Axiom of Completeness; Euclidean vector spaces; n-balls; infinite sets; cluster points.

1.1 LOGIC AND SETS

Much of mathematics is done in accordance with accepted tenets of the logic whose study was initiated by the Greek universal genius **Aristotle** (384–322 BCE). Fundamental in logic is the concept of a **proposition**, any statement about which it is meaningful to ask if it is true or false, such as

$$p: \text{"The integer 641 is prime."}$$

The negation of a proposition p is $\sim p$, and is such that if p is true (false), then $\sim p$ is false (true):

$$\sim p: \text{"The integer 641 is not prime."}$$

Two propositions can be combined so as to form compound propositions (Table 1.1).

Table 1.1 Some Compound Propositions

Name	Symbol	Meaning
1. **Conjunction**	$p \wedge q$	"p and q"
2. **Disjunction**	$p \vee q$	"p or q"
3. **Implication**	$p \to q$	"If p, then q"
4. **Double Implication**	$p \leftrightarrow q$	"If p, then q, and vice versa"[a]

[a] *More concisely read as "p, if and only if, q," or even more tersely as "p iff q."*

In logic, compound propositions have definite truth values (T or F) that depend only upon the truth values of the component simple propositions. This information is conveniently summarized in **truth tables** (Exercises 1.1–1.3), which are deducible from certain basic axioms of logic, including those given in Table 1.2.

Table 1.2 Some Axioms for Aristotelian Logic

Name	Statement
L1. **Law of Negation**	A proposition and the negation of its negation have the same truth value.
L2. **Law of the Excluded Middle**	Either a proposition is true or its negation is true.
L3. **Law of Contradiction**	A proposition and its negation cannot both be true.
L4. **Law of the Syllogism**	If proposition p implies proposition q, and if q implies proposition r, then p implies r.

■ Example 1.1

If p is the proposition given previously and q is the proposition "$\sqrt{641}$ is rational," then $p \vee q$ is read "641 is prime or $\sqrt{641}$ is rational." This compound proposition is true. ■

■ Example 1.2

p : "The function $f(x) = x^2$ is differentiable on the interval $[0, 1]$";

q : "The function $f(x) = x^2$ is continuous on the interval $[0, 1]$";

r : "The function $f(x) = x^2$ is integrable on the interval $[0, 1]$."

It is known that $p \to q$ is true and $q \to r$ is true. By Axiom L4, $p \to q$ is also true. ■

An interesting aside is that other systems of logic than that studied by Aristotle exist (Eves, 1990).[1] In them the Law of the Excluded Middle, for example, might not hold.

Besides logic, extensive use of concepts from the theory of sets is made in all branches of mathematics (Lipschutz, 1998). A **set** $\mathbf{S}$ is any well-defined collection of objects.[2] Membership in $\mathbf{S}$ (or the negation thereof) is written as $x \in \mathbf{S}$ (or as $x \notin \mathbf{S}$). Two sets $\mathbf{S}$ and $\mathbf{T}$ are **equal**, $\mathbf{S} = \mathbf{T}$, if they contain the same elements. Some common sets needed in advanced calculus are indicated in Table 1.3.

Table 1.3 Symbols for Some Important Sets

Set	Symbol	Illustrative Elements
Natural numbers	$\mathbf{N}$	$1, 2, 3, 4, 5, \ldots$
Integers	$\mathbf{Z}^a$	$\ldots, -2, -1, 0, 1, 2, \ldots$
Rational numbers	$\mathbf{Q}^b$	$-7/3, -1, 0, 2/31$
Real numbers	$\mathbf{R}$	$-\sqrt[3]{4}, 0, 1, 2/31, \pi$
Irrational numbers	$\mathbf{R}\backslash\mathbf{Q}^c$	$-\sqrt{\pi}, 1+\sqrt{2}, 2+2\pi$
Complex numbers	$\mathbf{C}$	$-1+i, 0, 2/31, e^{i\pi/3}$

[a] *From (die) Zahl (number* (Germ.)).
[b] *From the word "quotient."*
[c] *See entry 3 in Table 1.4.*

---

[1] Citations (annotated) and other references appear at the end of each chapter. These are intended for your enrichment.

[2] The German **Georg Cantor** (1845–1918) was the founder of the modern theory of sets. However, "set" (*die Menge* (Germ.)) first appeared in a technical context in an 1851 book (posthumous) by the Bohemian **Bernhard Bolzano** (1781–1848), a free thinker who was years ahead of his time.

Table 1.4 Four Standard Set Operations

1.	**Union** of S and T	$S \cup T = \{x : x \in S \text{ or } x \in T\}$
2.	**Intersection** of S and T	$S \cap T = \{x : x \in S \text{ and } x \in T\}$
3.	**Complement** of T Relative to S	$S \backslash T = \{x : x \in S \text{ and } x \notin T\}$
4.	S as a **Subset** of T	$S \subseteq T = \{x : (x \in S) \rightarrow (x \in T)\}$[a]

[a] *The notation indicates that* $S = T$ *is permitted; the more restricted notation* $S \subset T$ *means that there is at least one element* $x \in T$ *that is not a member of* S.

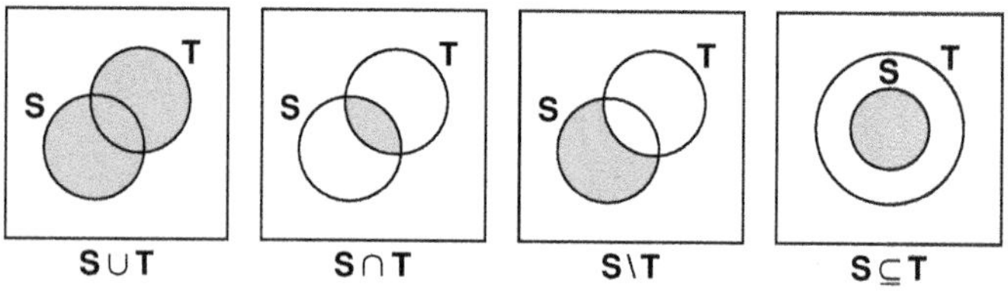

FIGURE 1.1

Construction of new sets from given sets.

New sets can be constructed from given sets in a number of ways. Four standard ways are indicated symbolically in Table 1.4 and pictorially in Figure 1.1.

It is convenient to introduce a set that has 0 elements, the **empty set**, which is denoted universally by the Scandinavian letter Ø. For any set $S \neq \emptyset$, the inclusion $\emptyset \subset S$ holds in a negative sense. Finally, two sets **S**, **T** are **disjoint** if $S \cap T = \emptyset$.

1.2 REAL NUMBERS

An area of frontline research during the period from 1880 to 1920 was the securing of the foundations of the system of real numbers (Kline, 1972). Amazingly, it proved possible to construct the entire real number system from just a few simple axioms for the very basic system of natural numbers.

All of this is very exciting, but we have not the luxury here to work through the large number of details needed to accomplish the program (Landau, 1966). Instead, following the German mathematician **David Hilbert** (1862–1943), we shall regard **R** itself as an axiomatic system. The first six axioms for **R** will certainly sound familiar to you (Table 1.5).

Axioms R1 through R6 are called the **field axioms** because statements analogous to them characterize any mathematical structure known as a **field**. The six axioms justify all of the arithmetic operations done in **R**.

Table 1.5 The Field Axioms for $\mathbf{R}$

R1.	**(Closure)** Addition and multiplication of any two members of $\mathbf{R}$ always produce members of $\mathbf{R}$.
R2.	**(Commutativity)** If $x, y \in \mathbf{R}$, then (a) $x + y = y + x$, (b) $xy = yx$.
R3.	**(Associativity)** If $x, y, z \in \mathbf{R}$, then (a) $(x + y) + z = x + (y + z)$, (b) $(xy)z = x(yz)$.
R4.	**(Left Distributive Law)** If $x, y, z \in \mathbf{R}$, then $x(y + z) = xy + xz$.
R5.	**(Identities)** There are elements $0, 1 \in \mathbf{R}$, the **additive identity** and the **multiplicative identity**, respectively, such that (a) $0 \neq 1$, and (b) for each $x \in \mathbf{R}, x + 0 = x$ and $x \cdot 1 = x$.
R6.	**(Inverses)** For each nonzero $x \in \mathbf{R}$ there are elements $-x, x^{-1} \in \mathbf{R}$, the **additive inverse** and the **multiplicative inverse** of x, respectively, such that (a) $x + (-x) = 0$, and (b) $x\left(x^{-1}\right) = 1$.

There is a seventh axiom for $\mathbf{R}$, which is not a field axiom:

R7. (Simple Ordering) There is a subset $\mathbf{P} \subset \mathbf{R}$, called the **positive numbers**, such that (a) $0 \notin \mathbf{P}$, (b) for each nonzero $x \in \mathbf{R}$, exactly one of $x, -x$ is in $\mathbf{P}$, and (c) for each $x, y \in \mathbf{P}, x + y \in \mathbf{P}$ and $xy \in \mathbf{P}$.

The field $\mathbf{Q}$ contains positive numbers, but the complex numbers $\mathbf{C}$ do not (why?). There are infinitely many fields that do not contain any positive numbers (Exercise 1.6). Focusing now on $\mathbf{R}$, however, we need the following definition in order to appreciate axiom R7 (Wilder, 1983):

Definition. *A set $\mathbf{S}$ is said to be **simply**, or **linearly**, **ordered** iff there is a binary relation $>$ defined on $\mathbf{S}$ and with these properties:*

1. *(Comparability) If $x, y \in \mathbf{S}$ and $x \neq y$, then either $x > y$ or $y > x$ holds, but not both.*
2. *(Nonreflexivity) For no $x \in \mathbf{S}$ is $x > x$ true.*
3. *(Transitivity) If $x, y, z \in \mathbf{S}$, and $x > y$ and $y > z$ hold, then $x > z$ also holds.*

We now use axiom R7 as the basis for defining a binary relation $>$ on $\mathbf{R}$. All three parts of R7 are needed for this.

Definition. *In $\mathbf{R}$ the symbol $>$ is defined by the correspondence*

$$x > y \leftrightarrow \{[x + (-y)] \in \mathbf{P}\}.$$

The notation $x < y$ means the same thing as $y > x$.

Suppose that $x, y \in \mathbf{R}$ and $x \neq y$; then we can establish that

$$x + (-y) \neq 0, y + (-x) \neq 0,$$

since (otherwise) postaddition of y or x, respectively, would lead to the contradiction that $x = y$ (Exercise 1.8). Next, using the elementary theorem that if $a, b, c \in \mathbf{R}$ and $a + b = a + c$, then $b = c$ (Exercise 1.9(a)), we can establish that for any $x, y \in \mathbf{R}$ we have (Exercise 1.11)

$$-[x + (-y)] = (-x) + y. \tag{*}$$

By axiom R7(b), exactly one of $x + (-y)$ and $(-x) + y$ in (*) then lies in $\mathbf{P}$. If it is the former, then $x > y$ by definition, and if it is the latter, then $y > x$. This establishes the Comparability aspect for $>$ in the definition of simple ordering.

The other two aspects are left to you to verify (Exercise 1.12). It follows that $\mathbf{R}$ is a simply-ordered field. This permits us to establish numerous theorems on inequalities (Exercises 1.13,1.19).

■ Example 1.3

Let $x, y \in \mathbf{R}$; if $x + z > y + z$, then $x > y$. To show this, note that the hypothesis (the given information) is equivalent to $(x + z) + [-(y + z)] = (x + z) + [-(z + y)] \in \mathbf{P}$. However, $-(z + y) = (-z) + (-y)$ by the same reasoning used to obtain (*). Hence, we have

$$\begin{aligned}(x + z) + [-(z) + (-y)] &= [(x + z) + (-z)] + (-y)\\ &= [x + \{z + (-z)\}] + (-y)\\ &= (x + 0) + (-y) = x + (-y) \in \mathbf{P},\end{aligned}$$

and this says that $x > y$. ■

1.3 THE AXIOM OF COMPLETENESS

Building further on axiom R7, we observe that the notion of boundedness of sets of real numbers can be introduced.

Definition. *Suppose that* $\mathbf{S} \subset \mathbf{R}$ *is nonempty.*

(a) *Then* $\mathbf{S}$ *is* ***bounded from above*** *iff there is a number* $u \in \mathbf{R}$*, called an* ***upper bound*** *for* $\mathbf{S}$*, such that* $x \le u$ *for all* $x \in \mathbf{S}$*; the* u *may, but need not, belong to* $\mathbf{S}$.

(b) *The set* $\mathbf{S}$ *is* ***bounded from below*** *iff there is a number* $l \in \mathbf{R}$*, called a* ***lower bound*** *for* $\mathbf{S}$*, such that* $x \ge l$ *for all* $x \in \mathbf{S}$*; the* l *may, but need not, belong to* $\mathbf{S}$.

(c) *The set* $\mathbf{S}$ *is* ***bounded*** *iff* l, u *both exist.*

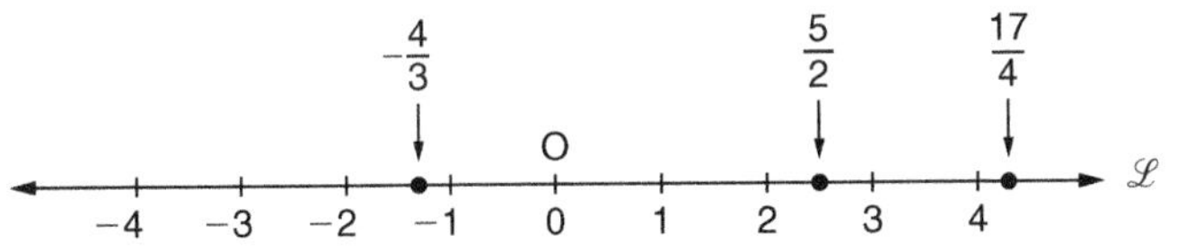

FIGURE 1.2
The line as a model for **R**.

It is now convenient to use the imagery of geometry. Following the Greeks of antiquity, we use a line $\mathcal{L}$ as a model for **R** (Figure 1.2); an arbitrary point O is chosen to represent the number 0. Elements of **P** lie to the right of O, and all other nonzero elements of **R** lie to the left of O. Most crucially, as the Greeks once believed, every point of $\mathcal{L}$ represents a rational number.

Let us next define two sets on $\mathcal{L}$:

$$\mathbf{S}_1 = \{x \in \mathbf{Q} : x \geq 0, 0 \leq x^2 < 5\}$$

$$\mathbf{S}_2 = \{x \in \mathbf{Q} : x \geq 0, 5 \leq x^2\}.$$

The set $\mathbf{S}_1 \cup \mathbf{S}_2$ should cover the entire right half of $\mathcal{L}$. We observe that $\mathbf{S}_1$ is certainly bounded from above, for example, by $u = 3, 3 \in \mathbf{S}_2$; also $\mathbf{S}_2$ is bounded from below, for example, by $l = 2, 2 \in \mathbf{S}_1$. Of course, if a set in **R** has one upper (lower) bound, then it has several. Our attention then is directed to the largest and/or smallest of the bounds.

Definition. *Suppose that* $\mathbf{S} \subset \mathbf{R}$ *is nonempty. Then a number* U *is called a* ***supremum*** *(or* ***least upper bound****) of* **S** *iff:*

(a) *S is bounded from above;*

(b) *U is an upper bound of* **S***;*

(c) *If u is any upper bound of* **S***, then* $U \leq u$.

Similarly, a number L is called an ***infimum*** *(or* ***greatest lower*** *bound) of* **S** *iff:*

(a′) *S is bounded from below;*

(b′) *L is a lower bound of* **S***;*

(c′) *If l is any lower bound of* **S***, then* $l \leq L$.

We denote the supremum and the infimum of **S** *by* sup **S**, *inf* **S**, *respectively, if they exist.*

If our set $\mathbf{S}_1$ has a supremum U, it must certainly be less than $u = 3$. Let us assume that $U \in \mathbf{S}_1$; as U is nonnegative, we compute the positive number

$$x = \frac{3U + 5}{3 + U} = 3 - \frac{4}{3 + U}. \qquad (*)$$

Clearly, x is rational if U is rational; squaring both sides of (*) gives

$$x^2 = 9 - \frac{24}{3+U} + \frac{16}{(3+U)^2}$$
$$= 5 + \frac{4}{(3+U)^2}(U^2 - 5).$$

But $U^2 < 5$, so $0 < x^2 < 5$ and $x \in \mathbf{S}_1$. Further, a rearrangement of (*) produces

$$x = U + \frac{5 - U^2}{3+U}, \qquad (**)$$

and $U^2 < 5$ now implies that $x > U$. Hence, U cannot be the supremum of $\mathbf{S}_1$; the supremum must be x or greater. But, this argument can be repeated indefinitely, beginning now with x and inserting this into the right-hand side of (*). Perhaps the supremum U of $\mathbf{S}_1$ is an element of $\mathbf{S}_2$. Substitution of U into (*) again carries us down to (**); but now because $U^2 > 5$, we have $x < U, x \in \mathbf{S}_2$, so U cannot be the *least* upper bound of $\mathbf{S}_1$ because x is smaller than U.

A candidate for U could be the left endpoint of $\mathbf{S}_2$, where $x^2 = 5$. But a simple divisibility argument (Exercise 1.25) shows that there is no element of **Q** whose square is 5. We appear to be left with the conclusion that $\mathbf{S}_1$ does not have a supremum U. Further, all the earlier reasoning can be adapted to yield the allied conclusion that $\mathbf{S}_2$ does not have an infimum L. So there is a "gap" between $\mathbf{S}_1$ and $\mathbf{S}_2$, and $\mathbf{S}_1 \cup \mathbf{S}_2$ apparently does not cover all of the right half of $\mathcal{L}$ (Figure 1.3).

There is nothing special about the way that sets $\mathbf{S}_1, \mathbf{S}_2$ were defined. Many choices other than 5 could have been made and would have led to similar difficulties. The line literally would be peppered with a great many "gaps." Such a line would be too perverse to be trustworthy in geometry, and in turn, **R** itself would be too perverse to be acceptable. The way out is provided by the recognition that in accepting the existence of all of **R**, we automatically admit also, as points on $\mathcal{L}$, numbers (like $\sqrt{5}$) that are not rational (**irrational**

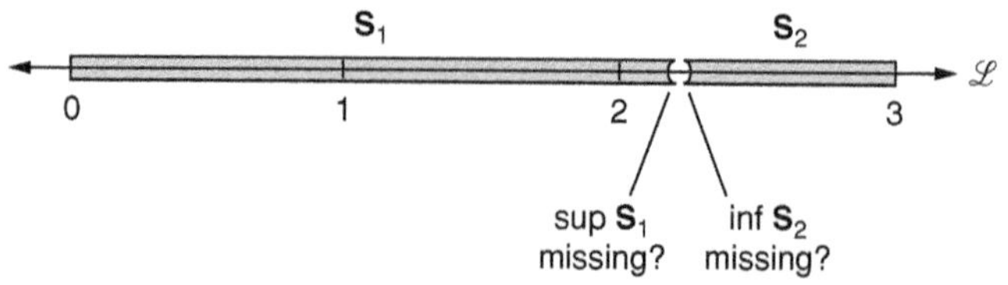

FIGURE 1.3
A "gap" on the real line?

numbers). These numbers completely fill all alleged "gaps." Set $\mathbf{S}_1$, because it is bounded from above, must then have a least upper bound, which in this case is $\sup \mathbf{S}_1 = \sqrt{5}$. The final axiom for **R** is, therefore:

R8 (**Axiom of Completeness**) Any nonempty set **S** of real numbers that is bounded from above has a supremum.

It is appropriate to say that R8 completes the description of the system of real numbers and that **R** is a **complete, simply-ordered field**. The axiom justifies the existence of all real irrational numbers, and *most of the theorems in this book would fail without it*. And, incidentally, there is a dual statement about infima that is derivable from axiom R8 and, as expected, $\inf \mathbf{S}_2 = \sqrt{5}$. Example 2.16 (and Exercise 2.30) show an explicit construction of an irrational number, ultimately from the Axiom of Completeness.

There have been many axiomatic treatments of **R**. A recent, interesting one is described in Oman (2009).

1.4 PROOF

The jewels of mathematics are its **theorems** ($\theta\varepsilon\omega\rho\eta\mu\alpha$ (Gk.), a product of contemplation), and it is the main business of mathematicians to establish them in a convincing way. We use standard techniques of proof in this book (Cupillari, 2005; Solow, 2004). A theorem usually is stated in the form of a single or double implication (Table 1.1):

If $\mathbf{S}, \mathbf{S}'$ are bounded, nonempty subsets of **R** and $\mathbf{S}' \subseteq \mathbf{S}$, then $\sup \mathbf{S}' \leq \sup \mathbf{S}$;

If x, y are real, then $xy = 0$ iff either $x = 0$ or $y = 0$.

The if-clause contains the **hypotheses** (or **premises**) of the theorem, and the then-clause is the **conclusion**. In **direct proof** we proceed logically from the hypotheses to the conclusion; definitions, axioms, previously proved theorems may all be used as justifications for individual steps. Proofs are written out in paragraph format with sufficient prose to make them readable.

Theorem 1.1.[3] *If* $\mathbf{S}, \mathbf{T}, \mathbf{W}$ *are sets, then* $\mathbf{S}\backslash(\mathbf{T} \cap \mathbf{W}) \subset [(\mathbf{S}\backslash\mathbf{T}) \cup (\mathbf{S}\backslash\mathbf{W})]$.

Proof. We need to show that any element of the set on the left-hand side is automatically an element of the set on the right-hand side (Table 1.4). Let $x \in \mathbf{S}\backslash(\mathbf{T} \cap \mathbf{W})$; then $x \in \mathbf{S}$ and $x \notin \mathbf{T} \cap \mathbf{W}$. The latter implies (by definition of intersection) that either $x \notin \mathbf{T}$ or $x \notin \mathbf{W}$. If $x \notin \mathbf{T}$ then $x \in \mathbf{S}\backslash\mathbf{T}$, and if $x \notin \mathbf{W}$, then $x \in \mathbf{S}\backslash\mathbf{W}$. In either case, by definition of union, $x \in [(\mathbf{S}\backslash\mathbf{T}) \cup (\mathbf{S}\backslash\mathbf{W})]$ and the desired set inclusion follows. ■

[3]Theorems in the book are enumerated as follows: THEOREM (chapter no.).(theorem no. within the chapter).

p	q	$\sim p$	$\sim q$	$p \to q$	$\sim q \to \sim p$
T	T	F	F	T	T
T	F	F	T	F	F
F	T	T	F	T	T
F	F	T	T	T	T

FIGURE 1.4
Truth table for a contrapositive.

You will find a sketch to be a helpful guide in the construction of the previous proof. In general, this is a useful habit but not a rigorous maneuver.

Theorem 1.2. *If* S, S′ *are bounded, nonempty subsets of* **R** *and* $S' \subseteq S$, *then* $\sup S' \leq \sup S$.

Proof. By axiom R8, both $\sup S'$ and $\sup S$ exist. For each $y \in S$, we have $y \leq \sup S$. But $S' \subseteq S$, so each $x \in S'$ is a $y \in S$; hence, S′ is bounded from above by $U = \sup S$. By the definition of supremum (Section 1.3), $\sup S' \leq U$ then follows. ■

We revisit Exercise 1.3. The proposition $\sim q \to \sim p$ is called the **contrapositive** of the proposition $p \to q$. The truth table shown in Figure 1.4 emerges.

The identity of the last two columns means that $p \to q$ and $\sim q \to \sim p$ are equivalent propositions; this is what the **Law of Contraposition** in Exercise 1.3 really means. Often it is easier to prove the contrapositive of a proposition than to prove the proposition directly. Thus, we have the technique of **proof by contrapositive**.

Theorem 1.3. *If a set* $S \subset \mathbf{R}$ *contains one of its upper bounds, then this upper bound is the supremum of* S.

Proof. Let $U = \sup S$ and let u be an upper bound of S; assume, contrapositively, that $u \neq U$. Then by the definition of a supremum, $U < u$. As U is itself an upper bound of S, $s \leq U$ for all $s \in S$, and therefore each s is strictly less than u. This says that u is not a member of S. ■

Somewhat similar in flavor to proof by contrapositive is **proof by contradiction**. In it we assume (a) the hypothesis p, (b) some proposition r (this could just be p, or an axiom, or a previously proved theorem), and (c) the negation $\sim q$ of the conclusion. Then we show that these assumptions imply $\sim r$. Since r and $\sim r$ cannot both be true (axiom L3), it follows that $\sim q$ is false, so q is true (axiom L1).

Theorem 1.4 (Euclid's Theorem).[4] *If the sequence of primes is written in ascending order, then the sequence has no end.*

Proof. Suppose that the primes have been written in ascending order (proposition p), and assume that the sequence terminates (proposition $\sim q$). Then the sequence has a largest member, $N_{\max}$ (proposition r):

$$2, 3, 5, 7, 11, 13, \ldots, N_{\max}.$$

Now consider the integer $N = 2 \cdot 3 \cdot 5 \cdot 7 \cdot 11 \cdot 13 \cdots N_{\max} + 1$. From the Fundamental Theorem of Arithmetic N is uniquely factorable (except for the order of listing) into powers of primes (see footnote 11):

$$N = p_1^{a_1} p_2^{a_2} \cdots p_n^{a_n}.$$

No prime p_i in this factorization can be found in the set of primes from 2 to $N_{\max}$, for if it could, then p_i would divide each term of the difference

$$N - (2 \cdot 3 \cdot 5 \cdot 7 \cdot 11 \cdot 13 \cdots N_{\max}),$$

that is, p_i would divide 1. This is impossible, so either (a) N is itself a prime larger than $N_{\max}$, or (b) N is not prime and its prime factors are larger than $N_{\max}$. Either case implies $\sim r$, so $\sim q$ is false and the sequence has no end. ■

Proof by contrapositive and proof by contradiction are two forms of **indirect proof**. In them, as we see, we do not proceed directly from premises to conclusion.

The next example is a consequence of the Axiom of Completeness and is an important characterization of $\mathbf{R}$. In order to proceed, we need a *definition* of the natural numbers.

Definition. *The **natural numbers** are the elements of the smallest subset $\mathbf{N} \subset \mathbf{R}$ with the property that $c + 1$ is an element of $\mathbf{N}$ iff $c = 0$ or $c \in \mathbf{N}$.*

Theorem 1.5 (Archimedean Property).[5] *If $x, y \in \mathbf{R}$ and $y > x > 0$, then there is a natural number n such that $nx > y$.*

Proof. Assume that $nx \leq y$ for all $n \in \mathbf{N}$. For fixed x, let $\mathbf{S}$ be

$$\mathbf{S} = \{nx : n \in \mathbf{N}\}.$$

---

[4] Proposition 20 in Book IX of Euclid's *Elements*.

[5] After the Greek genius **Archimedes of Syracuse** (287–212 BCE), who attributed a related statement to **Eudoxus** (ca. 408–355 BCE).

The assumption implies that **S** is bounded from above, so by axiom R8 it has a supremum U: $nx \leq U$ for all $n \in \mathbf{N}$. Any other upper bound for **S** is necessarily larger than U (proposition r). By the nature of $\mathbf{N}, n+1 \in \mathbf{N}$ whenever $n \in \mathbf{N}$. It follows that $(n+1)x = nx + x \leq U$, or $nx \leq U - x$ for all $n \in \mathbf{N}$. Since $x > 0$, then $U - x < U$ and we now have an upper bound for **S** that is smaller than U (proposition $\sim r$). It follows that the original assumption that $nx \leq y$ for all $n \in \mathbf{N}$ is false, and there must be some n for which $nx > y$. ∎

Combination of the definition of **N** with a few of the field axioms for **R** in Table 1.5 leads to the following statement:

Finite Induction

If **S** is a set of natural numbers such that (1) **S** contains the natural number 1, and (2) **S** contains the natural number $n+1$ whenever it contains the natural number n, then **S** is all of **N**.

If we begin the construction of **R** with an axiomatic treatment of **N**, then Finite Induction appears as one of the axioms of **N**. But, if we accept **R** at the outset (as we are doing) and then proceed to axiomatize it, then Finite Induction is a theorem. The Principle of Mathematical Induction (PMI) is a consequence of Finite Induction.

Theorem 1.6 (Principle of Mathematical Induction). *Suppose that $P(n)$ is a proposition that is defined for each $n \in \mathbf{N}$. If $P(1)$ is true and if, for each $k \in \mathbf{N}$, $P(k+1)$ is true whenever $P(k)$ is true, then $P(n)$ is true for all $n \in \mathbf{N}$.*

Proof. Let **S** be the set of natural numbers for which $P(n)$ is true. By hypothesis, $1 \in \mathbf{S}$ and whenever $k \in \mathbf{S}$, then $k+1 \in \mathbf{S}$. By Finite Induction, **S** is then all of **N**, so $P(n)$ is true for all $n \in \mathbf{N}$. ∎

A common use of the PMI is the verification of formulas that contain an integer variable. But arithmetical formulas are not the only objects of application of Theorem 1.6. The following two examples (theorems) are an interesting contrast.

Theorem 1.7 (Bernoulli's Inequality).[6] *If $x > -1, x \neq 0$, and $n \in \mathbf{N}$, then $(1+x)^{n+1} > 1 + (n+1)x$.*

Proof. The inequality is true for $n = 1$, since $(1+x)^2 = 1 + 2x + x^2 > 1 + 2x$, as $x^2 > 0$. Suppose that the inequality holds for $n = k$, that is, suppose that

[6] After the Swiss mathematician **Jakob Bernoulli** (1654–1705), who presented the result in a 1689 paper on infinite series. Extensions of Theorem 1.7 to nonintegral rational and irrational exponents will appear in Chapter 5.

$(1+x)^{k+1} > 1+(k+1)x$. Then $(1+x)^{(k+1)+1} = (1+x)(1+x)^{k+1}$,

$$\begin{aligned} \text{so} \quad (1+x)^{(k+1)+1} &> (1+x)[1+(k+1)x] \quad \text{because } 1+x>0 \\ &= 1+(k+2)x+(k+1)x^2 \\ &> 1+[(k+1)+1]x \quad \text{because } x^2>0. \end{aligned}$$

Hence, the inequality holds for $n = k+1$ whenever it holds for $n = k$, so by the PMI it is true for all $n \in \mathbf{N}$. ■

Theorem 1.8 (Well-Ordering Principle). *Any nonempty* $\mathbf{S} \subset \mathbf{N}$ *has a least element.*

Proof. Suppose, to the contrary, that some nonempty set $\mathbf{S} \subset \mathbf{N}$ has no least element. Let $\mathbf{T}$ be the set of all elements of $\mathbf{N}$ that are less than every element in $\mathbf{S}$; clearly, $\mathbf{S} \cap \mathbf{T} = \emptyset$. Then $1 \in \mathbf{T}$, for otherwise, it would be the least element of $\mathbf{S}$.

Now assume that $k \in \mathbf{T}$. Then every element of $\mathbf{S}$ is greater than k, that is (by the definition of $\mathbf{N}$), equal to or greater than $k+1$. But, again, if $k+1$ were in $\mathbf{S}$, then this would be the smallest element of $\mathbf{S}$, a contradiction to the initial assumption. Hence, $k+1 \in \mathbf{T}$ whenever $k \in \mathbf{T}$.

It follows by the PMI that every $n \in \mathbf{N}$ is in $\mathbf{T}$ and $\mathbf{S} \cap \mathbf{T} = \emptyset$ implies $\mathbf{S} = \emptyset$; that is, there is no nonempty set $\mathbf{S} \subset \mathbf{N}$ that has no least element. ■

The following comments on the two preceding theorems can be made: (a) When n is small, it is easy to prove Theorem 1.7 by the use of Newton's Binomial Theorem, but the PMI permits the proof of an entire family of theorems with a minimum of work. Could you, for example, handle the case when $x = -0.9$ and $n = 611$ without use of the PMI? (b) Theorem 1.8 is actually logically equivalent to Finite Induction; that is, the latter can be proved from the former. This shows that the natural numbers have a latent structure that is both interesting and powerful. Despite the equivalence of Theorem 1.8 and Finite Induction, however, either one of these might be a more natural tool for some kinds of problems than the other (MacHale, 2008).

1.5 EUCLIDEAN VECTOR SPACES

The focus of calculus is on the behavior of functions in various settings; the usual settings are the **real Euclidean vector spaces**, notably $\mathbf{R}^1$, $\mathbf{R}^2$, and $\mathbf{R}^3$ (some authors write $\mathbf{E}_1$, $\mathbf{E}_2$, and $\mathbf{E}_3$). What this means is summarized in the following paragraphs:

1. The elements of $\mathbf{R}^n$, $n \in \mathbf{N}$, are the set of all n-tuples of real numbers, $\mathbf{x} = (x_1, x_2, \ldots, x_n)$, on which the operations of addition ($\oplus$) and scalar multiplication ($\cdot$) are defined as follows:

$$\begin{aligned} \mathbf{x} \oplus \mathbf{y} &= (x_1, x_2, \ldots, x_n) \oplus (y_1, y_2, \ldots, y_n) \\ &= (x_1+y_1, x_2+y_2, \ldots, x_n+y_n) \end{aligned}$$

$$a \cdot \mathbf{x} = a \cdot (x_1, x_2, \ldots, x_n), \text{ where } a \in \mathbf{R}$$
$$= (ax_1, ax_2, \ldots, ax_n).$$

The constant a is referred to as a **scalar**; it is usually drawn from some field, such as **R** or **C**. An n-tuple is called a **vector**, or sometimes a **point with coordinates** $x_1, x_2, \ldots, x_n$. We shall indicate vectors in $\mathbf{R}^n$, $n \geq 2$, by boldface type and scalars and coordinates by italics. The vectors obey "desirable" rules of arithmetic. For example, addition is **closed** (the sum of two elements in $\mathbf{R}^n$ is an element in $\mathbf{R}^n$), commutative, and associative, there is an **additive identity** $\mathbf{0} = (0, 0, \ldots, 0)$ such that for any $x \in \mathbf{R}^n$ we have $\mathbf{0} \oplus \mathbf{x} = \mathbf{x}$, and each vector $\mathbf{x}$ has an **additive inverse** $(-\mathbf{x})$ such that $\mathbf{x} \oplus (-\mathbf{x}) = \mathbf{0}$. Multiplication by any scalar $a \in \mathbf{R}$ is distributive over vector addition, and for any $x \in \mathbf{R}^n$ we have $1 \cdot \mathbf{x} = \mathbf{x}$, where $1 \in \mathbf{R}$.

2. An **inner product** $(*)$ of two vectors in $\mathbf{R}^n$ yields a real number. The usual inner product for any $\mathbf{x}, \mathbf{y} \in \mathbf{R}^n$ is

$$\mathbf{x} * \mathbf{y} = \sum_{k=1}^{n} x_k y_k.$$

The following properties[7] of $*$ are consequences of this definition:
(a) (**Symmetry**) For any $\mathbf{x}, \mathbf{y} \in \mathbf{R}^n$, $\mathbf{x} * \mathbf{y} = \mathbf{y} * \mathbf{x}$;
(b) (**Positivity**) For any nonzero $\mathbf{x} \in \mathbf{R}^n$ one has $\mathbf{x} * \mathbf{x} > 0$; otherwise, we have $\mathbf{0} * \mathbf{0} = 0$;
(c) (**Linearity**)
  (i) For any $\mathbf{x}, \mathbf{y} \in \mathbf{R}^n$ and any $k \in \mathbf{R}$ we have $(k \cdot \mathbf{x}) * \mathbf{y} = k(\mathbf{x} * \mathbf{y})$;
  (ii) For any $\mathbf{x}, \mathbf{y}, \mathbf{z} \in \mathbf{R}^n$, we have $(\mathbf{x} \oplus \mathbf{y}) * \mathbf{z} = \mathbf{x} * \mathbf{z} + \mathbf{y} * \mathbf{z}$.

This completes the standard definition of a real Euclidean vector space (Birkhoff and MacLane, 1953). However, more is desired for practical aspects of calculus, and we shall always assume this of the symbol $\mathbf{R}^n$.

3. The **Euclidean norm** of a vector $\mathbf{x} \in \mathbf{R}^n$, which is a measure of the "length" of the vector, is an application of the Pythagorean Theorem to $\mathbf{R}^n$:[8]

$$||\mathbf{x}|| = \left[\sum_{k=1}^{n} |x_k|^p\right]^{1/p} = (\mathbf{x} * \mathbf{x})^{1/2} \quad (p = 2).$$

[7] In a more abstract setting, it is preferable to *define* a real inner product by the conditions of symmetry, positivity, and linearity, and then to show that the indicated formula actually meets these requirements.
[8] A generalization of the n-dimensional Pythagorean formula, in which $1 \leq p < \infty$ is arbitrary, is a valid norm for any $\mathbf{R}^n$ (or $\mathbf{C}^n$). The corresponding Triangle Inequality (see later) holds and is known as **Minkowski's Inequality**, a new proof of which was given recently (Kantrowitz and Neumann, 2008).

The norm has the following properties, the last two of which are considered later (Exercises 1.35, 1.36):

(a) For all $\mathbf{x} \in \mathbf{R}^n$ and all $k \in \mathbf{R}$, $||k \cdot \mathbf{x}|| = |k|\ ||\mathbf{x}||$, where the absolute value function for any $k \in \mathbf{R}$ is defined as

$$|k| = \begin{cases} k & k > 0 \\ 0 & k = 0; \\ -k & k < 0 \end{cases}$$

(b) For all nonzero $\mathbf{x} \in \mathbf{R}^n$, $||\mathbf{x}|| > 0$;
(c) (**Cauchy-Schwarz Inequality**) For all $\mathbf{x}, \mathbf{y} \in \mathbf{R}^n$, $|\mathbf{x} * \mathbf{y}| \le ||\mathbf{x}||\ ||\mathbf{y}||$;
(d) (**Triangle Inequality**) For all $\mathbf{x}, \mathbf{y} \in \mathbf{R}^n$, $||\mathbf{x} \oplus \mathbf{y}|| \le ||\mathbf{x}|| + ||\mathbf{y}||$.
Property (d), particularly for $n = 1$, is very useful in calculus.

4. In order to discuss the distance between two points in $\mathbf{R}^n$, we desire $\mathbf{R}^n$ to possess a **metric**, or distance function d, so that $\mathbf{R}^n$ together with the metric becomes a **metric space**. This metric space is carefully denoted $<\mathbf{R}^n, d>$. The Euclidean norm for $\mathbf{R}^n$ automatically induces a particular metric, written $d_n(\mathbf{x}, \mathbf{y})$, for the distance between $\mathbf{x}$ and $\mathbf{y}$:

$$\begin{aligned} d_n(\mathbf{x}, \mathbf{y}) &= ||\mathbf{x} \oplus (-\mathbf{y})|| \\ &= ||\mathbf{x} - \mathbf{y}|| \\ &= \left[\sum_{k=1}^{n} (x_k - y_k)^2\right]^{1/2}. \end{aligned}$$

Other metrics on $\mathbf{R}^n$ are conceivable. We shall always assume that $\mathbf{R}^n$ has the Euclidean metric, and in an abuse of language sometimes refer to $\mathbf{R}^n$ (instead of $<\mathbf{R}^n, d_n>$) as a metric space. Any metric d_n always has these characteristics:

(a) For any $\mathbf{x} \neq \mathbf{y}$, $d_n(\mathbf{x}, \mathbf{y}) > 0$; if $\mathbf{x} = \mathbf{y}$, then $d_n(\mathbf{x}, \mathbf{y}) = 0$;
(b) For any $\mathbf{x}, \mathbf{y} \in \mathbf{R}^n$, $d_n(\mathbf{x}, \mathbf{y}) = d_n(\mathbf{y}, \mathbf{x})$;
(c) For any $\mathbf{x}, \mathbf{y}, \mathbf{z} \in \mathbf{R}^n$, $d_n(\mathbf{x}, \mathbf{z}) \le d_n(\mathbf{x}, \mathbf{y}) + d_n(\mathbf{y}, \mathbf{z})$.
Observe that for $\mathbf{R}^1$, the usual metric is just $d_1(x, y) = |x - y|$.

■ Example 1.4

In $\mathbf{R}^2$ let $\mathbf{x} = (2, 3), \mathbf{y} = (-3, -1)$. Then $\mathbf{x} * \mathbf{y} = 2(-3) + 3(-1) = -9$, $\mathbf{x} * \mathbf{x} = 2^2 + 3^2 = 13, \mathbf{y} * \mathbf{y} = (-3)^2 + (-1)^2 = 10$, and $|\mathbf{x} * \mathbf{x}| = 9 \le ||\mathbf{x}||\ ||\mathbf{y}|| = \sqrt{13}\sqrt{10} \approx 11.40$. ■

■ Example 1.5

In $\mathbf{R}^3$ let $\mathbf{x} = (2, 3, 3), \mathbf{y} = (3, 1, 4), \mathbf{z} = (-2, -1, 4)$. Then

$$\mathbf{x} \oplus (-\mathbf{z}) = (4, 4, -1) \rightarrow d_3(\mathbf{x}, \mathbf{z}) = \left(4^2 + 4^2 + (-1)^2\right)^{1/2} = \sqrt{33}$$

$$\mathbf{x} \oplus (-\mathbf{y}) = (-1, 2, -1) \to d_3(\mathbf{x}, \mathbf{y}) = \left((-1)^2 + 2^2 + (-1)^2\right)^{1/2} = \sqrt{6}$$

$$\mathbf{y} \oplus (-\mathbf{z}) = (5, 2, 0) \to d_3(\mathbf{y}, \mathbf{z}) = \left(5^2 + 2^2 + 0^2\right)^{1/2} = \sqrt{29},$$

and

$$d_3(\mathbf{x}, \mathbf{z}) \approx 5.74 \le d_3(\mathbf{x}, \mathbf{y}) + d_3(\mathbf{y}, \mathbf{z})$$

$$\approx 2.45 + 5.39 = 7.84.$$

■

A metric in $\mathbf{R}^n$ is useful when we wish to refer to sets of points $\mathbf{p}$ that are nearby some other fixed point $\mathbf{p}_0$. The language here is purposely geometric.

Definition. *If $\mathbf{p}_0 \in \mathbf{R}^n$ and $r > 0$, then the* **open ball** *($n = 1$), or the* **open n-ball** *($n > 1$), of radius r and center at $\mathbf{p}_0$ is the set $\mathbf{B}_n(\mathbf{p}_0; r) = \{\mathbf{p} \in \mathbf{R}^n : d_n(\mathbf{p}, \mathbf{p}_0) < r\}$.*

In $\mathbf{R}^1$ an open ball has the same meaning as an open interval (Figure 1.5). We recall the four types of intervals in $\mathbf{R}^1$:

1. CLOSED
   $[a, b] = \{x : a \le x \le b\}$
2. LEFT HALF-OPEN
   $(a, b] = \{x : a < x \le b\}$
3. RIGHT HALF-OPEN
   $[a, b) = \{x : a \le x < b\}$
4. OPEN
   $(a, b) = \{x : a < x < b\}$.

n-balls have various shapes in different dimensional spaces; for $n \ge 4$ the n-balls cannot even be visualized. Even in a given space, different metrics will lead to different shapes for the balls (Exercise 4.24), and this invariably leads to interesting geometric consequences in the space (Exercise 4.25).

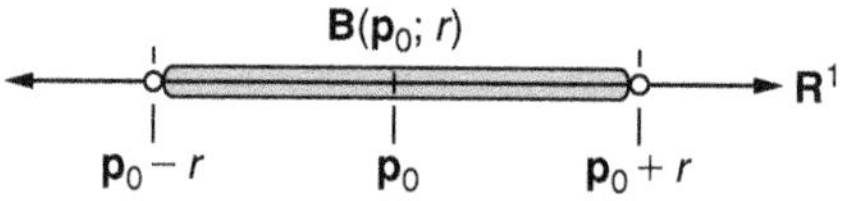

FIGURE 1.5
In $\mathbf{R}^1$ an open ball is an open interval.

1.6 GENERAL ASPECTS OF FUNCTIONS

The **Cartesian product** of two sets, $\mathbf{D} \times \mathbf{S}$, is the set of all ordered pairs (x, y) such that $x \in \mathbf{D}, y \in \mathbf{S}$. We use the notion of an ordered pair to give a modern definition of function.

Definition 1. *A* ***function*** *f from a set $\mathbf{D}$ into a set $\mathbf{S}$ is a set of ordered pairs (x, y) in $\mathbf{D} \times \mathbf{S}$ such that if (x, y), $(x, y') \in f$, then $y = y'$. If $(x, y) \in f$, then we say that f is* ***defined*** *(or has* **value***) at x, and we write commonly $y = f(x)$.*

This definition of function has the merit that it is phrased entirely in terms of sets. The definition presumes that equality has been defined on **S**.

Definition 2. *If* **D**, **S** *are sets, then by a* ***function*** f *we shall mean the operation by which to each element* $x \in \mathbf{D}$ *there is assigned a unique element* $f(x)$ *in* **S**.

This older definition employs the undefined "operation ... assigned," but it has the merit that vitality is breathed into the concept. Other synonyms for "function" that also connote something that is *done* are **mapping** and **transformation**. We shall feel free to use either Definition 1 or Definition 2.

The set of all first elements x (Definition 1) of a function f is the **domain D** of f (or, more completely, $\mathbf{D}(f)$). The set of all second elements y of f is the **range** of f (or, more completely, $\mathbf{R}(f)$). A very general notation for a function from **D** into **S** is $f : \mathbf{D} \to \mathbf{S}$. The set **S** is called the **codomain** of f; in general, $\mathbf{R}(f) \subseteq \mathbf{S}$.

Definition 1 and 2 are so general that the sets **D**, **S** could be sets of any sort of mathematical object. Consequently, the modern conception of function is unlike any that Newton or Leibniz possessed (Kleiner, 1989). The following is typical of the flexibility of the modern idea.

■ Example 1.6

Let f be the set of all ordered pairs (x, y), in which y is given by the formula $y = 1 + \ln x$, and let $\mathbf{I} = \{x : x \in [1, e]\}$. We define the **direct image** of any set **I** under f (Figure 1.6) to be the set

$$f(\mathbf{I}) = \{y : (x, y) \in f,\ x \in \mathbf{I}\}.$$

If **I** were the empty set, then we define $f(\varnothing) = \varnothing$. Since the domain of f had not been specified, the conventional assumption is that it is the largest or natural domain of the function. In the present case, $\mathbf{I} \subset \mathbf{D}(f) = (0, \infty)$, so

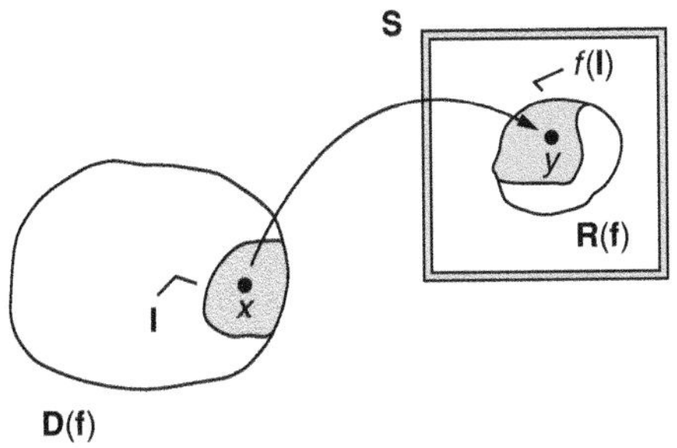

FIGURE 1.6
A direct image under f.

the set $f(\mathbf{I})$ makes sense. The set $\{(x, y) : x \in \mathbf{I}, y \in f(\mathbf{I})\}$ is itself a function; it is called a **restriction** of f, one notation for which is $f \mid \mathbf{I}$. ■

From among the numerous mappings of a set $\mathbf{D}(f)$ into a set $\mathbf{S}$, we pick out a few important types. The following terminology is standard.

Definition. *Let $f : \mathbf{D} \to \mathbf{S}$ be a mapping.*

(a) *The mapping is an **injection** (or is a **one-one mapping**) iff whenever $(x_1, y_1), (x_2, y_2) \in f$ and $x_1 \neq x_2$, then $y_1 \neq y_2$.*

(b) *The mapping is a **surjection** (or is a mapping of $\mathbf{D}(f)$ **onto** $\mathbf{S}$) iff $\mathbf{R}(f) = \mathbf{S}$.*

(c) *The mapping is a **bijection** from $\mathbf{D}(f)$ to $\mathbf{S}$ iff it is an injection and a surjection.*

Proof by contrapositive often is used to show that a mapping is an injection. To prove that a mapping is a surjection, we try to show that for any $y \in \mathbf{S}$ there is at least one $x \in \mathbf{D}(f)$ such that $(x, y) \in f$.

■ Example 1.7

Let $f : \mathbf{R}^1 \backslash \{2\} \to \mathbf{R}^1$ be defined by $y = x^2/(x - 2)$. Choose $y_1 = y_2 = -2$; then $x^2 + 2x - 4 = 0$ and $x_1 = -1 + \sqrt{5}, x_2 = -1 - \sqrt{5}$, for example. Hence, f is not an injection. ■

■ Example 1.8

For the function of the previous example, choose $y = 1$. Then $x^2 - x + 2 = 0$, and neither root of this lies in $\mathbf{D}(f)$. Hence, f is not a surjection. ■

■ Example 1.9

The function of Example 1.7 becomes a surjection if the codomain is restricted to $\mathbf{R}^1 \backslash (0, 8)$. [Hint: Differentiate the function or, alternatively, prepare a graph.] ■

Other important general aspects of functions include the composition of functions and the inverses of functions. These are reviewed briefly in Exercises 1.42 and 1.43.

1.7 INFINITE SETS

An important application of bijections, in particular, is to the cardinality of sets, a vast subject. The empty set, which has no members, is said to have cardinality 0. A set $\mathbf{S}$ is **finite** either if it is empty or if there exists a bijection between $\mathbf{S}$ and the set $\mathbf{N}_n = \{1, 2, 3, \ldots, n\}$ of consecutive natural numbers; $\mathbf{S}$ then is said to have cardinality n. Thus, the Greek alphabet has cardinality 24, and the set of Platonic solids has cardinality 5.

	n	$\phi(n+1)$

Row 1 $\quad \frac{1}{1}$ $\qquad$ 1

Row 2 $\quad \frac{2}{1}\ \frac{1}{2}$ $\qquad$ 2

Row 3 $\quad \frac{3}{1}\ \frac{1}{3}$ $\qquad$ 2

Row 4 $\quad \frac{4}{1}\ \frac{3}{2}\ \frac{2}{3}\ \frac{1}{4}$ $\qquad$ 4

Row 5 $\quad \frac{5}{1}\ \frac{1}{5}$ $\qquad$ 2

Row 6 $\quad \frac{6}{1}\ \frac{5}{2}\ \frac{4}{3}\ \frac{3}{4}\ \frac{2}{5}\ \frac{1}{6}$ $\qquad$ 6

Row 7 $\quad \frac{7}{1}\ \frac{5}{3}\ \frac{3}{5}\ \frac{1}{7}$ $\qquad$ 4

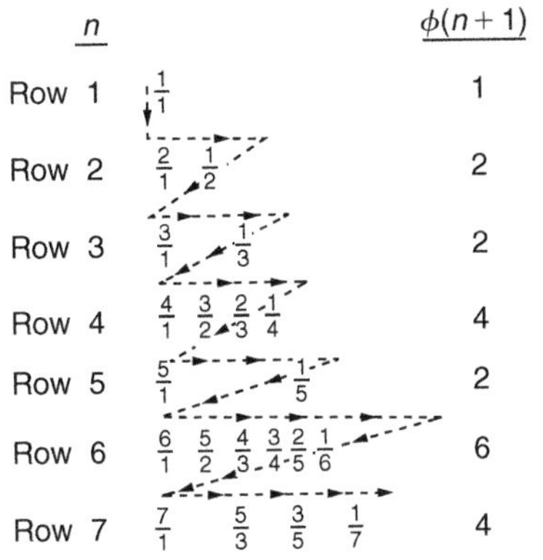

FIGURE 1.7

A small portion of the table for the enumeration of the positive rationals.

A set is **infinite** iff it is not finite. Such sets do exist.[9] We distinguish two classes. A **countably infinite** (or **denumerable**) set S is one for which there exists a bijection between S and N. Such an infinite set has cardinality $\aleph_0$ (Cantor's choice of symbol!). An infinite set S is **uncountably infinite** (or **nondenumerable**) iff there is no bijection possible between S and N. Obviously, N itself is a countably infinite set. Less obvious, and more interesting, is the fact that the set of positive rational numbers is countably infinite. The proof to be presented avoids redundancy in the counting process. The following idea from arithmetic is needed:

Definition. *Two unequal natural numbers m, n are **relatively prime** iff the largest natural number d that divides both of them is 1. The integer d is called the **greatest common divisor** of m, n.*

We take it for granted that any two natural numbers have a greatest common divisor, and that this can always be calculated (in principle) by the Euclidean algorithm (Andrews, 1994; Dence and Dence, 1999).

Theorem 1.9. *The positive members of* $\mathbf{Q}$ *are a denumerable set.*

Proof. The positive rationals, $\mathbf{Q}^+$, are arranged in rows as shown in Figure 1.7. Each ROW n contains rationals in descending order whose numerator + denominator $= n + 1$. Additionally, the denominators are those natural

---

[9] How glibly we take the existence of infinite sets for granted. Both Gauss and Cauchy denied their existence, although Bolzano (a lone voice in the wilderness, here also) defended their existence. But it was **Georg Cantor** who single-handedly created the mathematics of infinite sets. His work met with stiff opposition from older mathematicians (Kline, 1972), but is regarded today as pioneering, brilliant, and fundamental.

numbers less than and relatively prime to $n+1$. The number of rationals in ROW n is given in number theory by the **Euler ϕ-function**; that is, $\phi(n+1)$ is the number of natural numbers less than $n+1$ that are relatively prime to $n+1$ (by convention, $\phi(1)=1$). For example, $\phi(4)=2$ (see ROW 3) and $\phi(7)=6$ (see ROW 6). For any $n \in \mathbf{N}$, we have $1 \leq \phi(n+1) \leq n$.

Numbered rows of the table can be continued indefinitely; every positive rational will appear somewhere in the table, and only once, because the reduction to lowest terms always gives only one representation of a rational number. If counting of the rationals is begun at the top of the zigzag path, then we have an injection from $\mathbf{Q}^+$ into $\mathbf{N}$. If $K \in \mathbf{N}$ is specified, then to find the Kth rational, first find the smallest integer M such that

$$\sum_{n=2}^{M+1} \phi(n) \geq K.$$

Then read across row M, from left to right, until the $\left[K - \sum_{n=2}^{M} \phi(n)\right]$-th entry is encountered. Thus, we have a surjection of $\mathbf{Q}^+$ onto $\mathbf{N}$; hence, there is a bijection from $\mathbf{Q}^+$ onto $\mathbf{N}$, and $\mathbf{Q}^+$ is a denumerable set. ■

■ Example 1.10

In the scheme of Theorem 1.9, what is the $K=25$th rational?

We first determine $\phi(8)=4$ and $\phi(9)=6$. Then we find that $\sum_{n=2}^{7+1} \phi(n) = 21 < 25$, but $\sum_{n=2}^{8+1} \phi(n) = 27 > 25$, so the 25th positive rational is in row $M=8$; there, the third element from the left is $5/4$. ■

■ Example 1.11

What natural number corresponds to the rational $22/9$?

Since $22+9+1$ is relatively prime to 9, then the rational lies in ROW 30. From the data in Table 1.6 we compute

$$\sum_{n=2}^{30} \phi(n) = \sum_{n=2}^{9} \phi(n) + \sum_{n=10}^{30} \phi(n) = 27 + 250 = 277.$$

ROW 30 begins $\frac{30}{1}, \frac{29}{2}, \frac{28}{3}, \frac{27}{4}, \frac{26}{5}, \frac{25}{6}, \frac{24}{7}, \frac{23}{8}, \frac{22}{9}$. Hence, the natural number $K = 277 + 9 = 286$ corresponds to $22/9$. ■

We have not yet exhibited an uncountably infinite set, yet there exists an abundance of them. This will be dealt with in Exercise 1.47.

Table 1.6 Some Values of the Euler ϕ-function

n	$\phi(n+1)$	n	$\phi(n+1)$	n	$\phi(n+1)$
8	6	16	16	24	20
9	4	17	6	25	12
10	10	18	18	26	18
11	4	19	8	27	12
12	12	20	12	28	28
13	6	21	10	29	8
14	8	22	22	30	30
15	8	23	8	31	16

1.8 CLUSTER AND OTHER POINTS

In the next section we will review and expand the concept of the limit of a function. For this purpose we shall need the important idea of a cluster point.

Definition. *A* ***cluster point*** *of a set* $\mathbf{S} \subset \mathbf{R}^n$ *is any point* $\mathbf{a}$ *such that* every *n-ball about* $\mathbf{a}$ *contains an infinite subset of* $\mathbf{S}$.

A careful reading of the definition furnishes these four conclusions: (a) the set $\mathbf{S}$ has to be an infinite set, (b) the point $\mathbf{a}$ could belong to $\mathbf{S}$ but it may not, (c) all the points in the n-ball about $\mathbf{a}$ may belong to $\mathbf{S}$, but some points in the n-ball could lie outside of $\mathbf{S}$, (d) every n-ball about $\mathbf{a}$, regardless of radius, must be considered.

Some sets in a metric space, even some infinite sets, do not have any cluster points. For example, the set $\mathbf{S} \subset \mathbf{R}^3$, where $\mathbf{S} = \{\mathbf{x}_k : \mathbf{x}_k = (k, -k, k), k \in \mathbf{N}\}$, has none (why?), and neither does the set of integers, $\mathbf{S} = \mathbf{Z}$ or the set $\mathbf{S} \subset \mathbf{R}^2$ of prime lattice points in the plane, where $\mathbf{S} = \{(x, y) : x, y \text{ are primes}\}$.

A cluster point $\mathbf{a}$ of a set $\mathbf{S}$ is especially important if it is not a member of $\mathbf{S}$, but is merely a boundary point of $\mathbf{S}$ just outside of $\mathbf{S}$. To appreciate this, we define $\mathbf{p}$ to be an **interior point** of a set $\mathbf{S}$ iff there is a $\delta > 0$ such that $\mathbf{B}(\mathbf{p}; \delta) \subset \mathbf{S}$. The set of all interior points of a set $\mathbf{S}$ is denoted by $\text{Int}(\mathbf{S})$; a set $\mathbf{S}$ is said to be **open** iff $\text{Int}(\mathbf{S}) = \mathbf{S}$. Next, a point $\mathbf{p}$ is an **exterior point** of a set $\mathbf{S}$ iff there is a $\delta > 0$ such that $\mathbf{B}(\mathbf{p}; \delta) \cap \mathbf{S} = \emptyset$. The set of all exterior points of a set $\mathbf{S}$ is denoted $\text{Ext}(\mathbf{S})$; clearly, $\text{Int}(\mathbf{S}) \cap \text{Ext}(\mathbf{S}) = \emptyset$. We note that if $\mathbf{S}$ were the interval $[1, 2]$, then $\text{Int}(\mathbf{S}) = (1, 2)$, and $\mathbf{p} = 3$ would be an exterior point. In fact, $\text{Ext}(\mathbf{S}) = (-\infty, 1) \cup (2, \infty)$.

Finally, a point $\mathbf{p}$ is a **boundary point** of $\mathbf{S}$ if every ball about $\mathbf{p}$ contains at least one point from $\mathbf{S}$ and at least one point external to $\mathbf{S}$. The set of all boundary points of $\mathbf{S}$ is denoted by $\text{Bd}(\mathbf{S})$; $\mathbf{S} \cup \text{Bd}(\mathbf{S})$ is called the **closure** of $\mathbf{S}$ and is

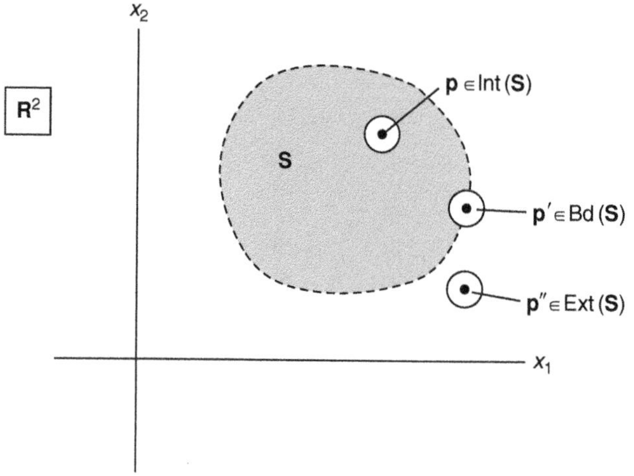

FIGURE 1.8
Three types of points in $\mathbf{R}^2$.

denoted by $\overline{\mathbf{S}}$. Interior point, exterior point, and boundary point are illustrated in Figure 1.8.

■ Example 1.12

Given a set **S**, is $\mathrm{Int}(\overline{\mathbf{S}})$ necessarily the same as $\overline{\mathrm{Int}(\mathbf{S})}$? Sketch some sets in $\mathbf{R}^1$ and see what you can discover. Can you produce a set $\mathbf{S} \subset \mathbf{R}^1$ such that zero to three combinations of the operations interior and closure, Int( ), $(\overline{\ })$ lead to five distinct sets? ■

The definition of boundary point (when this is not an isolated point) implies that any ball about such a point actually contains infinitely many points from **S**. Thus, a cluster point can be either a boundary point or an interior point of a set **S** but never an exterior point of **S**. If, for instance, **S** is the set of rational numbers contained in the unit interval, $\mathbf{S} = \mathbf{Q} \cap [0, 1]$, then no point in **S** is an interior point of **S** and $\overline{\mathbf{S}} = [0, 1]$.

■ Example 1.13

Let $\mathbf{S} \subset \mathbf{R}^2$ be the set of all points $\mathbf{x}_k = (x_{k1}, x_{k2})$, where $x_{k1} = k^{-1}, x_{k2} = (k^2 + 2)/(2k^2 - k + 1), k \in \mathbf{N}$. A rough sketch shows that the $\mathbf{x}_k$'s are discrete points that lie along a path from $(1, 3/2)$ (which belongs to **S**) and approach $(0, 1/2)$ (which does not belong to **S**). Near this latter terminus, and nowhere else along the path, the points crowd closely together. Hence, $(0, 1/2)$ is the only cluster point of **S**. It is a boundary point that does not belong to **S**. All the $\mathbf{x}_k$'s are also boundary points; they belong to **S**. There are no interior points in **S**; all other points in $\mathbf{R}^2$ are exterior points. ■

1.9 LIMITS OF FUNCTIONS

Limits play a central role in the structure of calculus (Figure 1.9). We recall that the limit of a function f of a single variable x, as x approaches some value a, is (loosely speaking) a description of the behavior of f when x is "close to a." We carry this idea over to broader functions $\mathbf{f}: \mathbf{D}(\mathbf{f}) \to \mathbf{R}^m, \mathbf{D}(\mathbf{f}) \subseteq \mathbf{R}^n$.

We build upon the ideas in Section 1.8. We wish to consider limits at cluster points, both those that are interior points and those that are boundary points.

Definition. *Let* $\mathbf{f}: \mathbf{D}(\mathbf{f}) \to \mathbf{R}^m, \mathbf{D}(\mathbf{f}) \subseteq \mathbf{R}^n$ *be a mapping and let* $\mathbf{a}$ *be a cluster point of* $\mathbf{D}(\mathbf{f})$. *Then* $\mathbf{L} \in \mathbf{R}^m$ *is a* ***limit*** *of* $\mathbf{f}$ *at* $\mathbf{a}$, *and we write* $\lim_{\mathbf{x}\to\mathbf{a}} \mathbf{f} = \mathbf{L}$ *iff, given any* $\varepsilon > 0$, *there exists a* $\delta > 0$ *such that whenever* $(\mathbf{x}, \mathbf{y}) \in \mathbf{f}$ *and* $\mathbf{x} \in [\mathbf{B}_n(\mathbf{a}; \delta)\backslash\{\mathbf{a}\}] \cap \mathbf{D}(\mathbf{f})$, *then* $\mathbf{y} \in \mathbf{B}_m(\mathbf{L}; \varepsilon)$.

The definition of limit is pictorialized in Figure 1.10 for an interior point $\mathbf{a}$ of $\mathbf{D}(\mathbf{f})$ and a boundary point $\mathbf{b}$ of $\mathbf{D}(\mathbf{f})$. Several pertinent comments now follow.

1. The n-ball $\mathbf{B}_n(\mathbf{a}; \delta)\backslash\{\mathbf{a}\}$ is a **deleted** n-ball; the point $\mathbf{x} = \mathbf{a}$ is not included in it so that we can consider the limit of $\mathbf{f}$ at $\mathbf{a}$, where $\mathbf{f}$ may not be defined.
2. The symbolism $[\mathbf{B}_n(\mathbf{a}; \delta)\backslash\{\mathbf{a}\}] \cap \mathbf{D}(\mathbf{f})$ reinforces the idea that we can consider "sided" limits at cluster points $\mathbf{a}$ that are boundary points, as well as limits at interior points of $\mathbf{D}(\mathbf{f})$. In $\mathbf{R}^1$ this leads to consideration of the familiar right-hand and left-hand limits of $\mathbf{f}$ at the left and right endpoints, respectively (Exercises 1.57, 1.58).

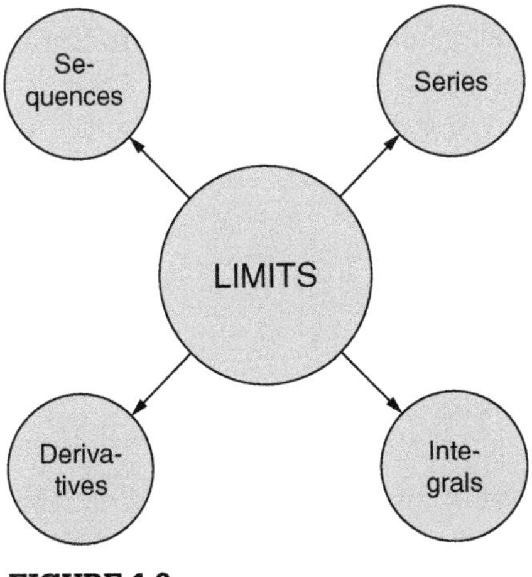

FIGURE 1.9
Limits are the entry point into derivatives, sequences, series, and integrals.

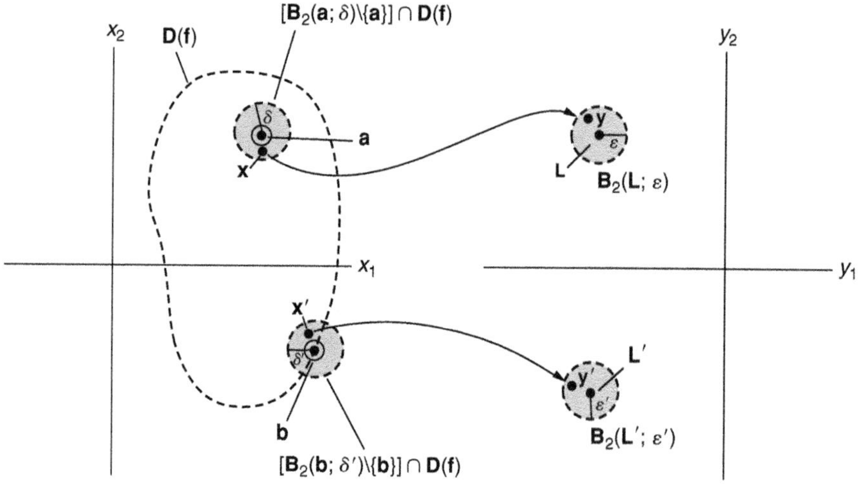

FIGURE 1.10

Limits of a function $\mathbf{f}: \mathbf{D}(\mathbf{f}) \to \mathbf{R}^2, \mathbf{D}(\mathbf{f}) \subset \mathbf{R}^2$, *at two cluster points* $\mathbf{a}, \mathbf{b}$.

3. The positive number ε is stipulated first; a $\delta > 0$ is then sought. In general, δ will depend upon ε, and we could write, more precisely, $\delta(\varepsilon)$.
4. By requiring $\mathbf{a}$ to be a cluster point of $\mathbf{D}(\mathbf{f})$, we guarantee that $\mathbf{B}_n(\mathbf{a}; \delta)\backslash\{\mathbf{a}\}$ will contain points of $\mathbf{D}(\mathbf{f})$, no matter how small is δ.
5. The definition refers to $\mathbf{L}$ as "a" limit of $\mathbf{f}$ at $\mathbf{a}$. In fact, a proof by contradiction will show that the limit is unique (Exercise 1.51).
6. The definition gives no guidance on how to determine $\mathbf{L}$. Thus, whereas the definition of limit is useful for theoretical purposes, the practicalities of limits are more easily handled by the use of various limit theorems.

Theorem 1.10. *Let f, g be mappings from a common domain $\mathbf{D} \subseteq \mathbf{R}^n$ into $\mathbf{R}^1$, and let $\mathbf{a}$ be a cluster point of $\mathbf{D}$. Suppose that $\lim_{\mathbf{x}\to\mathbf{a}} f = F$, $\lim_{\mathbf{x}\to\mathbf{a}} g = G$, and that $k \in \mathbf{R}^1$. Then*

(i) $\lim_{\mathbf{x}\to\mathbf{a}} k \cdot f = kF$

(ii) $\lim_{\mathbf{x}\to\mathbf{a}} (f + g) = F + G$

(iii) $\lim_{\mathbf{x}\to\mathbf{a}} fg = FG.$

Additionally, if $F \neq 0$ and if f is nonzero everywhere in the modified n-ball $[\mathbf{B}_n(\mathbf{a}; \delta)\backslash\{\mathbf{a}\}] \cap \mathbf{D}(f)$, then

(iv) $\lim_{\mathbf{x}\to\mathbf{a}} 1/f = F^{-1}.$

Proof.

(ii) Let $\varepsilon > 0$ be given. Then there exist numbers $\delta_1, \delta_2 > 0$ such that $\mathbf{x} \in [\mathbf{B}_n(\mathbf{a}; \delta_1)\backslash\{\mathbf{a}\}] \cap \mathbf{D}$ implies $y_1 \in \mathbf{B}(F; \varepsilon/2), (\mathbf{x}, y_1) \in f$, and $\mathbf{x} \in [\mathbf{B}_n(\mathbf{a}; \delta_2)\backslash\{\mathbf{a}\}] \cap \mathbf{D}$ implies $y_2 \in \mathbf{B}(G; \varepsilon/2), (\mathbf{x}, y_2) \in g$. Let $\delta = \min\{\delta_1, \delta_2\}$; use the Triangle Inequality in Section 1.5 to conclude what $\mathbf{x} \in [\mathbf{B}_n(\mathbf{a}; \delta)\backslash\{\mathbf{a}\}] \cap \mathbf{D}$ implies.

(iii) Let $\varepsilon > 0$ be given, and note that $|f(\mathbf{x}) - F| < 1$ if $\delta > 0$ is small enough. Then write $fg - FG$ as

$$|f(\mathbf{x})g(\mathbf{x}) - FG| = |f(\mathbf{x})\{g(\mathbf{x}) - G\} + G\{f(\mathbf{x}) - F\}|,$$

and make use of the Triangle Inequality.

(iv) Let $\varepsilon > 0$ be given. Then by hypothesis there is a $\delta > 0$ such that $\mathbf{x} \in [\mathbf{B}_n(\mathbf{a}; \delta)\backslash\{\mathbf{a}\}] \cap \mathbf{D}$ implies $f(\mathbf{x}) \neq 0$ and $|f(\mathbf{x}) - F| < \min\left\{\frac{|F|}{2}, \frac{|F|^2}{2}\varepsilon\right\}$. Now consider

$$\left|\frac{1}{f(\mathbf{x})} - \frac{1}{F}\right|,$$

and make use of the Triangle Inequality and Exercise 1.33.

The proof of part (i) and the completions of the proofs of the other three parts are left to you. ■

Theorem 1.11. *Let* $\mathbf{f}: \mathbf{D}(\mathbf{f}) \to \mathbf{R}^m, \mathbf{D}(\mathbf{f}) \subseteq \mathbf{R}^n$ *be a mapping, and denote the value of* $\mathbf{f}$ *at any point* $\mathbf{x} \in \mathbf{D}(\mathbf{f})$ *by* $(f_1(\mathbf{x}), f_2(\mathbf{x}), \ldots, f_m(\mathbf{x}))$. *Let* $\mathbf{a}$ *be a cluster point of* $\mathbf{D}(\mathbf{f})$. *Then* $\mathbf{L} = (L_1, L_2, \ldots, L_m)$ *is the limit of* $\mathbf{f}$ *at* $\mathbf{a}$ *iff*

$$\lim_{\mathbf{x}\to\mathbf{a}} f_i(x) = L_i, \quad 1 \le i \le m.$$

Proof. ($\leftarrow$) We prove this half and leave the other half as an exercise. Suppose that $\lim_{\mathbf{x}\to\mathbf{a}} f_i(x) = L_i$ for each natural number $i \in [1, m]$. Let $\varepsilon > 0$ be given; then for each i there is a $\delta_i > 0$ such that $\mathbf{x} \in [\mathbf{B}_n(\mathbf{a}; \delta_i)\backslash\{\mathbf{a}\}] \cap \mathbf{D}(\mathbf{f})$ implies that $|L_i - f_i(\mathbf{x})| = [(L_i - f_i(\mathbf{x}))^2]^{1/2} < \varepsilon/m$. Let $\delta = \min\{\delta_1, \delta_2, \ldots, \delta_m\}$; then whenever $\mathbf{x} \in [\mathbf{B}_n(\mathbf{a}; \delta)\backslash\{\mathbf{a}\}] \cap \mathbf{D}(\mathbf{f})$, the previous inequality holds for all i.

For any set $\{c_1, c_2, \ldots, c_m\}$ of real numbers, we have

$$c_1^2 + c_2^2 + \cdots + c_m^2 \le \sum_{k=1}^{m} |c_k|^2 + 2 \sum_{1\le i\le j\le m} |c_i||c_j| = [|c_1| + |c_2| + \cdots |c_m|]^2$$

or

$$[c_1^2 + c_2^2 + \cdots + c_m^2]^{1/2} \le |c_1| + |c_2| + \cdots + |c_m|.$$

Applying this to the present problem, we then obtain

$$\left\{\sum_{i=1}^{m}(L_i - f_i(\mathbf{x}))^2\right\}^{1/2} \le \sum_{i=1}^{m}\{(L_i - f_i(\mathbf{x}))^2\}^{1/2} = \sum_{i=1}^{m}|L_i - f_i(\mathbf{x})|$$
$$< m\left(\frac{\varepsilon}{m}\right) = \varepsilon,$$

so $\mathbf{L} = (L_1, L_2, \ldots, L_m)$ is $\lim_{\mathbf{x}\to\mathbf{a}} \mathbf{f}$. ■

■ Example 1.14

Let $f\colon \mathbf{D}(f) \to \mathbf{R}^2, \mathbf{D}(f) \subseteq \mathbf{R}^3$, be defined by $\mathbf{x} = (x_1, x_2, x_3), f(\mathbf{x}) = \left(x_1^2 - 3, (2x_2 + x_3)/x_1\right)$. Find $\lim_{\mathbf{x}\to\mathbf{a}} f$ at $\mathbf{a} = (2, 1, 0)$.

Using all parts of Theorem 1.10, we determine

$$\lim_{\mathbf{x}\to\mathbf{a}}(x_1^2 - 3) = 1, \ \lim_{\mathbf{x}\to\mathbf{a}} \frac{2x_2 + x_3}{x_1} = 1.$$

Hence, by Theorem 1.11, $\lim_{\mathbf{x}\to\mathbf{a}} f = (1, 1)$. ■

The two preceding theorems presume that all quantities are real or are tuples of real numbers. Some extension is required for common cases where the domain or the range of a function is unbounded. The **extended real number** system, **Re**, is the system obtained by adjoining to **R** the two fictitious points $+\infty$ (or, more commonly, just ∞) and $-\infty$, and then defining on **Re** the order relationships $-\infty < x < \infty$ for any $x \in \mathbf{R}$. For a set $\mathbf{S} \subseteq \mathbf{R}$ that is unbounded from above (below) we then write $\sup \mathbf{S} = \infty$ ($\inf \mathbf{S} = -\infty$).

Definition. *Let $f\colon \mathbf{D}(f) \to \mathbf{R}^1$ and suppose that $\mathbf{D}(f) \subseteq \mathbf{R}^1$ is unbounded from above. Then $L \in \mathbf{R}^1$ is the* **limit** *of f as $x \to \infty$ and we write $\lim_{x\to\infty} f = L$ iff, given any $\varepsilon > 0$, there exists an $M > 0$ such that for all $x \in [M, \infty) \cap \mathbf{D}(f)$ we have $f(x) \in \mathbf{B}(L; \varepsilon)$.*

We shall also say that f has limit ∞ (in **Re***) as $x \to \infty$ and we will write $\lim_{x\to\infty} f = \infty$ iff, given any $r \in \mathbf{R}^1$, there exists an $M > 0$ such that for all $x \in [M, \infty) \cap \mathbf{D}(f)$ we have $f(x) > r$.*

■ Example 1.15

Let $f\colon \mathbf{D}(f) \to \mathbf{R}^1$ be the mapping defined by $f(x) = x^{-1}, \mathbf{D}(f) = \mathbf{R}^+ = (0, \infty)$. Show that $\lim_{x\to\infty} f = L = 0$.

For no $x \in \mathbf{D}(f)$ does $x^{-1} = 0$ or $x^{-1} < 0$ hold, since multiplications by x lead to contradictions; hence, $\mathbf{R}(f) \subseteq \mathbf{R}^+$. Then if $\varepsilon > 0$ be given, $f(x) \in \mathbf{B}(L; \varepsilon)$ iff $0 < x^{-1} < \varepsilon$ iff $x \in [M, \infty)$, where $M = \varepsilon^{-1}$. The result follows. ■

Some other limiting situations are dealt with shortly (Exercises 1.56–1.59), and some additional examples to work are given in Exercises 1.60 and 1.61.

EXERCISES[10]

Section 1.1

1.1. The truth table for conjunction is shown in Figure 1.11(a). Would the proposition in Example 1.1 still be true if "or" was replaced by "and"?

1.2. The truth table for disjunction is shown in Figure 1.11(b). Would the proposition in Example 1.1 still be true if "641" were replaced by "441"? What connection exists, in general, between the truth values for $\sim (p \wedge q)$ and those of $(\sim p \vee \sim q)$?

1.3. In logic $p \rightarrow q$ is defined to mean $\sim p \vee q$; this is somewhat controversial among logicians.

(a) Nevertheless, show that we obtain the truth table in Figure 1.12.

p	q	$p \wedge q$
T	T	T
T	F	F
F	T	F
F	F	F

(a)

p	q	$p \vee q$
T	T	T
T	F	T
F	T	T
F	F	F

(b)

FIGURE 1.11
Truth tables (a) for conjunction, and (b) for disjunction.

p	q	$p \rightarrow q$
T	T	T
T	F	F
F	T	T
F	F	T

FIGURE 1.12
Truth table for implication.

[10]Throughout your use of this book, don't forget about Appendix A at the back of the book.

(b) Define $p \leftrightarrow q$ to mean $(p \rightarrow q) \wedge (q \rightarrow p)$. Work out the truth table for this double implication.

(c) Finally, work out the truth table for the proposition $(p \rightarrow q) \leftrightarrow (\sim q \rightarrow \sim p)$. This is the **Law of Contraposition**.

1.4. Pictures do not constitute proofs in mathematics; they may point the way to correct ideas. Let $\mathbf{S}, \mathbf{T}, \mathbf{W}$ be sets; conjecture from pictures which of these may hold:

(a) $\mathbf{S} \subseteq \mathbf{T} \rightarrow (\mathbf{S} \cap \mathbf{T} = \mathbf{S})$

(b) $(\mathbf{S} \cap \mathbf{T} = \mathbf{S}) \rightarrow \mathbf{S} \subseteq \mathbf{T}$

(c) $\mathbf{W} \backslash (\mathbf{S} \cup \mathbf{T}) = [(\mathbf{W} \backslash \mathbf{S}) \cup (\mathbf{W} \backslash \mathbf{T})]$

(d) $(\mathbf{S} \cap \mathbf{T}) \subseteq [(\mathbf{S} \cap \mathbf{T}) \cap \mathbf{W}]$

(e) $[(\mathbf{S} \backslash \mathbf{T}) \cup (\mathbf{T} \backslash \mathbf{S})] = \mathbf{S} \cup \mathbf{T}$

(f) $(\mathbf{S} \backslash \mathbf{T}) \backslash \mathbf{W} = \mathbf{S} \backslash (\mathbf{T} \cup \mathbf{W})$

Section 1.2

1.5. Why is no Right Distributive Law, $(y + z)x = yx + zx$, given in Table 1.5? Which of the axioms in Table 1.5 would be false if they were reworded so as to apply to 3×3 real matrices instead of to elements of $\mathbf{R}$?

1.6. On the set $\mathbf{Z}_5 = \{0, 1, 2, 3, 4\}$ the operations $\oplus$ and $\otimes$ are defined as follows:

$\oplus$	0	1	2	3	4
0	0	1	2	3	4
1	1	2	3	4	0
2	2	3	4	0	1
3	3	4	0	1	2
4	4	0	1	2	3

$\otimes$	0	1	2	3	4
0	0	0	0	0	0
1	0	1	2	3	4
2	0	2	4	1	3
3	0	3	1	4	2
4	0	4	3	2	1

(a) Confirm that the algebraic structure $< \mathbf{Z}_5, \oplus, \otimes >$ is a field.

(b) After referring to axiom R7, explain how you know that $< \mathbf{Z}_5, \oplus, \otimes >$ has no positive elements.

(c) Try to construct some other finite fields. See if you can frame any conjectures.

1.7. It might be supposed that there could be two distinct additive identities (0 and $0', 0 \neq 0'$) in $\mathbf{R}$, and also two distinct multiplicative identities (1 and $1', 1 \neq 1'$).

(a) Present an informal argument (i.e., not a formal proof) against distinct $0, 0'$.

(b) Do the same for distinct $1, 1'$.

1.8. Suppose that $x, y \in \mathbf{R}$ and $x \neq y$. Explain, from the axioms, how it follows that $x + (-y) \neq 0$ and $y + (-x) \neq 0$.

1.9. **(a)** Suppose that $a, b, c \in \mathbf{R}$ and $a + b = a + c$. Explain, from the axioms, how it follows that $b = c$.

(b) Begin with $y + 0 = y, y \in \mathbf{R}$, and premultiply both sides by any $x \in \mathbf{R}$. Explain, from the axioms, what you can conclude about $x \cdot 0$.

1.10. Begin with $1 + (-1) = 0$, and premultiply both sides by any nonzero $x \in \mathbf{R}$. Explain, from the axioms and Exercise 1.9(b), how you can conclude that $(-1)x = -x$. As a corollary, deduce $(-1)(-1) = 1$.

1.11. Begin with $x + [(-x) + y] = x + [(-x) + y]$, for any $x, y \in \mathbf{R}$. Use the axioms and Exercises 1.9(a) and 1.10, and explain how you conclude that $-[x + (-y)] = (-x) + y$.

1.12. Explain how you know that Nonreflexivity and Transitivity hold for $>$ in $\mathbf{R}$.

1.13. Suppose that $a, b, c \in \mathbf{R}$ and that $b > a, c > 0$. Explain, from the axioms, how it follows that $cb > ca$.

Section 1.3

1.14. Two sets, $\mathbf{S}_1$ and $\mathbf{S}_2$, are defined similarly to those in the text, but with "3" in place of "5." Let $x_{k+1} = 4 - [13/(4 + x_k)]$, and $x_1 = 1$.
- **(a)** Write a short program and tabulate $x_2, x_3, \ldots, x_{18}$, and their squares.
- **(b)** Verify algebraically that $x_k < x_{k+1}$ and $x_k^2 < 3$ for all $k \in \mathbf{N}$.
- **(c)** Prepare a second table of results for $\mathbf{S}_2$; start with $x_1 = 2$.
- **(d)** What seems to be $\sup \mathbf{S}_1$, $\inf \mathbf{S}_2$, and how do you know these numbers exist?

1.15. A set $\mathbf{S} \subset \mathbf{R}$ has a supremum. Explain how you know that there is only one supremum.

1.16. A set $\mathbf{S} \subset \mathbf{R}$ is bounded from below by l. Use axiom R8 to explain how you know that $\inf \mathbf{S}$ exists.

1.17. A set $\mathbf{S} \subset \mathbf{R}$, $\mathbf{S} = \{x_0, x_1, x_2, \ldots\}$, is defined by
$$x_k = \begin{cases} 0 & k = 0 \\ 1 & k = 1 \\ (4x_{k-1} - x_{k-2})/3 & k \geq 2. \end{cases}$$
- **(a)** Write a short program and tabulate x_0 to x_{12}.
- **(b)** Conjecture a value for $\sup \mathbf{S}$.
- **(c)** Extra credit if you can give a plausibility argument for the existence of $\sup \mathbf{S}$. This is not a rigorous proof.

Section 1.4

1.18. Prove that if $x, y \in \mathbf{R}$, then $x \cdot (-y) = -(x \cdot y)$. You may assume all the results in Exercises 1.9 through 1.13.

1.19. Suppose that $x, y, z \in \mathbf{R}$. Establish:
- **(a)** If $x > y$ and $z < 0$, then $zy > zx$;
- **(b)** If $xy < 0$, then either $x > 0$ AND $y < 0$, or $x < 0$ AND $y > 0$;
- **(c)** If $x \neq 0$, then $x^4 > 0$.

1.20. Prove that if a set $\mathbf{S} \subset \mathbf{R}$ contains one of its lower bounds, then this lower bound is $\inf \mathbf{S}$.

1.21. Prove that in $\mathbf{R}$ there is no smallest positive number.

1.22. If $x, y \in \mathbf{R}$ and are both positive, then their **geometric mean** is $\sqrt{xy}$ and their **arithmetic mean** is $(x + y)/2$. Prove that if the two means are unequal, then $x \neq y$.

1.23. Let $U = \sup \mathbf{S}$. Prove that if $x < U$, then there is an $s \in \mathbf{S}$ such that $x < s \leq U$. Draw a picture of this result.

1.24. Let $U = \sup \mathbf{S}$. Prove that if $\varepsilon > 0$, then there is an $s \in \mathbf{S}$ such that $U - \varepsilon < s \leq U$. Draw a picture of this result.

1.25. Prove, as asserted in the discussion surrounding Figure 1.3, that there is no $x \in \mathbf{Q}^+$ such that $x^2 = 5$.[11]

1.26. Suppose that $\mathbf{S}, \mathbf{T}$ are nonempty, bounded subsets of $\mathbf{R}$, and that $\mathbf{S} \subseteq \mathbf{T}$. Prove that $\inf \mathbf{T} \leq \inf \mathbf{S}$.

1.27. Suppose that $x, y \in \mathbf{R}$ and $y > x$. Use Theorem 1.5 to prove that there is a rational number strictly between x, y.

1.28. Prove that for all $k \in \mathbf{N}$, the number of diagonals in a convex polygon of $k+2$ sides is equal to $\frac{1}{2}(k-1)(k+2)$.

1.29. The union of n sets is defined as

$$\bigcup_{k=1}^{n} \mathbf{S}_k = \begin{cases} \mathbf{S}_1 & n = 1 \\ \mathbf{S}_1 \cup \mathbf{S}_2 & n = 2 \\ \{x : x \in \bigcup_{k=1}^{n-1} \mathbf{S}_k \text{ or } x \in \mathbf{S}_n\} & n > 2. \end{cases}$$

Now suppose that for each $k \in \mathbf{N}$, the set $\mathbf{S}_k$ is a bounded subset of $\mathbf{R}$. Prove that for each $n \in \mathbf{N}$, $\sup \bigcup_{k=1}^{n} \mathbf{S}_k = \max\{\sup \mathbf{S}_1, \sup \mathbf{S}_2, \ldots, \sup \mathbf{S}_n\}$.

Section 1.5

1.30. **(a)** If the underlying set $\mathbf{V}$ is the set of all real-valued solutions of $f''(x) + f(x) = 0$, $\oplus$ is pointwise addition of functions, and $\bullet$ is pointwise multiplication of a function by a real number, is $<\mathbf{V}, \oplus, \bullet>$ a vector space?

(b) If the underlying set $\mathbf{V}$ is the set of all invertible 3×3 matrices with real entries, $\oplus$ is matrix addition, and $\bullet$ is ordinary multiplication of a matrix by a real number, is $<\mathbf{V}, \oplus, \bullet>$ a vector space?

1.31. If $\mathbf{x}, \mathbf{y} \in \mathbf{R}^3$ and (*) is defined by $\mathbf{x} * \mathbf{y} = \sum_{k=1}^{3} x_k y_k$, prove that (*) has the properties of an inner product.

1.32. Two nonzero vectors $\mathbf{x}, \mathbf{y} \in \mathbf{R}^n$ are **orthogonal** if $\mathbf{x} * \mathbf{y} = 0$. In $\mathbf{R}^3$ let $\mathbf{x} = (1, 2, -1)$ and $\mathbf{y} = (-2, 0, 3)$. Find a vector $\mathbf{z}$ that has norm 1 and is orthogonal to both $\mathbf{x}$ and $\mathbf{y}$.

1.33. If $\mathbf{x}, \mathbf{y} \in \mathbf{R}^1$, prove that

(a) $||x| - |y|| \leq |x - y|$;

(b) $|xy| = |x||y|$.

1.34. Let $\mathbf{S} \subset \mathbf{R}$ consist of $x_1, x_2, \ldots, x_n$. Prove that $\left|\sum_{k=1}^{n} x_k\right| \leq \sum_{k=1}^{n} |x_k|$.

1.35. Here we consider the Cauchy-Schwarz Inequality, one of the properties of the Euclidean norm on $\mathbf{R}^n$.

(a) Show that the Inequality holds if either $\mathbf{x} = \mathbf{0}$ or $\mathbf{y} = \mathbf{0}$.

(b) If neither $\mathbf{x}$ nor $\mathbf{y}$ is $\mathbf{0}$, define the point $\mathbf{P} = \mathbf{x} \oplus c\mathbf{y}$, where $c \in \mathbf{R}$ is arbitrary. Why is $\mathbf{P} * \mathbf{P} > 0$? Expand this inner product, and write it as a quadratic equation in c.

---

[11] This requires the **Fundamental Theorem of Arithmetic**; see Andrews (1994) or Dence and Dence (1999), or ask your instructor.

(c) Since $P * P$ is greater than 0, what must be true about the discriminant of the quadratic? Deduce, from this, the Inequality.

1.36. Expand $||\mathbf{x} \oplus \mathbf{y}||^2, \mathbf{x}, \mathbf{y} \in \mathbf{R}^n$, make use of Exercise 1.35, and deduce the Triangle Inequality.

1.37. Suppose that for a fixed $\mathbf{p}$ and a fixed $\mathbf{a}$, $\mathbf{p} \in \mathbf{B}_n(\mathbf{a}; \varepsilon)$ for every $\varepsilon > 0$. Prove that $\mathbf{p} = \mathbf{a}$. Draw a picture for the one-dimensional case.

Section 1.6

1.38. Let $\mathbf{S} = \{1, 2, 3\}, \mathbf{T} = \{4, 5, 6, 7\}$, and $\mathbf{W} = \{8, 9, 10, 11, 12\}$. How many elements are in $\mathbf{S} \times \mathbf{T} \times \mathbf{W}$? How many subsets of $\mathbf{S} \times \mathbf{T} \times \mathbf{W}$ are there?

1.39. Definition. *Let $f\colon \mathbf{D}(f) \to \mathbf{S}$ be a mapping. If $\mathbf{H} \subseteq \mathbf{S}$, then the **inverse image** of $\mathbf{H}$ under f is the subset of $\mathbf{D}(f)$*

$$f^{-1}(\mathbf{H}) = \{x : (x, y) \in f,\ y \in \mathbf{H}\}.$$

(a) Construct an example of a function f and a proper subset $\mathbf{I} \subset \mathbf{D}(f)$ such that $f^{-1}(f(\mathbf{I})) = \mathbf{I}$.
(b) Prove that if $f\colon \mathbf{D}(f) \to \mathbf{S}$ is an injection and $\mathbf{I} \subseteq \mathbf{D}(f)$, then $f^{-1}(f(\mathbf{I})) = \mathbf{I}$.
(c) Construct another function f and a proper subset $\mathbf{I} \subset \mathbf{D}(f)$ such that $f^{-1}(f(\mathbf{I})) \neq \mathbf{I}$. In view of (b), what can you conclude?
(d) When $f^{-1}(f(\mathbf{I})) \neq \mathbf{I}$, is $f^{-1}(f(\mathbf{I})) \supset \mathbf{I}$ true or is $f^{-1}(f(\mathbf{I})) \subset \mathbf{I}$ true?

1.40. (a) Prove that if $f\colon \mathbf{D}(f) \to \mathbf{S}$ is a surjection and $\mathbf{H} \subseteq \mathbf{S}$, then $f(f^{-1}(\mathbf{H})) = \mathbf{H}$.
(b) Construct a function f and a proper subset $\mathbf{H} \subset \mathbf{S}$ such that $f(f^{-1}(\mathbf{H})) \neq \mathbf{H}$. In view of (a), what can you conclude?
(c) When $f(f^{-1}(\mathbf{H})) \neq \mathbf{H}$, is $f(f^{-1}(\mathbf{H})) \supset \mathbf{H}$ true or is $f(f^{-1}(\mathbf{H})) \subset \mathbf{H}$ true?

1.41. Suppose that $f\colon \mathbf{D}(f) \to \mathbf{S}$ is a function and that $\mathbf{I}, \mathbf{J} \subseteq \mathbf{D}(f)$. Prove that $f(\mathbf{I} \cup \mathbf{J}) = f(\mathbf{I}) \cup f(\mathbf{J})$, but that for intersections we can write only $f(\mathbf{I} \cap \mathbf{J}) \subseteq f(\mathbf{I}) \cap f(\mathbf{J})$. Give a specific example of this latter relation.

1.42. Recall the definition of the composition of two functions:

Definition. *Suppose that $f\colon \mathbf{D}(f) \to \mathbf{S}$ and $g\colon \mathbf{D}(g) \to \mathbf{T}$ are functions and that $\mathbf{R}(f) \subseteq \mathbf{D}(g)$. The **composition** of g on f is the new function $g[f]\colon \mathbf{D}(f) \to \mathbf{T}, g[f] = \{(x, y)\colon x \in \mathbf{D}(f), y \in \mathbf{T}, y = g[f(x)]\}$.*

In each case determine if neither, only one, or both of $g[f], f[g]$ make sense:
(a) $f = \{(x, y)\colon x \geq 0, y = x^2 - 2x + 1\}, g = \{(x, y)\colon x \geq 0, y = \sqrt{x}\}$.
(b) $f = \{(x, \mathbf{y})\colon -4 \leq x \leq 4, \mathbf{y} = (2x^2 - 1, e^{-x}/3)\}, g = \{(\mathbf{x}, y)\colon \mathbf{x} = (x_1, x_2), -1 \leq x_1, x_2 \leq 1, y = x_1 + x_2\}$.
(c) $f = \{(\mathbf{x}, \mathbf{y})\colon \mathbf{x} = (x_1, x_2), x_1, x_2 \geq 0, \mathbf{y} = (x_1 - 1, \sin x_2)\}, g = \{(\mathbf{x}, \mathbf{y})\colon \mathbf{x} = (x_1, x_2), x_1, x_2 \geq \frac{1}{2}, \mathbf{y} = (\sin x_1, \sin x_2)\}$.

1.43. Recall the definition of the inverse function of a given function:

Definition. *If $f\colon \mathbf{D}(f) \to \mathbf{S}$ is a function, then $f^{-1} = \{(y, x)\colon (x, y) \in f\}$ is the **inverse function** of f iff $(y, x_1) \in f^{-1}$ and $(y, x_2) \in f^{-1}$ imply $x_1 = x_2$.*
(a) Show that if $f\colon \mathbf{D}(f) \to \mathbf{S}$ is an injection, then the inverse function f^{-1} exists.
(b) Prove that for all $x \in \mathbf{D}(f)$ for the function of part (a), $f^{-1}[f(x)] = x$, and that for all $y \in \mathbf{R}(f) \subseteq \mathbf{S}, f[f^{-1}(y)] = y$.

(c) Show that if $f\colon \mathbf{D}(f) \to \mathbf{S}$ is a bijection of $\mathbf{D}(f)$ onto $\mathbf{S}$, then f^{-1} is a bijection of $\mathbf{S}$ onto $\mathbf{D}(f)$. Thus, we have a bijection *between* $\mathbf{D}(f)$ and $\mathbf{S}$.

Section 1.7

1.44. Which of the following are infinite sets?

(a) The set of all $q \in \mathbf{Q}^+$ such that $|q - \sqrt{5}| < 10^{-6}$;

(b) The set of all $n \in \mathbf{N}$ such that $n/\ln(n) > 100$;

(c) The set of all irrational numbers in (2, 4) whose base-10 expansions are repeating decimals;

(d) The set of all f such that at any $x \in [0, 1], f'(x) = 3x^2$;

(e) The set of all $n \in \mathbf{N}$ of the form $n = p_1p_2p_3$, where p_1, p_2, p_3 are distinct, positive, odd primes and $p_1 + p_2 + p_3 = 100,000,001$.

1.45. Prove that all of $\mathbf{Q}$ is a countably infinite set.

1.46. Write a short computer program to compile values of the Euler ϕ-function. Then use it to compute, in the algorithmic scheme of Theorem 1.9,

(a) the 1685th rational

(b) the natural number that corresponds to the rational 171/250.

1.47. During 1873 to 1874, and again during 1890 to 1891, Cantor addressed the issue of whether a bijection between $\mathbf{N}$ and *all* the real numbers in $[0, 1]$ could be defined. Let us agree to write each such real number as a decimal. We also stipulate that any nonzero number that has an infinite tail of 0's, such as 123/1000, be written in its equivalent form, like 0.1229999....

Suppose that to each real number in $[0, 1]$, a natural number k could be assigned. All the real numbers a_k in $[0, 1]$ can then be tabulated:

k	a_k
1	$0.a_{1,1}a_{1,2}a_{1,3}a_{1,4}\ldots$
2	$0.a_{2,1}a_{2,2}a_{2,3}a_{2,4}\ldots$
3	$0.a_{3,1}a_{3,2}a_{3,3}a_{3,4}\ldots$
$\vdots$	$\vdots$

Now define the following real number in $[0, 1]$:

$$N = 0.c_1c_2c_3c_4\ldots, \qquad \text{where } c_k = \begin{cases} 9 & a_{k,k} = 1 \\ 1 & a_{k,k} \neq 1. \end{cases}$$

What can you conclude about N, and what broader conclusion do we draw from this?

1.48. Members of the uncountably infinite set of irrational numbers fall into two geometrically significant subsets. On the one hand, if we are given a line segment 1 unit in length, it is impossible to classically construct a segment π or $\sqrt[3]{2}$ units in length. On the other hand, tell how we could construct, using only Euclidean tools (compass and unmarked straightedge), a line segment $\sqrt[4]{2}$ units long from one that is 1 unit in length.

Section 1.8

1.49. Identify all the cluster points of each set $\mathbf{S}$, if there are any:

(a) $\mathbf{S} = (0, 1)$;

(b) $\mathbf{S} = \{n : n \in \mathbf{N}, n > 1000\}$;

(c) $\mathbf{S} = \{(x, y) : 0 < x < 1, 0 < y < 1\}$;

(d) $\mathbf{S} = \{x : x = 1/n, n \in \mathbf{N}\}$;

(e) $\mathbf{S} = \{(n^2 - 1)/(n^2 + 2) : n \in \mathbf{N}\}$;

(f) $\mathbf{S} = \mathbf{Q} \cap (0, 1)$.

Classify each cluster point as an interior point or a boundary point.

1.50. Prove that $\mathbf{p}$ is a cluster point of $\mathbf{S} \subseteq \mathbf{R}^n$ iff any n-ball about $\mathbf{p}$ contains at least one point of $\mathbf{S}$ different from $\mathbf{p}$.

Section 1.9

1.51. Let $\mathbf{f}: \mathbf{R}^n \to \mathbf{R}^m$ be a mapping, and suppose that $\lim_{\mathbf{x} \to \mathbf{a}} \mathbf{f}$ exists. Prove that the limit is unique.

1.52. **(a)** Give an example of a function $f: D(f) \to \mathbf{R}^1, D(f) \subset \mathbf{R}^1$, and of a cluster point a not in $D(f)$, such that $\lim_{x \to a} f$ exists (in $\mathbf{R}^1$).

(b) Give examples of functions $\mathbf{f}: \mathbf{D(f)} \to \mathbf{R}^2, \mathbf{g}: \mathbf{D(g)} \to \mathbf{R}^2, \mathbf{D(f)}, \mathbf{D(g)} \subset \mathbf{R}^2$, and of a common cluster point $\mathbf{a}$, such that neither $\lim_{\mathbf{x} \to \mathbf{a}} \mathbf{f}$ nor $\lim_{\mathbf{x} \to \mathbf{a}} \mathbf{g}$ exists (in $\mathbf{R}^2$), but $\lim_{\mathbf{x} \to \mathbf{a}} (\mathbf{f} \oplus \mathbf{g})$ does exist (in $\mathbf{R}^2$).

1.53. **(a)** Prove part (i), and complete the proof of part (ii) in Theorem 1.10.

(b) Complete the proof of part (iii). The indicated hint in the outline of the proof will allow you to put a bound on $|f(\mathbf{x})|$.

(c) Complete the proof of part (iv).

1.54. Prove the other half of Theorem 1.11.

1.55. **(a)** Supply all the details in Example 1.14.

(b) Let $\mathbf{f}: \mathbf{R}^2 \to \mathbf{R}^2$, where $\mathbf{x} = (x_1, x_2)$ and $\mathbf{f}(\mathbf{x}) = (x_1(x_1 + x_2), x_2(x_2 - x_1))$. Determine $\lim_{\mathbf{x} \to (2, 1)} \mathbf{f}$, justifying each step.

1.56. **(a)** Write a plausible definition for what it means for a function $f: D(f) \to \mathbf{R}^1, D(f) \subseteq \mathbf{R}^1$ to have limit $-\infty$ (in **Re**) as $x \to a$.

(b) Prove that the function $f: D(f) \to \mathbf{R}^1$, defined by

$$f(x) = \frac{1}{1 - e^{|x|}}, \quad x \neq 0,$$

has limit $-\infty$ (in **Re**) as $x \to 0$.

1.57. **Definition.** *Let $f: D(f) \to \mathbf{R}^1, D(f) \subseteq \mathbf{R}^1$ be a function and let a be a cluster point of the set $\{x : x \in D(f), x > a\}$. Then $L \in \mathbf{R}^1$ is a* ***right-hand limit*** *of f at a and we write $\lim_{x \to a^+} f = L$ iff, given any $\varepsilon > 0$, there exists a $\delta > 0$ such that whenever $x \in D(f)$ and $0 < x - a < \delta$, then $|f(x) - L| < \varepsilon$.*

(a) Prove that if f has a right-hand limit at a, then it is unique.

(b) Prove that f has a right-hand limit at a if $\lim_{x \to a} f$ exists (in $\mathbf{R}^1$).

1.58. A **left-hand limit**, notated $\lim_{x \to a^-} f = L$, is analogous to a right-hand limit. Write out the formal definition. Give an example:

(a) of a function that has a left-hand limit but no right-hand limit (in $\mathbf{R}^1$).

(b) of a function that has both a right-hand limit and a left-hand limit, but $\lim_{x \to a} f$ does not exist (in $\mathbf{R}^1$).

1.59. Extensions of Theorem 1.10 to **Re** are possible.

(a) Let f: $\mathbf{D}(f) \to \mathbf{R}^1$ and g: $\mathbf{D}(g) \to \mathbf{R}^1$, $\mathbf{D}(f), \mathbf{D}(g) \subseteq \mathbf{R}^n$, and suppose that $\mathbf{a}$ is a cluster point of both $\mathbf{D}(f)$ and $\mathbf{D}(g)$. Further, suppose that $\lim_{\mathbf{x} \to \mathbf{a}} f = L \in \mathbf{R}^1$ and $\lim_{\mathbf{x} \to \mathbf{a}} g = \infty$ (in **Re**). Prove that $\lim_{\mathbf{x} \to \mathbf{a}}(f + g) = \lim_{\mathbf{x} \to \mathbf{a}} f + \lim_{\mathbf{x} \to \mathbf{a}} g$, if we make the definition that $y + \infty = \infty$ (in **Re**) for any $y > -\infty$.

(b) Let g be as in part (a), and suppose k is real and negative. Prove that $\lim_{\mathbf{x} \to \mathbf{a}} kg = k \lim_{\mathbf{x} \to \mathbf{a}} g$, if we make the definition that $k \cdot \infty = -\infty$ (in **Re**) for any real $k < 0$.

1.60. Determine the limits, if they exist, of the indicated functions, and justify your answers rigorously. Use any definitions and results that have been given.

(a) $\lim_{x \to 1^-} \sqrt{1 - x^3}$;

(b) $\lim_{x \to -3} \frac{|x^2-9|}{x+3}$;

(c) $\lim_{x \to 0} \frac{x^3}{x}$;

(d) $\lim_{x \to 2}\{(x-2)\cos[(2-x)^{-1}]\}$;

(e) $\lim_{x \to \infty} \left[\frac{3x}{x+2} - \frac{x^2}{x+2}\right]$;

(f) $\lim_{x \to 3^+} f(x)$, $f(x) = \left(2^{1/(3-x)}, \frac{1}{1+\left(\frac{1}{3-x}\right)^2}\right)$.

1.61. A function f: $\mathbf{R}^2 \to \mathbf{R}^1$ is defined by $\mathbf{x} = (x_1, x_2)$,

$$f(\mathbf{x}) = \begin{cases} \frac{x_1 x_2}{x_1^2 + x_2^2} & \mathbf{x} \neq 0 \\ 0 & \mathbf{x} = 0. \end{cases}$$

Explain how you know that $\lim_{\mathbf{x} \to \mathbf{0}} f$ does not exist.

1.62. Consult Landau (1966) and read a selection of pages in that classic work. Prepare a summary of what you have learned. Ask if your professor will give you extra credit for this.

1.63. Consult Halmos (1974), which like Landau (1966) is a very famous work. It reads something like a sequence of short, mathematical essays. Select two or three of them and prepare a summary of what you have learned.

REFERENCES

Cited Literature

Andrews, G.E., *Number Theory*, Dover Publications, NY, 1994, pp. 12–29. These pages give a thorough introduction to the Fundamental Theorem of Arithmetic and prior material, including the Euclidean algorithm.

Birkhoff, G. and MacLane, S., *A Survey of Modern Algebra*, rev. ed., Macmillan, NY, 1953, pp. 189–191. See the indicated pages in this classic for a brief discussion of abstract Euclidean vector spaces.

Cupillari, A., *The Nuts and Bolts of Proofs*, 3rd ed., Academic Press, Orlando, 2005. A short paperback on the spirit and mechanics of how to set up proofs.

Dence, J.B. and Dence, T.P., *Elements of the Theory of Numbers*, Harcourt/Academic Press, San Diego, 1999, pp. 39–46, 65–67, 243–244, 265–267. The first citation is to the Fundamental Theorem of Arithmetic, the second to the Euclidean algorithm, and the others are to aspects of the Euler ϕ-function.

Eves, H., *Foundations and Fundamental Concepts of Mathematics*, 3rd ed., PWS-Kent Pub. Co., Boston, 1990, pp. 257–262. Very interesting discussion of "other logics."

Halmos, P.R., *Naive Set Theory*, Springer-Verlag, NY, 1974, pp. 22–25, 59–61. Another mathematical classic; these pages discuss, insightfully, ordered pairs and the Axiom of Choice, respectively. Recommended.

Kantrowitz, R., and Neumann, M.M., "Yet Another Proof of Minkowski's Inequality," *Amer. Math. Monthly*, **115**, 445–447 (2008). This proof uses ideas of convex sets and concave functions. The paper could be used in senior seminar.

Kleiner, I., "Evolution of the Function Concept: A Brief Survey," *Coll. Math. J.*, **20**, 282–300 (1989). Well-written, as are all of Kleiner's articles.

Kline, M., *Mathematical Thought from Ancient to Modern Times*, Oxford University Press, NY, 1972, pp. 979–992; 992–1004. Well-written thumbnail summaries of the history of the rigorization of **R** and of Cantor's work on infinite sets.

Landau, E.G., *Foundations of Analysis*, 3rd ed., Chelsea Publishing Co., NY, 1966. In 134 beautiful pages Landau deduces the real and complex fields of numbers and many of their properties.

Lipschutz, S., *Set Theory and Related Topics*, 2nd ed., Schaum's Outline Series, McGraw-Hill, NY, 1998. Well-written general coverage of the theory of sets; recommended.

MacHale, D., "The Well-Ordering Principle for **N**," *Math. Gaz.*, **92**, 257–259 (2008). A nifty, short article that illustrates some situations where the Well-Ordering Principle is a more natural tool than Finite Induction.

Oman, G., "An Independent Axiom System for the Real Numbers," *Coll. Math. J.*, **40**, 78–86 (2009). This well-written article will be of interest to the reader who is attracted to the foundations of mathematics.

Solow, D., *How to Read and Do Proofs*, 4th ed., John Wiley & Sons, NY, 2004. A paperback (288 pp.) that is a must on the personal bookshelf of every serious student of mathematics.

Wilder, R.L., *Introduction to the Foundations of Mathematics*, 2nd ed., Krieger Publishing Co., Malabar, FL, 1983, pp. 23–53; 63–76. Contains discussions of the axiomatic method, simple ordering, and the Axiom of Choice.

Additional Literature

Dauben, J.W., *Georg Cantor: His Mathematics and Philosophy of the Infinite*, Princeton University Press, Princeton, 1990.

Devlin, K., "2003: Mathematicians Face Uncertainty," *Discover*, **25** (1), 36 (2004) (the changing face of mathematical proof).

Hardy, G.H., "Mathematical Proof," *Mind*, **38**, 1–25 (1929).

Hardy, G.H., *A Course of Pure Mathematics*, 10th ed., Cambridge University Press, Cambridge, 1967, pp. 3–19, 28–32 (Dedekind cuts).

Kac, M. and Ulam, S.M., *Mathematics and Logic*, Dover Publications, NY, 1992.

Kemeny, J.G., Snell, J.L. and Thompson, G.L., *Introduction to Finite Mathematics*, Prentice-Hall, Englewood Cliffs, NJ, 1957, pp. 1–78. (logic; sets and subsets).

Maligranda, L., "Simple Norm Inequalities," *Amer. Math. Monthly*, **113**, 256–260 (2006).

Smith, W.K., *Inverse Functions*, Macmillan Co., NY, 1966.

Smith, D., Eggen, M. and St. Andre, R., *A Transition to Advanced Mathematics*, 3rd ed., Brooks/Cole Publishing Co., Pacific Grove, CA, 1990, pp. 135–164; 165–193 (functions; cardinality).

Sominsky, I.S., *The Method of Mathematical Induction*, Mir Publishers, Moscow, 1975.

Wang, X., "Volumes of Generalized Unit Balls," *Math. Mag.*, **78**, 390–395 (2005).

CHAPTER 2

Sequences

"Of the two concepts *sequence* and *series*, the former is the simpler and more primitive one."
Konrad Knopp

CONTENTS

Reviewed in this chapter	Elementary features of sequences; subsequences.
New in this chapter	Convergence of sequences; Bolzano-Weierstrass Theorem and cluster points; limit superior (inferior); Cauchy sequences and complete metric spaces.

2.1 GENERAL PROPERTIES OF SEQUENCES

Sequences of real numbers occur everywhere in mathematics, and you are already intuitively familiar with them. They are especially fundamental in any study of calculus.

Definition. *A **sequence** is a function f in which $\mathbf{D}(f)$ is $\mathbf{N}$; it is conventionally indicated by $\{x_k\}_{k=1}^{\infty}$. Any x_k is called a **term** of the sequence, or when x_k is a number, a **value** of the sequence.*

The symbolism $\{x_k\}_{k=1}^{\infty}$, which indicates that $\mathbf{D}(f)$ and the set of terms are countably infinite sets, is incomplete until the manner in which x_k is determined from k has been indicated. Simply listing the first few terms will not unambiguously specify the sequence. Finally, note that sequences can be sequences of any sort of mathematical object.

■ Example 2.1

The set $\mathbf{S}$ in Example 1.13 is a set of terms of a sequence. ■

■ Example 2.2

We can't say what is the next term in "the sequence" $\{2, 4, 8, 16, 32, \ldots\}$. If x_k is given, however, by the formula

$$x_k = (k^4 - 6k^3 + 23k^2 - 18k + 24)/12,$$

then the next term is not 64 (Exercise 2.1). The integer sequence here is representative of more than 80,000 such sequences accumulated in a database by (Sloane, N.J.A., 2005). We urge you to consult this fascinating material. ■

■ Example 2.3

The sequence of unnormalized **Hermite polynomials**,[1] $\{H_k(x)\}_{k=1}^{\infty}$, important in physics, is defined by

$$H_k(x) = (-1)^k e^{x^2} D^{(k)}(e^{-x^2}),$$

where $D^{(k)}(f)$ denotes the kth derivative of f. We find $H_1(x) = 2x$, $H_2(x) = 4x^2 - 2$, $H_3(x) = 8x^3 - 12x$, $H_4(x) = 16x^4 - 48x^2 + 12$ (verify!). Some properties of $\{H_k(x)\}_{k=1}^{\infty}$ are dealt with in Exercise 2.2. ■

Two important concepts that are applicable to many kinds of sequences are boundedness (in any metric space) and subsequence (of any sequence, whatsoever). We let $< \mathbf{M}, d >$ denote an arbitrary metric space (Section 1.5).

---

[1] After the French mathematician **Charles Hermite** (1822–1901), famous for having proved in 1873 that e is transcendental.

Definition. *Let* $\mathbf{a} \in \mathbf{M}$ *be a fixed point in some metric space* $< \mathbf{M}, d >$. *A sequence defined in* $\mathbf{M}$ *is said to be* **bounded** *iff there is a real number* $r > 0$ *such that all terms of the sequence lie inside the ball* $\mathbf{B}(\mathbf{a}; r) = \{\mathbf{p} \in \mathbf{M} : d(\mathbf{p}, \mathbf{a}) < r\}$.

This definition generalizes that given at the start of Section 1.3. A common case involves sequences in $\mathbf{R}^n$. Let $\mathbf{x}_k = (x_{k1}, x_{k2}, \ldots, x_{kn})$ denote a general term of a sequence in $\mathbf{R}^n$, and choose arbitrarily (but, commonly) the fixed point $\mathbf{a} = \mathbf{0} = (0, 0, \ldots, 0)$. Then since we use the Euclidean metric on $\mathbf{R}^n$, the sequence $\{\mathbf{x}_k\}_{k=1}^{\infty}$ will be bounded in $\mathbf{R}^n$ iff there is some real $r > 0$ such that $\left[\sum_{i=1}^{n} x_{ki}^2\right]^{1/2} < r$ for all $\mathbf{x}_k$. It follows fairly directly (Exercise 2.4) that:

Theorem 2.1. *A sequence* $\{\mathbf{x}_k\}_{k=1}^{\infty}$ *in* $\mathbf{R}^n$ *is bounded iff the sequence* $\{x_{ki}\}_{k=1}^{\infty}$, $i = 1, 2, \ldots, n$, *of values of each coordinate in* $\mathbf{R}^n$ *is bounded in* $\mathbf{R}^1$.

Proof. The proof is left to you. Note that there are two separate theorems to prove here, since the proposition is an iff-statement. ■

■ Example 2.4

The sequence in Example 1.13 is bounded, in view of Theorem 2.1. ■

■ Example 2.5

The sequence $\{x_k\}_{k=1}^{\infty}$ where $x_k = k\,[1 + (-1)^k]$, is bounded from below in $\mathbf{R}^1$, but is not bounded from above.

The very important sequence $\{y_k\}_{k=1}^{\infty}$ where $y_k = (1 + k^{-1})^k$, is a bounded sequence in $\mathbf{R}^1$, although this is not obvious. An upper bound is 3; more on this in Exercise 2.6 and Example 2.7. ■

Turning now to subsequences, we see that the idea is a simple one.

Definition. *Let* $k_1 < k_2 < k_3 < \ldots$ *be an arbitrary increasing sequence of natural numbers. Then* $\{\mathbf{x}_{k_n}\}_{n=1}^{\infty}$ *is called a* **subsequence** *of* $\{\mathbf{x}_k\}_{k=1}^{\infty}$.

A subsequence is a choice of an infinite subset of the terms of a given sequence. Thus, if in the first sequence of Example 2.5 only the even-indexed terms are chosen, then the following subsequence is obtained:

$$\{x_{k_n}\}_{n=1}^{\infty} = \{x_{2n}\}_{n=1}^{\infty} = \{4n\}_{n=1}^{\infty} = \{4, 8, 12, 16, \ldots\}.$$

Whereas many sequences $\{x_k\}_{k=1}^{\infty}$ are presented initially by an indication of how x_k is determined from k, many other sequences are defined **recursively**, that is, the kth term is obtained from one or more prior terms according to some prescription. An example is the well-known **Fibonacci sequence**, $\{F_k\}_{k=1}^{\infty}$, where

$$F_k = \begin{cases} 1 & k = 1, 2 \\ F_{k-1} + F_{k-2} & k > 2. \end{cases}$$

This innocent-looking sequence has many intriguing properties (Hoggatt, Jr., 1972; Vajda, 1989).

Sequences of real numbers, such as $\{F_k\}_{k=1}^{\infty}$, may possess one feature that other classes of sequences cannot possess. A sequence $\{x_k\}_{k=1}^{\infty}$ of real numbers is **increasing (decreasing)** iff for each $k \in \mathbf{N}$ we have $x_k \leq x_{k+1}$ $(x_k \geq x_{k+1})$. Many sequences, in their entirety, are neither increasing nor decreasing, but become so for all n beyond a certain value. Other sequences never become increasing or decreasing. A sequence that is (or becomes) increasing or decreasing is said to **be** (or to **become**) **monotonic**. The feature of being (becoming) monotonic is clearly a consequence of Axiom R7 (Section 1.2).

■ Example 2.6

The sequence $\{F_n\}_{n=1}^{\infty}$ of Fibonacci numbers is a monotonic sequence.

The sequence $\{x_n\}_{n=1}^{\infty}$, where $x_n = 4^n/n!$, becomes a decreasing sequence for all $n \geq 3$ because

$$\frac{x_{n+1}}{x_n} = \frac{4^{n+1}/(n+1)!}{4^n/n!} = \frac{4}{n+1} \leq 1 \quad \text{if } n \geq 3.$$

■

■ Example 2.7

The sequence $\{y_k\}_{k=1}^{\infty}$ in Example 2.5 is an increasing sequence. We have, for $k > 1$,

$$\begin{aligned}\frac{y_{k+1}}{y_k} &= \frac{[1+(k+1)^{-1}]^{k+1}}{(1+k^{-1})^k} = \left[\frac{1+(k+1)^{-1}}{1+k^{-1}}\right]^k [1+(k+1)^{-1}] \\ &= \left[1 - \frac{1}{(k+1)^2}\right]^k \left(1 + \frac{1}{k+1}\right) \\ &> \left(1 - \frac{k}{(k+1)^2}\right)\left(1 + \frac{1}{k+1}\right),\end{aligned}$$

from Bernoulli's Inequality (Theorem 1.7). Multiplication of the binomials then yields (verify!)

$$\frac{y_{k+1}}{y_k} > 1 + \frac{1}{(k+1)^3} > 1.$$

■

■ Example 2.8

The sequence of ratios of consecutive Fibonacci numbers $\{F_{n+1}/F_n\}_{n=1}^{\infty}$ never becomes increasing or decreasing. It oscillates, forever (Exercise 2.8). ■

2.2 CONVERGENCE OF SEQUENCES

In order to ascertain deeper properties of sequences, we should investigate their limits (as hinted in Figure 1.9). We adapt the definition of $\lim_{x\to\infty} f(x)$ that was given in Section 1.9 immediately after Example 1.14. A convenient class of sequences with which to begin are those defined in $\mathbf{R}^1$.

Definition. *A sequence* $\{x_n\}_{n=1}^{\infty}$ *in* $\mathbf{R}^1$ ***converges*** *to* $L \in \mathbf{R}^1$ *iff, given any* $\varepsilon > 0$, *there exists an* $N \in \mathbf{N}$ *such that for all integers* $n > N$, $x_n \in \mathbf{B}(L;\varepsilon)$.

A crucial part of the definition is the requirement of *all* integers $n > N$. If $\{x_n\}_{n=1}^{\infty}$ converges to L, then L is termed the **limit** of the sequence. However, if the limit is a point at infinity (in **Re**), then use of the word "limit" is permitted, but the sequence is not said to converge. A sequence that does not converge is said to **diverge** (or be **divergent**). Also note that the definition carries the usual burden of providing no assistance on how to ascertain L. The determination of L may often be quite difficult.

Theorem 2.2. *A bounded, monotonic sequence* $\{x_n\}_{n=1}^{\infty}$ *in* $\mathbf{R}^1$ *that does not become constant converges.*

Proof. Suppose that $\{x_n\}_{n=1}^{\infty}$ is increasing, and let $\mathbf{S}$ be the set of terms, $\{x_1, x_2, x_3, \ldots\}$. Then by the Axiom of Completeness (Axiom R8; Section 1.3), $U = \sup \mathbf{S}$ exists. Let $\varepsilon > 0$ be given; then from Exercise 1.24 there is a term x_N such that $U - \varepsilon < x_N < U + \varepsilon$. As the sequence is increasing, then $U - \varepsilon < x_n < U + \varepsilon$ for all $n \geq N$, that is, $x_n \in \mathbf{B}(U;\varepsilon)$. This says that $\{x_n\}_{n=1}^{\infty}$ converges to U. The proof is analogous if the sequence is decreasing. ∎

We actually have proved a bit more than was required, namely, that for a bounded, increasing sequence, $\sup \mathbf{S}$ has the properties of a limit. This is not true, in general; that is, the limit of a sequence in $\mathbf{R}^1$ (if it has one) need not equal the supremum of the set of terms.

■ Example 2.9

The sequence $\{y_k\}_{k=1}^{\infty}, y_k = (1 + k^{-1})^k$, converges, in view of Exercise 2.6, Example 2.7, and Theorem 2.2. In fact, $\lim_{k\to\infty} y_k = e \approx 2.71828$. ∎

■ Example 2.10

The monotonic sequence $\{x_n\}_{n=1}^{\infty}, x_n = a^n, a > 1$, in $\mathbf{R}^1$ diverges to ∞. For any $n \in \mathbf{N}, x_{n+1}/x_n = a^{n+1}/a^n = a > 1$, so the sequence is increasing. Let $r > 1$ be given; if $a \geq r$, then for any $n \in \mathbf{N}, a^n > a \geq r$ and by the definition in Section 1.9 this says that $\lim_{n\to\infty} a^n = \infty$. If $r > a$, then $\ln r > \ln a$ and by the Archimedean Property (Theorem 1.5) there is an $N \in \mathbf{N}$ such that $N \ln a > \ln r$ or, equivalently, $a^N > r$. Hence, $a^n > r$ for all $n \geq N$ and this again says $\lim_{n\to\infty} a^n = \infty$. ∎

2.3 GENERAL THEOREMS ON CONVERGENT SEQUENCES

Convergent sequences possess several nice properties. We restrict our discussion to sequences in $\mathbf{R}^m$, although some of the properties apply (as stated) even more generally. The definition of convergence in Section 2.2 carries over automatically to sequences in $\mathbf{R}^m$, if $\mathbf{B}(L;\varepsilon)$ is replaced by $\mathbf{B}_m(\mathbf{L};\varepsilon)$.

Theorem 2.3. *If a sequence $\{x_n\}_{n=1}^{\infty}$ in $\mathbf{R}^m$ converges, then the limit is unique.*

Proof. Suppose that $\mathbf{x}, \mathbf{x}'$ are two distinct limits; let $\varepsilon = \|\mathbf{x} - \mathbf{x}'\| > 0$ and use the Triangle Inequality (Section 1.5). The completion of the proof is left to you. ■

Theorem 2.4. *If a sequence $\{\mathbf{x}_n\}_{n=1}^{\infty}$ in $\mathbf{R}^m$ is convergent, then it is bounded.*

Proof. Let $\varepsilon > 0$ be given, denote $\lim_{n\to\infty} \mathbf{x}_n = \mathbf{x} = (x_1, x_2, \ldots, x_m)$, and let $\mathbf{a} = (a_1, a_2, \ldots, a_m)$ be an arbitrary point in $\mathbf{R}^m$. Denote $D = \max\{|x_1 - a_1|, |x_2 - a_2|, \ldots, |x_m - a_m|\}$; then from the Triangle Inequality we have for any $n \in \mathbf{N}$,

$$||\mathbf{x}_n - \mathbf{a}|| = ||\{\mathbf{x}_n - \mathbf{x}\} \oplus \{\mathbf{x} - \mathbf{a}\}|| \le ||\mathbf{x}_n - \mathbf{x}|| + ||\mathbf{x} - \mathbf{a}||.$$

There is an $N \in \mathbf{N}$ such that for all $n > N$, $||\mathbf{x}_n - \mathbf{x}|| < 1$ (since the sequence is convergent). Hence, for all $n > N$,

$$||\mathbf{x}_n - \mathbf{a}|| < 1 + ||\mathbf{x} - \mathbf{a}|| = 1 + \left[\sum_{i=1}^{m} (x_i - a_i)^2\right]^{1/2} \le 1 + \sqrt{m}\, D.$$

If $M = \max\{||\mathbf{x}_1 - \mathbf{a}||, ||\mathbf{x}_2 - \mathbf{a}||, \ldots, ||\mathbf{x}_N - \mathbf{a}||\}$, then for all $n \in \mathbf{N}$, $||\mathbf{x}_n - \mathbf{a}|| \le \max\{1 + \sqrt{m}\, D, M\}$. This says that $\{\mathbf{x}_n\}_{n=1}^{\infty}$ is bounded. ■

Theorem 2.5. *Let the general term of a sequence $\{\mathbf{x}_n\}_{n=1}^{\infty}$ in $\mathbf{R}^m$ be denoted $\mathbf{x}_n = (x_{n1}, x_{n2}, \ldots, x_{nm})$. Then $\{\mathbf{x}_n\}_{n=1}^{\infty}$ converges to $\mathbf{L} = (L_1, L_2, \ldots, L_m) \in \mathbf{R}^m$ iff each component sequence $\{x_{nk}\}_{n=1}^{\infty}$, $k = 1, 2, \ldots, m$, converges to $L_k \in \mathbf{R}^1$.*

Proof. ($\rightarrow$) Suppose that $\{\mathbf{x}_n\}_{n=1}^{\infty}$ converges to $\mathbf{L} \in \mathbf{R}^m$. Then, given any $\varepsilon > 0$, there is an $N \in \mathbf{N}$ such that for all $n > N$, $\mathbf{x}_n \in \mathbf{B}_m(\mathbf{L};\varepsilon)$, that is,

$$||\mathbf{x}_n \oplus (-\mathbf{L})|| = \left[\sum_{j=1}^{m} (x_{nj} - L_j)^2\right]^{1/2} < \varepsilon.$$

Choose an arbitrary natural number $k \in [1, m]$; then

$$(x_{nk} - L_k)^2 \le \sum_{j=1}^{m} (x_{nj} - L_j)^2 < \varepsilon^2,$$

so $|x_{nk} - L_k| < \varepsilon$ for all $n > N$. Thus, $\{x_{nk}\}_{n=1}^{\infty}$ converges to $L_k \in \mathbf{R}^1$. The other direction of the proof is left to you. ■

Theorem 2.6. *If in* $\mathbf{R}^m$, $\lim_{n\to\infty} \mathbf{x}_n = \mathbf{x}$ *and* $\lim_{n\to\infty} \mathbf{y}_n = \mathbf{y}$, *and* $k \in \mathbf{R}^1$, *then*

(i) $\lim_{n\to\infty} (k \cdot \mathbf{x}_n) = k \cdot \mathbf{x}$;

(ii) $\lim_{n\to\infty} (\mathbf{x}_n \oplus \mathbf{y}_n) = \mathbf{x} \oplus \mathbf{y}$.

Further, if $m = 1$, *then*

(iii) $\lim_{n\to\infty} (x_n y_n) = xy$;

(iv) *if* $y_n \neq 0$ *for all n and* $y \neq 0$, *then* $\lim_{n\to\infty} (x_n/y_n) = x/y$.

Proof. Pattern your work after the proof of Theorem 1.10. The completion of the proof is left to you. ■

Theorems 2.3 through 2.6 have the look of familiarity, based upon our prior work with functions, in general. The next two theorems are not quite so obvious. In order to prepare for them, we require two short lemmas, which are interesting in their own right. The first lemma, in particular, is an analog of the result in Exercise 1.33(a).

Lemma 2.3.1.[2] *If* $\mathbf{x}, \mathbf{z} \in \mathbf{R}^m$, *then* $|\, ||\mathbf{x}|| - ||\mathbf{z}||\, | \leq ||\mathbf{x} - \mathbf{z}||$.

Proof. Let $\mathbf{x}, \mathbf{y}, \mathbf{z}, \mathbf{w} \in \mathbf{R}^m$; then from the Triangle Inequality,

$$||\mathbf{x} - \mathbf{y}|| = ||\{\mathbf{x} - \mathbf{z}\} \oplus \{\mathbf{z} - \mathbf{w}\} \oplus \{\mathbf{w} - \mathbf{y}\}||$$
$$\leq ||\mathbf{x} - \mathbf{z}|| + ||\mathbf{z} - \mathbf{w}|| + ||\mathbf{w} - \mathbf{y}||$$

or

$$||\mathbf{x} - \mathbf{y}|| - ||\mathbf{z} - \mathbf{w}|| \leq ||\mathbf{x} - \mathbf{z}|| + ||\mathbf{w} - \mathbf{y}||. \tag{*}$$

Throughout (*) interchange **y** with **w** and **x** with **z**:

$$||\mathbf{z} - \mathbf{w}|| - ||\mathbf{x} - \mathbf{y}|| \leq ||\mathbf{z} - \mathbf{x}|| + ||\mathbf{y} - \mathbf{w}||. \tag{**}$$

The right-hand sides of (*) and (**) are identical, but the left-hand side of (*) is the negative of that of (**). Hence, the two inequalities are equivalent to $|\, ||\mathbf{x} - \mathbf{y}|| - ||\mathbf{z} - \mathbf{w}||\, | \leq ||\mathbf{x} - \mathbf{z}|| + ||\mathbf{w} - \mathbf{y}||$.[3] The Lemma follows by setting $\mathbf{y} = \mathbf{w} = 0$. ■

Lemma 2.3.2. *If a, b are fixed real numbers where for all* $\varepsilon > 0, b \leq a + \varepsilon$, *then* $b \leq a$.

---

[2]Lemmas in the book are enumerated as follows:
LEMMA (chapter no.).(section no.).(lemma no. within the section).

[3]When viewed as a statement about distances in a metric space, this has been dubbed the **Quadrilateral Inequality** (Shilov, 1996).

Proof. Suppose, to the contrary, that $b > a$. Then choose $\varepsilon = \frac{b-a}{2}$. The completion of the proof is left to you. ■

Theorem 2.7. *If* $\{\mathbf{x}_n\}_{n=1}^{\infty}$ *is a sequence in* $\mathbf{R}^m$ *and* $\lim_{n\to\infty} \mathbf{x}_n = \mathbf{x} \in \mathbf{R}^m$, *then* $\lim_{n\to\infty} ||\mathbf{x}_n|| = ||\mathbf{x}||$.

Proof. The result follows immediately from Lemma 2.3.1. The completion of the proof is left to you. ■

Theorem 2.8 (Comparison Theorem). *If* $\{\mathbf{x}_n\}_{n=1}^{\infty}, \{\mathbf{y}_n\}_{n=1}^{\infty}$ *converge to* $\mathbf{x}, \mathbf{y}$, *respectively, in* $\mathbf{R}^m$, *and if for all natural numbers n larger than some natural number* N_1 *we have* $||\mathbf{x}_n|| \geq ||\mathbf{y}_n||$, *then* $||\mathbf{x}|| \geq ||\mathbf{y}||$.

Proof. Let $\varepsilon > 0$ be given. Then from Theorem 2.7 there is an $N_2 \in \mathbf{N}$ such that for all $n > N_2$

$$||\mathbf{x}_n|| \leq ||\mathbf{x}|| + \frac{\varepsilon}{2}. \qquad (*)$$

Similarly, there is an $N_3 \in \mathbf{N}$ such that for all $n > N_3$

$$|\,||\mathbf{y}_n|| - ||\mathbf{y}||\,| \leq \frac{\varepsilon}{2},$$

or equivalently,

$$-||\mathbf{y}_n|| \leq -||\mathbf{y}|| + \frac{\varepsilon}{2}. \qquad (**)$$

Let $N = \max\{N_1, N_2, N_3\}$; then combination of $(*)$ and $(**)$ with the hypothesis yields, for all $n > N$,

$$0 \leq ||\mathbf{x}_n|| - ||\mathbf{y}_n|| \leq ||\mathbf{x}|| - ||\mathbf{y}|| + \varepsilon,$$

or

$$||\mathbf{y}|| \leq ||\mathbf{x}|| + \varepsilon$$

for all $\varepsilon > 0$. The theorem then follows from Lemma 2.3.2. ■

■ Example 2.11

In $\mathbf{R}^2$ the sequence $\{\mathbf{x}_n\}_{n=1}^{\infty}$ is defined by $\mathbf{x}_n = (x_{n1}, x_{n2}) = (1/n, 2 - 4^{-n/3})$. We have $\lim_{n\to\infty} x_{n1} = 0$ and $\lim_{n\to\infty} x_{n2} = 2$; Theorem 2.5 then says that $\{\mathbf{x}_n\}_{n=1}^{\infty}$ converges in $\mathbf{R}^2$ to $\mathbf{x} = (0, 2)$. ■

■ Example 2.12

In $\mathbf{R}^1$ the sequence $\{\mathbf{x}_n\}_{n=1}^{\infty}, x_n = \cos n$, is certainly bounded. We cannot conclude from Theorem 2.4 that the sequence converges because the converse of Theorem 2.4 is not true, in general. ■

■ Example 2.13

(a) $\lim_{n\to\infty} \frac{1}{n} = 0$ (Exercise 2.24(a)).

(b) If $\{\mathbf{x}_n\}_{n=1}^{\infty}$, $\{\mathbf{y}_n\}_{n=1}^{\infty}$ agree at all natural numbers n and if $\lim_{n\to\infty} \mathbf{x}_n = \mathbf{x}$, then $\lim_{n\to\infty} \mathbf{y}_n = \mathbf{x}$ (Exercise 2.24(a)).

(c) In $\mathbf{R}^1$ the sequence $\{y_n\}_{n=1}^{\infty}$ is defined by $y_n = \frac{n^2+n-1}{2n^2-n+3}$. Then define the sequence $\{x_n\}_{n=1}^{\infty}$, where $x_n = \frac{1+n^{-1}-n^{-2}}{2-n^{-1}+3n^{-2}} = \frac{f_n}{g_n}$. From part (a) and Theorem 2.6 (i, ii, iii), we have $\lim_{n\to\infty} f_n = 1$ and $\lim_{n\to\infty} g_n = 2$. From Theorem 2.6 (iv), $\lim_{n\to\infty} x_n = 1/2$, and finally, from part (b), $\lim_{n\to\infty} y_n = 1/2$. ■

■ Example 2.14

For the sequence in Example 2.11, we have

$$||\mathbf{x}_n|| = (\mathbf{x}_n * \mathbf{x}_n)^{1/2} = \left[\frac{1}{n^2} + 4 - 4^{1-(n/3)} + 4^{-2n/3}\right]^{1/2}.$$

By Theorem 2.7 we obtain

$$\lim_{n\to\infty} ||\mathbf{x}_n|| = ||\mathbf{x}|| = ||(0,2)|| = [0^2 + 2^2]^{1/2} = 2.$$

■

2.4 CLUSTER POINTS OF SEQUENCES

The definition of a cluster point[4] of a set that was given in Section 1.8 is easily adapted for use with sequences.

Definition. *A* ***cluster point*** *of a sequence* $\{\mathbf{x}_n\}_{n=1}^{\infty}$ *in* $\mathbf{R}^m$ *is a point* $\mathbf{p} \in \mathbf{R}^m$ *iff, given any* $\varepsilon > 0$, *there are infinitely many* $n \in \mathbf{N}$ *such that* $\mathbf{x}_n \in \mathbf{B}_m(\mathbf{p}; \varepsilon)$.

Pictorially, we can say that infinitely many terms of a sequence *cluster* (arbitrarily closely) about a cluster point. Note, however, the difference between a limit and a cluster point. If $\mathbf{x}$ is the limit of $\{\mathbf{x}_n\}_{n=1}^{\infty}$, then *all* terms with n larger than some N approach $\mathbf{x}$ arbitrarily closely; if $\mathbf{p}$ is merely a cluster point of $\{\mathbf{x}_n\}_{n=1}^{\infty}$, then "only" infinitely many terms approach $\mathbf{p}$ arbitrarily closely.

■ Example 2.15

The sequence $\{x_n\}_{n=1}^{\infty}$ defined in $\mathbf{R}^1$ by

$$x_n = \begin{cases} 1 - \frac{1}{n} & n = 2k \\ -1 + \frac{1}{n} & n = 2k-1 \end{cases}$$

has two cluster points, namely, 1 and -1, but no limit.

---

[4]Commonly employed synonyms are **limit point** and **accumulation point**.

The sequence $\{x_n\}_{n=1}^{\infty}$ in $\mathbf{R}^1$ defined by $x_n = 4/n^2$ has one cluster point and a limit, both at $x = 0$.

The sequence $\{\mathbf{x}_n\}_{n=1}^{\infty}$ in $\mathbf{R}^2$ defined by $\mathbf{x}_n = (1, n)$ has neither a cluster point nor a limit in $\mathbf{R}^2$. ■

Cluster point and boundedness of a sequence intersect in a famous theorem. We will state the higher-dimensional Euclidean version of it in Theorem 2.10; the standard version, usually given for sequences in $\mathbf{R}^1$, is a special case. For this latter, we shall use the imagery of geometry. We shall say that a sequence of intervals $\{\mathbf{I}_n\}_{n=1}^{\infty}$ along $\mathbf{R}^1$ is called a **nested sequence** iff for each $n \in \mathbf{N}$, $\mathbf{I}_{n+1} \subseteq \mathbf{I}_n$. When each interval is a closed, bounded interval, $\mathbf{I}_n = [a_n, b_n]$, $a_n < b_n$, then the nesting implies that for each $n \in \mathbf{N}$, $a_n \le a_{n+1} < b_{n+1} \le b_n$.

Theorem 2.9 (Nested Intervals Theorem). *Let $\{\mathbf{I}_n\}_{n=1}^{\infty}$ be a nested sequence of closed, bounded intervals, $\mathbf{I}_n = [a_n, b_n]$. Then there exists at least one number M common to each interval, that is, $\bigcap_{n=1}^{\infty} \mathbf{I}_n \neq \emptyset$.*

Proof. For each $m, n \in \mathbf{N}$ we have $a_m < b_n$ because (a) if $m > n$, then $a_m < b_m < b_n$ holds, and (b) if $m \le n$, then $a_m \le a_n < b_n$ holds. Let $\mathbf{S} = \{a_1, a_2, \ldots\}$ be the set of left-hand endpoints; $\mathbf{S}$ is bounded above by b_1. Hence, by the Completeness Axiom, $M = \sup \mathbf{S}$ exists (in $\mathbf{R}^1$), so for each $n \in \mathbf{N}$, $a_n \le M$. Further, for each $n \in \mathbf{N}$, $M \le b_n$ because M is the *least* upper bound of $\mathbf{S}$. Hence, for each $n \in \mathbf{N}$, $a_n \le M \le b_n$. ■

Corollary 2.9.1.[5] *If in the sequence $\{\mathbf{I}_n\}_{n=1}^{\infty}$ we have $\lim_{n\to\infty}(b_n - a_n) = 0$, then $\bigcap_{n=1}^{\infty} \mathbf{I}_n$ reduces to a singleton set.*

Proof. By Theorem 2.9, $\bigcap_{n=1}^{\infty} \mathbf{I}_n$ is nonempty. Suppose that $M_1, M_2 \in \bigcap_{n=1}^{\infty} \mathbf{I}_n$, $M_2 - M_1 = \delta > 0$. Let $\varepsilon = \delta/2$; by hypothesis there is a natural number N such that $b_N - a_N < \varepsilon$. Then

$$\begin{aligned} \delta/2 > b_N - a_N &= (b_N - M_2) + (M_2 - M_1) + (M_1 - a_N) \\ &= (b_N - M_2) + \delta + (M_1 - a_N) > \delta, \end{aligned}$$

a contradiction. Hence, only one of M_1, M_2 is in $\bigcap_{n=1}^{\infty} \mathbf{I}_n$, and it is the M of Theorem 2.9. ■

---

[5] Corollaries in the book are enumerated as follows:
COROLLARY (theorem no.).(corollary no. of the theorem).

■ Example 2.16

A sequence of intervals $\{\mathbf{I}_n\}_{n=1}^{\infty}$ is defined by

$$\mathbf{I}_n = \begin{cases} [x_{n+1}, x_n] & n = 2k-1 \\ [x_n, x_{n+1}] & n = 2k, \end{cases}$$

where

$$x_n = \begin{cases} 3 & n = 1 \\ \frac{x_{n-1}+2}{x_{n-1}+1} & n > 1. \end{cases}$$

The first five intervals are found to be (verify!)

$$\mathbf{I}_1 = [1.25, 3] \qquad \mathbf{I}_2 = [1.25, 1.\overline{4}] \qquad \mathbf{I}_3 = [1.4\overline{09}, 1.\overline{4}]$$

$$\mathbf{I}_4 = [1.4\overline{09}, 1.4150943] \qquad \mathbf{I}_5 = [1.4140625, 1.4150943],$$

and are shown (greatly exaggerated) in Figure 2.1.

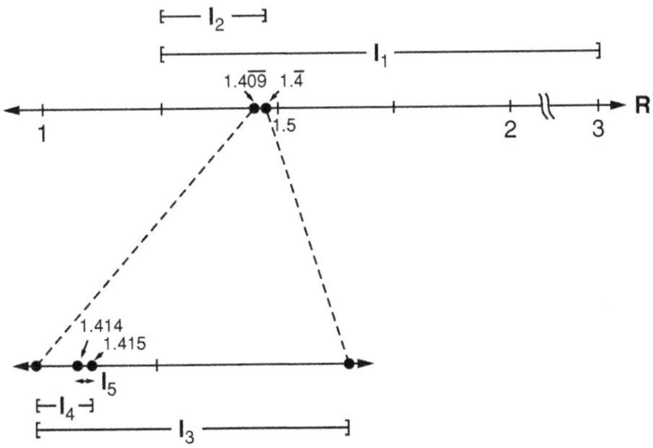

FIGURE 2.1
A nested sequence of intervals?

The widths of the intervals $\mathbf{I}_n$ approach 0 (Exercise 2.30), so by Corollary 2.9.1 there is a unique number common to all the $\mathbf{I}_n$'s. ■

We generalize Theorem 2.9 and Corollary 2.9.1 to $\mathbf{R}^m$. Let $\mathbf{a} = (a_1, a_2, \ldots, a_m) \in \mathbf{R}^m$ and let $r_1, r_2, \ldots, r_m > 0$. The set of all points $\mathbf{x} = (x_1, x_2, \ldots, x_m) \in \mathbf{R}^m$ that satisfy $|x_k - a_k| \le r_k, k = 1, 2, \ldots, m$, is a closed box $\mathfrak{B}$ about $\mathbf{a}$ of dimensions $2r_1 \times 2r_2 \times \cdots \times 2r_m$ (Fig. 2.2). It has $2m$ faces, which are the hyperplanes $0 \le x_k - a_k \le r_k, -r_k \le x_k - a_k \le 0$ parallel to the coordinate hyperplanes.

1. The box $\mathfrak{B}$ is bounded, since if $r = \max\{r_1, r_2, \ldots, r_m\}$, then the m-ball $\mathbf{B}_m(\mathbf{a}; \sqrt{m}r)$ contains the box (Figure 2.2).

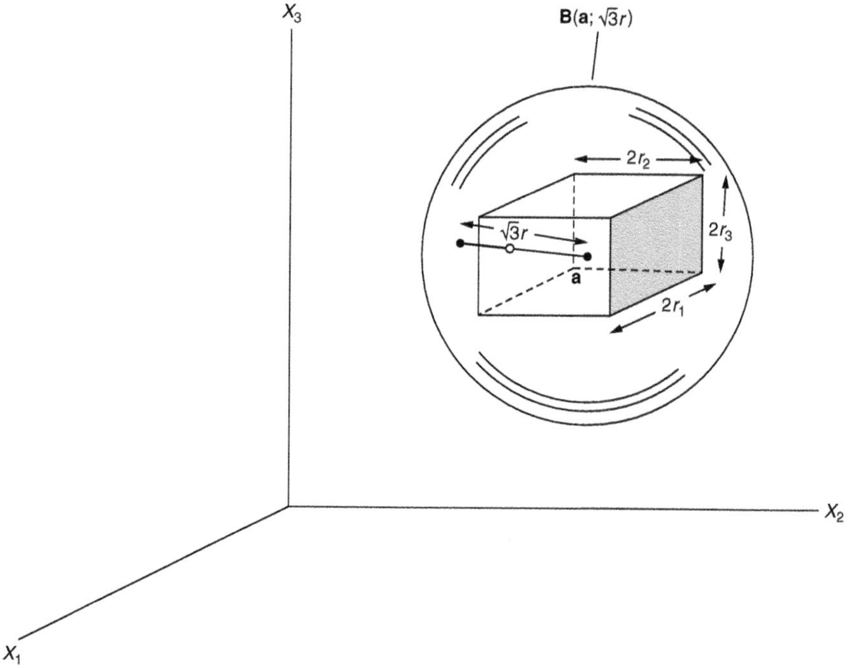

FIGURE 2.2
A closed box in $\mathbf{R}^3$ *and contained in a 3-ball* $\mathbf{B}_3(\mathbf{a}; \sqrt{3}r)$.

2. The box $\mathfrak{B}$ is divisible into 2^m congruent, closed, bounded subboxes by the m hyperplanes that pass through **a** and are parallel to the coordinate hyperplanes.

3. An upper bound to the maximum distance between two points in the box $\mathfrak{B}$ is

$$\left\{\sum_{k=1}^{m}[(a_k + r_k) - (a_k - r_k)]^2\right\}^{1/2} \leq 2\sqrt{m}\,r = D.$$

   We call D the diameter of the box $\mathfrak{B}$.

4. A sequence of closed, bounded boxes in $\mathbf{R}^m$, $\{\mathfrak{B}_n\}_{n=1}^{\infty}$, is a **nested sequence** if for each $n \in \mathbf{N}$, $\mathfrak{B}_{n+1} \subseteq \mathfrak{B}_n$.

The analog of Theorem 2.9 for a box $\mathfrak{B}$ in $\mathbf{R}^m$ follows by application of that theorem to each of the dimensions of box $\mathfrak{B}$.[6] The analog of Corollary 2.9.1, which follows, is a consequence of that corollary.

---

[6]The analogs of Theorem 2.9 and Corollary 2.9.2, taken together, are sometimes referred to as **Cantor's Intersection Theorem** (Simmons, 1983).

Corollary 2.9.2. *If* $\{\mathfrak{B}_n\}_{n=1}^{\infty}$, *is a nested sequence of closed, bounded boxes in* $\mathbf{R}^m$ *and if* $\lim_{n\to\infty} D_n = 0$, *then* $\bigcap_{n=1}^{\infty} \mathfrak{B}_n$ *reduces to a singleton set.*

Now look again at the first two examples in Example 2.15. Those two sequences possessed a cluster point; they were also bounded sequences. This is no accident; here is where we put the concept of nesting to work.

Theorem 2.10 (Bolzano-Weierstrass Theorem). *Every bounded sequence in* $\mathbf{R}^m$ *has at least one cluster point.*

Proof. Suppose that the sequence $\{\mathbf{x}_n\}_{n=1}^{\infty}$ contains only finitely many distinct terms. Then there exists some subsequence $\{\mathbf{x}_{n_k}\}_{k=1}^{\infty}$, all of whose terms are identical, say to $\mathbf{x}$, so $\mathbf{x}$ is a cluster point of $\{\mathbf{x}_n\}_{n=1}^{\infty}$, and we are done.

Suppose now that $\{\mathbf{x}_n\}_{n=1}^{\infty}$ does not contain any term that occurs infinitely often. As the sequence is bounded, there is a closed, bounded box $\mathfrak{B}_1$ that contains all the terms. Partition $\mathfrak{B}_1$ into 2^m congruent, closed, bounded subboxes; one of these, possibly more than one, must contain infinitely many terms of $\{\mathbf{x}_n\}_{n=1}^{\infty}$. Denote this subbox by $\mathfrak{B}_2$. If D_1 is the diameter of $\mathfrak{B}_1$, then $D_1/2$ is the diameter of $\mathfrak{B}_2$.

The partitioning is continued; a nested sequence, $\mathfrak{B}_1 \supset \mathfrak{B}_2 \supset \mathfrak{B}_3 \supset \ldots$, in $\mathbf{R}^m$ is obtained. The diameter of box $\mathfrak{B}_n$ is $D_n = D_1/2^{n-1}$; since $\lim_{n\to\infty} D_n = 0$, then by Corollary 2.9.2 there is a unique point $\mathbf{p} \in \bigcap_{n=1}^{\infty} \mathfrak{B}_n$.

Let $\varepsilon > 0$ be given. The partitioning ensures that there is a $\mathfrak{B}_n$ such that $D_n < \varepsilon$, so $\mathfrak{B}_n \subset \mathbf{B}_m(\mathbf{p};\varepsilon)$. Thus, there are infinitely many terms in $\mathbf{B}_m(\mathbf{p};\varepsilon)$, and this says that $\mathbf{p}$ is a cluster point of $\{\mathbf{x}_n\}_{n=1}^{\infty}$. ■

The theorem may fail if the terms of the sequence are not bounded, for it may then happen that there are no convergent subsequences (see Theorem 2.11).

The Bolzano-Weierstrass Theorem for sequences or sets in $\mathbf{R}^1$ was known as early as 1817 to Bolzano (footnote 2 in Section 1.1). The first formal proof of it, however, is attributed to the German mathematician **Karl Weierstrass** (1815–1897), one of the pioneers in the rigorization of calculus.

■ Example 2.17

Suppose a sequence $\{\mathbf{x}_n\}_{n=1}^{\infty}$ in $\mathbf{R}^2$ is defined by $\mathbf{x}_n = \left(\frac{\sin n}{n}, 2(-1)^n - n^{-1}\right)$. This is bounded because each component is bounded for all $n \in \mathbf{N}$ (Theorem 2.1). A cluster point of the sequence is $\mathbf{p} = (0, 2)$. We observe that if $\varepsilon = 0.01$, then

$$d_2(\mathbf{x}_n, \mathbf{p}) = \left[\sum_{k=1}^{2}(x_k - p_k)^2\right]^{1/2}$$

$$= \left[\left(\frac{\sin n}{n} - 0 \right)^2 + \left(2 - \frac{1}{n} - 2 \right)^2 \right]^{1/2} \qquad (n = \text{even})$$

$$= \frac{\sqrt{1 + \sin^2 n}}{n} \leq \frac{\sqrt{2}}{n},$$

and this is less than ε if $n > 141$. Thus, infinitely many terms of the sequence lie inside the 2-ball $\mathbf{B}_2(\mathbf{p}; 0.01)$. ■

■ Example 2.18

In $\mathbf{R}^1$ a sequence $\{x_n\}_{n=1}^{\infty}$ is defined by

$$x_n = \cos n;$$

a partial graph is shown in Figure 2.3. The sequence is clearly bounded, so it possesses a cluster point. Picking one out, however, is not obvious. We consider this in Exercise 2.31. We can also investigate (in Exercise 2.32) the question of how to find a monotonic subsequence of $\{x_n\}_{n=1}^{\infty}$. ■

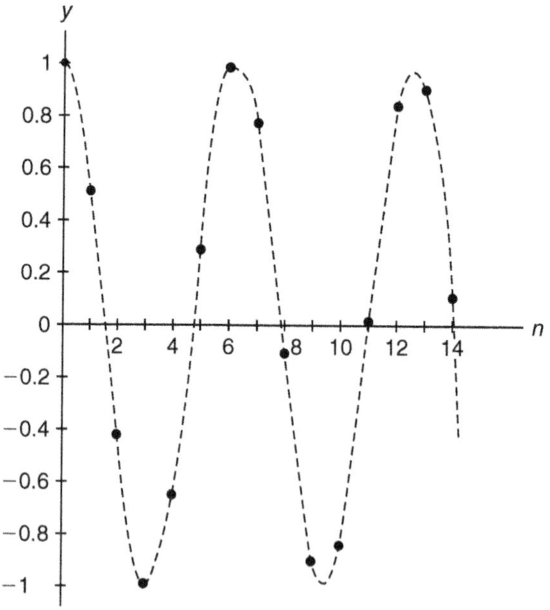

FIGURE 2.3
The graph of $y = \cos n$.

2.5 THE CONNECTION WITH SUBSEQUENCES

Although a sequence can have, at most, one limit, it can have many cluster points.

Theorem 2.11. *A point* $\mathbf{p} \in \mathbf{R}^m$ *is a cluster point of a sequence* $\{\mathbf{x}_n\}_{n=1}^{\infty}$ *iff there is a subsequence that converges to* $\mathbf{p}$.

Proof. ($\rightarrow$) Suppose $\mathbf{p} \in \mathbf{R}^m$ is a cluster point of $\{\mathbf{x}_n\}_{n=1}^{\infty}$. For each $k \in \mathbf{N}$, there is a term $\mathbf{x}_{n_k}$ such that $\mathbf{x}_{n_k} \in \mathbf{B}_m(\mathbf{p}; k^{-1})$. Let $\varepsilon > 0$ be given and choose $N = \lceil \varepsilon^{-1} \rceil$.[7] Then $k > N \geq \varepsilon^{-1}$ implies $k^{-1} < N^{-1} \leq \varepsilon$, so $\mathbf{x}_{n_k} \in \mathbf{B}_m(\mathbf{p}; \varepsilon)$. This says that $\mathbf{p}$ is the limit of the subsequence $\{\mathbf{x}_{n_k}\}_{k=1}^{\infty}$.

($\leftarrow$) This direction of the proof is left to you. ■

When a subsequence converges, we refer to its limit as a **subsequential limit** of the original sequence $\{\mathbf{x}_n\}_{n=1}^{\infty}$. Theorem 2.11 can be viewed either as an "interpretation" of a cluster point of a sequence, or as an alternative definition of a cluster point.

Corollary 2.11.1. *Any bounded sequence in* $\mathbf{R}^m$ *has a convergent subsequence.*

Proof. This follows immediately from Theorems 2.10 and 2.11. ■

■ Example 2.19

The corollary holds for the somewhat obvious case of Example 2.17 and for the far less obvious case of Example 2.18. ■

A sequence in $\mathbf{R}^m$ could have just one cluster point in $\mathbf{R}^m$; for lack of a better term, let us call such a sequence a **single-cluster sequence**. An example of a single-cluster sequence is $\{x_n\}_{n=1}^{\infty}$, where

$$x_n = \begin{cases} n^{-1} & n = 2k - 1,\ k \in \mathbf{N} \\ n & n = 2k. \end{cases}$$

This sequence diverges, but $x = 0$ is its sole cluster point in $\mathbf{R}^1$. Convergent sequences are a special class of single-cluster sequences. Here is a more complete statement.

Theorem 2.12. *A single-cluster sequence* $\{\mathbf{x}_n\}_{n=1}^{\infty}$ *in* $\mathbf{R}^m$ *converges to* $\mathbf{x}$ *iff every subsequence converges to* $\mathbf{x}$.

Proof. ($\leftarrow$) This direction is trivial because $\{\mathbf{x}_n\}_{n=1}^{\infty}$ is a particular subsequence of itself.

($\rightarrow$) Suppose that $\lim_{n\to\infty} \mathbf{x}_n = \mathbf{x}$; let $\varepsilon > 0$ be given. Then there is an $N \in \mathbf{N}$ such that for all $n > N$, $\mathbf{x}_n \in \mathbf{B}_m(\mathbf{x}; \varepsilon)$. Let $n_1, n_2, \ldots, n_k, \ldots$ be an arbitrary increasing

[7] $\lceil x \rceil$, known as the **ceiling function** of $x \in \mathbf{R}$, is the smallest integer that equals or exceeds x. Thus, $\lceil -3.4 \rceil = -3$, $\lceil 4.49 \rceil = 5$, and $\lceil 7 \rceil = 7$.

sequence of natural numbers; for each k we have $n_k \geq k$ (use the PMI). For those k that exceed N, we then have $n_k > N$, so for all $n_k > N, \mathbf{x}_{n_k} \in \mathbf{B}_m(\mathbf{x}; \varepsilon)$. Thus, the arbitrary subsequence $\{\mathbf{x}_{n_k}\}_{k=1}^{\infty}$ converges to $\mathbf{x}$. ■

■ Example 2.20

The sequence in Example 2.17 certainly cannot converge because the two subsequences

$$\{\mathbf{x}_{n_k}\}_{k=1}^{\infty}, \quad n_k = 2m$$

$$\{\mathbf{x}_{n_k}\}_{k=1}^{\infty}, \quad n_k = 2m - 1$$

converge to $(0, 2)$ and $(0, -2)$, respectively. This example illustrates one use of Theorem 2.12. ■

It is interesting that if no requirements at all are imposed upon a sequence in $\mathbf{R}^1$, it is still possible to find a monotonic subsequence (Bell, 1964). There is no guarantee, though, that this subsequence will converge. It follows that if boundedness is imposed upon $\{x_n\}_{n=1}^{\infty}$ in $\mathbf{R}^1$, then there must exist a convergent, monotonic subsequence. Boundedness is a powerful condition.

2.6 LIMIT SUPERIOR AND LIMIT INFERIOR

In this section we develop a concept for sequences in $\mathbf{R}^1$ that is similar in spirit to, but is weaker than, that of the limit of a sequence. If $\{x_n\}_{n=1}^{\infty}$ is a sequence in $\mathbf{R}^1$, then it has a subsequential limit in **Re** (see Section 1.9 for definition of symbol). This follows because either (a) $\{x_n\}_{n=1}^{\infty}$ is bounded in $\mathbf{R}^1$ and Corollary 2.11.1 then applies, or (b) $\{x_n\}_{n=1}^{\infty}$ is unbounded from above (from below) and there is a subsequence $\{x_{n_k}\}_{k=1}^{\infty}$ such that $\lim_{k\to\infty} x_{n_k} = \infty(-\infty)$ (Exercise 2.35). Hence, the set of subsequential limits in **Re** of a sequence in $\mathbf{R}^1$ is always nonempty. Accordingly, we can seek the supremum and the infimum of the set.

Definition. *If $\{x_n\}_{n=1}^{\infty}$ is a sequence in $\mathbf{R}^1$, then we denote the set of all subsequential limits of $\{x_n\}_{n=1}^{\infty}$ by*

$$\mathrm{E} = \{x \in \mathbf{Re} \colon \lim_{k\to\infty} x_{n_k} = x\}.$$

*The **limit superior (limit inferior)**[8] of $\{x_n\}_{n=1}^{\infty}$, written as $\limsup_{n\to\infty} x_n$ ($\liminf_{n\to\infty} x_n$), are defined as*

$$\limsup_{n\to\infty} x_n = \sup \mathrm{E}, \quad \liminf_{n\to\infty} x_n = \inf \mathrm{E}.$$

---

[8]Some authors (Shilov, 1996) use the terms **upper limit** and **lower limit**, respectively. These terms, however, sound suspiciously close to ordinary bounds.

For a sequence $\{x_n\}_{n=1}^{\infty}$ in $\mathbf{R}^1$, the set $\mathbf{E}$ consists of at least one element; it could have an unlimited number of elements. But for any such sequence, both $\limsup_{n\to\infty} x_n$ and $\liminf_{n\to\infty} x_n$ must exist and be unique, although in some cases they may be identical. Also, note from above that one or both of $\limsup_{n\to\infty} x_n$, $\liminf_{n\to\infty} x_n$ might be infinite.

■ Example 2.21

Define $\{x_n\}_{n=1}^{\infty}$ by $x_n = -n$. Then $\mathbf{E} = \{-\infty\}$, so $\limsup_{n\to\infty} x_n = \liminf_{n\to\infty} x_n = -\infty$. What is $\lim_{n\to\infty} x_n$?

Define $\{x_n\}_{n=1}^{\infty}$ by $x_n = (-1)^n - n^{-1}$. Then $\mathbf{E} = \{-1, 1\}$, so $\limsup_{n\to\infty} x_n = 1$ and $\liminf_{n\to\infty} x_n = -1$. This example shows that $\inf\{x_1, x_2, \ldots\}$ does not always equal $\liminf_{n\to\infty} x_n$; in fact, $\inf\{x_1, x_2, \ldots\} \leq \liminf_{n\to\infty} x_n$. When $\liminf_{n\to\infty} x_n$ is real, it is the smallest number to which infinitely many x_n's can approach to within any prescribed $\varepsilon > 0$. ■

■ Example 2.22

Define $\{x_n\}_{n=1}^{\infty}$ to be any sequence that has arranged all the rationals in the open interval $(0, 1)$ for enumeration (see Section 1.7). Then $\mathbf{E}$ is the closed interval $[0, 1]$, and $\limsup_{n\to\infty} x_n = 1$ and $\liminf_{n\to\infty} x_n = 0$. ■

If $\mathbf{E}$ is a finite set, then $\sup \mathbf{E}$ and $\inf \mathbf{E}$ belong to $\mathbf{E}$, since these are identical to the maximum element and the minimum element of $\mathbf{E}$, respectively (see Exercise 1.20). The case where $\mathbf{E}$ is an infinite set is intrinsically more interesting. We shall also restrict our consideration now to sets $\mathbf{E}$ that are bounded. If such is not the case, see Exercise 2.35.

Theorem 2.13. *Let the bounded, infinite set $\mathbf{E} \subset \mathbf{R}^1$ be the set of subsequential limits of the sequence $\{x_n\}_{n=1}^{\infty}$, and let $\{b_n\}_{n=1}^{\infty}$ be any convergent sequence with range in $\mathbf{E}$. Suppose that $\lim_{n\to\infty} b_n = b$. Then there is a subsequence $\{x_{n_k}\}_{k=1}^{\infty}$ such that $\lim_{k\to\infty} x_{n_k} = b$.*

Proof. Since $b_1 \in \mathbf{E}$, there is a subsequence of $\{x_n\}_{n=1}^{\infty}$ that converges to b_1. Hence, we can find a value of n, say n_1, such that $x_{n_1} \in \mathbf{B}(b_1; 1)$. Similarly, there is a subsequence of $\{x_n\}_{n=1}^{\infty}$ that converges to b_2, so from the set of natural numbers $\{n_1 + 1, n_1 + 2, n_1 + 3, \ldots\}$ choose one (call it n_2) such that $x_{n_2} \in \mathbf{B}(b_2; 1/2)$. For b_3 choose an integer from $\{n_2 + 1, n_2 + 2, n_2 + 3, \ldots\}$ (call it n_3) such that $x_{n_3} \in \mathbf{B}(b; 1/3)$. Continuing in this way, we arrive at an increasing sequence of natural numbers, $n_1 < n_2 < n_3 < \ldots$, such that for each $k \in \mathbf{N}$, we have $x_{n_k} \in \mathbf{B}(b_k; 1/k)$. Now let $\varepsilon > 0$ be given. Choose natural numbers N_1, N_2 such that

$$\frac{1}{N_1} < \frac{\varepsilon}{2}, \quad b_k \in \mathbf{B}(b; \varepsilon/2), \text{ for all } k > N_2.$$

Denote $N = \max\{N_1, N_2\}$, so that both of these conditions hold for all $k > N$. We obtain, for all $k > N$,

$$|x_{n_k} - b| = |x_{n_k} - b_k + b_k - b| \leq |x_{n_k} - b_k| + |b_k - b| < \varepsilon.$$

This says that $\lim_{k\to\infty} x_{n_k} = b$. ■

Let e_1 be an arbitrary element in the interior of **E**; we have $e_1 < \sup \mathbf{E} = \limsup_{n\to\infty} x_n$. By Exercise 1.23 there is an element e_2 such that $e_1 < e_2 < \limsup_{n\to\infty} x_n$. Continuing in this way, we obtain a sequence with range in **E** that converges to $\limsup_{n\to\infty} x_n$. Hence, $\limsup_{n\to\infty} x_n$ is a b as described in Theorem 2.13, and we have the following corollary:

Corollary 2.13.1. $\limsup_{n\to\infty} x_n$ *and (by parallel reasoning)* $\liminf_{n\to\infty} x_n$ *belong to* **E**. ■

Two basic properties of lim sup's now follow; the corresponding properties for lim inf's are somewhat parallel. One use for these properties occurs in Theorem 2.16 (see Section 2.7).

Theorem 2.14. *Let* $\{x_n\}_{n=1}^{\infty}$ *be a bounded sequence in* $\mathbf{R}^1$, *and suppose that* $\limsup_{n\to\infty} x_n = \nu \in \mathbf{R}^1$; *let* l, L *be two real numbers. Then*

(a) *if* $L > \nu$, *then there is an* $N \in \mathbf{N}$ *such that for all* $n > N$ *we have* $x_n < L$;

(b) *if* $l < \nu$, *then for infinitely many* $n \in \mathbf{N}$ *we have* $x_n > l$.

Proof.

(a) Suppose, to the contrary, that there is a number $L > \nu$ such that $x_n \geq L$ for infinitely many $n \in \mathbf{N}$. This set of x_n's is bounded by hypothesis, so by the Bolzano-Weierstrass Theorem (Theorem 2.10) it contains a subsequence $\left\{x_{n_k}\right\}_{k=1}^{\infty}$ that converges to some number $x \geq L$, which must be in **E** by definition. But ν, a supremum, is the largest element in **E**, so the contradiction implies that the initial assumption is false. If N is chosen, only finitely many $x_n \geq L$, and by choosing N sufficiently large we then have $x_n < L$ for all $n > N$.

(b) By Theorem 2.13 and Corollary 2.13.1, the point ν is a cluster point of $\{x_n\}_{n=1}^{\infty}$, that is, it is the limit of some subsequence $\left\{x_{n_k}\right\}_{k=1}^{\infty}$. Let ε satisfy $0 < \varepsilon < \nu - l$; then there are infinitely many k such that $x_{n_k} \in \mathbf{B}(\nu; \varepsilon)$. Each such x_{n_k} exceeds l because $x_{n_k} > \nu - \varepsilon > l$. ■

The theorem has a counterpart for lim inf's (Exercise 2.37). Theorem 2.14 can be reworded in the following useful manner:

(a′) If $\varepsilon > 0$ is given, there are only finitely $n \in \mathbf{N}$ such that $x_n \geq \varepsilon + \limsup_{n\to\infty} x_n$.

(b′) If $\varepsilon > 0$ is given, there are infinitely many $n \in \mathbf{N}$ such that $x_n > \limsup_{n\to\infty} x_n - \varepsilon$.

Statements (a′),(b′) are nicely illustrated by the sequence $x_n = 1 + (-1)^n/n$, with $\varepsilon = 0.01$.

Theorem 2.15. *If $\{x_n\}_{n=1}^{\infty}$ is a sequence in $\mathbf{R}^1$ and $x \in$ **Re**, then $\lim_{n\to\infty} x_n = x$ iff*

$$\limsup_{n\to\infty} x_n = \liminf_{n\to\infty} x_n = x.$$

Proof. ($\rightarrow$) This direction is trivial, in view of Theorems 2.12, 2.13.

($\leftarrow$) Suppose, first, that $\limsup_{n\to\infty} x_n = \liminf_{n\to\infty} x_n = x \in \mathbf{R}^1$. Let $\varepsilon > 0$ be given. From statement (a′), there is an $N_1 \in \mathbf{N}$ such that for all $n > N_1$ we have $x_n < \varepsilon + \limsup_{n\to\infty} x_n$. Similarly, the analogous statement for lim inf's is that there is an $N_2 \in \mathbf{N}$ such that for all $n > N_2$ we have $x_n > \liminf_{n\to\infty} x_n - \varepsilon$. Combination of the two inequalities then gives $x - \varepsilon < x_n < x + \varepsilon$ for all $n > N = \max\{N_1, N_2\}$, or equivalently, $|x_n - x| < \varepsilon$, for all $n > N$. This says that $\lim_{n\to\infty} x_n = x$.

The completion of the proof for the case where $\limsup_{n\to\infty} x_n = \liminf_{n\to\infty} x_n = \pm\infty$ is left to you. ■

■ Example 2.23

Any sequence constructed as in Example 2.22 cannot converge, since $\limsup_{n\to\infty} x_n \neq \liminf_{n\to\infty} x_n$. ■

2.7 CAUCHY SEQUENCES

The Axiom of Completeness has given us in Theorem 2.2 a criterion for the convergence of a sequence in $\mathbf{R}^1$ without our knowing the limit of the sequence. This criterion applies, however, only to monotonic sequences.

The French mathematician **Augustin-Louis Cauchy** (1789–1857), who gave early, solid definitions of limit of a function and convergence of a sequence (in $\mathbf{R}^1$) (Grabiner, 1981), gave without proof a convergence criterion that could be applied to nonmonotonic sequences. The criterion had been anticipated four years earlier by Bolzano, but Cauchy may have been unaware of this work. We will state the criterion in Theorem 2.16 after some crucial terms have been introduced.

Definition. *A sequence $\{x_n\}_{n=1}^{\infty}$ in $\mathbf{R}^1$ is termed a* ***Cauchy sequence***[9] *iff, given any $\varepsilon > 0$, there is an $N \in \mathbf{N}$ such that for any $n > m > N$ we have $|x_n - x_m| < \varepsilon$.*

---

[9] Less commonly called a **fundamental sequence** (by Cantor, for example). There are a number of equivalent formulations of the definition.

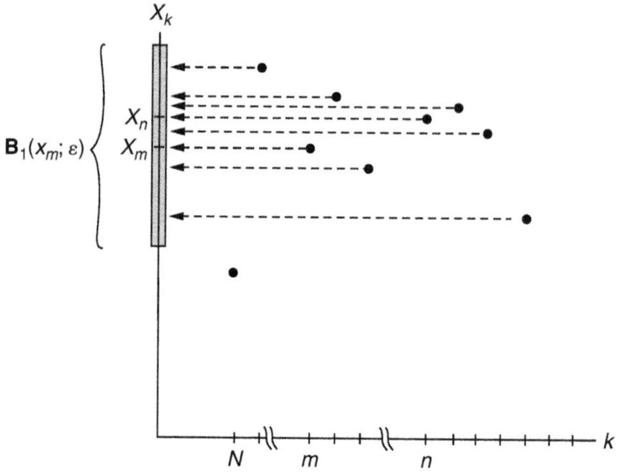

FIGURE 2.4

The idea of a Cauchy sequence.

The definition says that if a ball of radius ε is constructed about any term x_m $(m > N)$, then all the succeeding terms $x_n (n > m)$ will lie inside the ball (Figure 2.4). Also, the definition can be extended, in an obvious way, to sequences in more general metric spaces.

■ Example 2.24

A sequence $\{x_n\}_{n=1}^{\infty}$ in $\mathbf{R}^1$ is defined by $x_n = (n+1)/(n+2)$. Then

$$\begin{aligned}
|x_n - x_m| &= \left|\frac{n+1}{n+2} - \frac{m+1}{m+2}\right| \\
&= \left|\left(1 - \frac{1}{n+2}\right) - \left(1 - \frac{1}{m+2}\right)\right| \\
&= \frac{1}{m+2} - \frac{1}{n+2} \quad \text{since } n > m \\
&< \frac{1}{m+2}.
\end{aligned}$$

If $\varepsilon > 0$ is given, then choose $N = \left\lceil \frac{1}{\varepsilon} \right\rceil$. Then $m > N \rightarrow m + 2 > \left\lceil \frac{1}{\varepsilon} \right\rceil + 2 \geq \varepsilon^{-1} + 2 > \varepsilon^{-1} \rightarrow (m+2)^{-1} < \varepsilon \rightarrow |x_n - x_m| < \varepsilon$, as required. ■

We recognize that the Cauchy sequence in Example 2.24 is also a convergent sequence. The question of when this is true, in general, is fundamental.

Definition. *A space* **M** *equipped with a metric (such as the space* $\mathbf{R}^1$*, or anything more general) is termed* ***complete*** *iff every Cauchy sequence in* **M** *converges to an element of* **M**. *Otherwise,* **M** *is referred to as* ***incomplete.***

Note that this is a different use of the term "complete" from that in Section 1.3.

An example of an incomplete metric space is **Q**. The sequence $\{x_n\}_{n=1}^{\infty}$, where $x_n = \sum_{k=1}^{n}(-1)^{k+1}k^{-1}$, is a Cauchy sequence in **Q**, for given any $\varepsilon > 0$, the choice of $N = \lceil \varepsilon^{-1}$ is such that $n > m > N$ implies that

$$\left| \sum_{k=1}^{n}(-1)^{k+1}k^{-1} - \sum_{k=1}^{m}(-1)^{k+1}k^{-1} \right| < \varepsilon.$$ [10]

Further, the sequence converges (in $\mathbf{R}^1$), but to $\ln 2$, however, which is not rational. Thus, **Q** is not a complete metric space. A second example, and a simpler one, would be to consider the sequence $\{x_n\}_{n=1}^{\infty}$ where x_n is the rational number formed by the first n digits in the decimal expansion of π (Exercise 2.44).

Cauchy believed, without proof, that $\mathbf{R}^1$ is a complete metric space. The modern statement of his criterion for sequences in $\mathbf{R}^1$ is as follows:

Theorem 2.16 (Cauchy's Convergence Criterion). *Let* $\{x_n\}_{n=1}^{\infty}$ *be a sequence in* $\mathbf{R}^1$. *This converges iff the sequence is a Cauchy sequence.*

Proof. ($\rightarrow$) This direction is straightforward. Its proof is left to you.

($\leftarrow$) A number of proofs are known for this direction. We have chosen one that makes use of material from Section 2.6. Suppose that $\{x_n\}_{n=1}^{\infty}$ is a Cauchy sequence. We first show that it is bounded. Choose, initially, $\varepsilon = 1/2$. Then there is an $N \in \mathbf{N}$ such that for all $n > N$ we have $|x_n - x_N| < 0.5$, or equivalently, for $n > N$

$$x_N - 0.5 < x_n < x_N + 0.5. \qquad (*)$$

Next, let m, M be defined by

$$m = \min\{x_1, x_2, \ldots, x_{N-1}\}, \quad M = \max\{x_1, x_2, \ldots, x_{N-1}\},$$

and then set

$$A = \min\{m, x_N - 0.5\}, \quad B = \max\{M, x_N + 0.5\}.$$

Combination of this last line with (*) then yields $A \le x_n \le B$ for all $n \in \mathbf{N}$, so $\{x_n\}_{n=1}^{\infty}$ is bounded. In view of this, $\limsup_{n\to\infty} x_n$ and $\liminf_{n\to\infty} x_n$ are both real.

[10] A theorem due to Young, cited in Section 3.4, is sufficient to prove this.

Next, we *assume* that

$$\varepsilon = \frac{1}{2}\left(\lim_{n\to\infty} \sup x_n - \lim_{n\to\infty} \inf x_n\right) > 0.$$

From Theorem 2.14 (b′) there are infinitely many $n \in \mathbf{N}$ such that

$$x_n > \lim_{n\to\infty} \sup x_n - \frac{\varepsilon}{2}.$$

These n's define a sequence $\{x_{n_k}\}_{k=1}^{\infty}$ such that for each $k \in \mathbf{N}$

$$x_{n_k} > \lim_{n\to\infty} \sup x_n - \frac{\varepsilon}{2}.$$

Similarly, from the analog of Theorem 2.14 (b′) for infima (Exercise 2.38), we deduce the existence of a sequence $\{x_{m_k}\}_{k=1}^{\infty}$ such that for each $k \in \mathbf{N}$

$$x_{m_k} < \lim_{n\to\infty} \inf x_n + \frac{\varepsilon}{2}.$$

Subtraction of the two inequalities gives

$$\begin{aligned} x_{n_k} - x_{m_k} &> \lim_{n\to\infty} \sup x_n - \lim_{n\to\infty} \inf x_n - \varepsilon \\ &= 2\varepsilon - \varepsilon \\ &= \varepsilon. \end{aligned}$$

This says that no matter how large N is chosen, we can select k large enough so that $n_k > N$ and $m_k > N$, and then $|x_{n_k} - x_{m_k}|$ will not be less than ε. This contradicts the fact that $\{x_n\}_{n=1}^{\infty}$ is Cauchy. It follows that $\varepsilon > 0$ is false; since for any set $\mathbf{E}$ of real numbers we have $\inf \mathbf{E} \le \sup \mathbf{E}$, it follows that $\varepsilon = 0$ is the only possibility. From Theorem 2.15 this implies that $\lim_{n\to\infty} x_n = \lim_{n\to\infty} \sup x_n = \lim_{n\to\infty} \inf x_n = x$. ■

■ Example 2.25

Cauchy's Convergence Criterion allows an almost instantaneous answer to the question of the convergence of the sequence in Example 2.8. We have, for $n > m$,

$$\begin{aligned} \left|\frac{F_{n+1}}{F_n} - \frac{F_{m+1}}{F_m}\right| &= \left|\frac{F_m F_{n+1} - F_{m+1} F_n}{F_n F_m}\right| = \left|\frac{F_m(F_n + F_{n-1}) - (F_m + F_{m-1})F_n}{F_n F_m}\right| \\ &= \left|\frac{F_{m-1} F_n - F_m F_{n-1}}{F_n F_m}\right| \end{aligned}$$

after one reduction; all indices in the numerator have decreased by 1. A total of $m-1$ reductions then give

$$\left|\frac{F_{n+1}}{F_n}-\frac{F_{m+1}}{F_m}\right| = \left|\frac{F_1F_{n-m+2}-F_2F_{n-m+1}}{F_nF_m}\right| = \frac{F_{n-m}}{F_nF_m},$$

since $F_1 = F_2 = 1$. It follows that

$$\left|\frac{F_{n+1}}{F_n}-\frac{F_{m+1}}{F_m}\right| = \frac{F_{n-m}}{F_nF_m} < \frac{1}{F_m} < \varepsilon$$

if m is large enough. Hence, $\{F_{n+1}/F_n\}_{n=1}^{\infty}$ is a Cauchy sequence, and by Theorem 2.16 it converges.

To illustrate numerically, let $\varepsilon = 0.001$. Then $F_m^{-1} < \varepsilon$ implies $F_m > 1000$; choose $m = 17$, so that $F_{17} = 1597$. Choose arbitrarily $F_n = F_{19} = 4181$. We compute

$$\left|\frac{F_{20}}{F_{19}}-\frac{F_{18}}{F_{17}}\right| = \left|\frac{6765}{4181}-\frac{2584}{1597}\right| = 1.49\times 10^{-7} < \frac{1}{F_{17}}$$

$$= 6.26\times 10^{-4} < \varepsilon.$$

■

From Theorem 2.16 we can extend the concept of completeness to $\mathbf{R}^m$, in general. When this is combined with Theorem 2.5, we obtain

Theorem 2.17. *Any Euclidean vector space* $\mathbf{R}^m$ *is complete.*

Proof. Let $\{\mathbf{x}_n\}_{n=1}^{\infty}$, $\mathbf{x}_n = (x_{n1}, x_{n2}, \ldots, x_{nm})$, be a sequence in $\mathbf{R}^m$. Now consider the Cauchy nature of the separate sequences of the corresponding components of the vectors. The proof is left to you. ■

EXERCISES

Section 2.1

2.1. Refer to Example 2.2.
- **(a)** Compute x_6 to x_{10}. Formulate some conjectures.
- **(b)** Is x_{92305} a natural number?

2.2. Refer to Example 2.3.
- **(a)** Prove that for each k, $H_{k+1}(x) = 2xH_k(x) - H_k'(x)$.
- **(b)** Deduce $H_8(x), H_9(x)$.
- **(c)** Prove that $H_k(0) = 0$ if k is odd.

2.3. A sequence of 2×2 matrices, $\{\mathbf{M}_n\}_{n=1}^{\infty}$, is defined as follows:

$$m_{11} = 1/(2n+1)! \qquad m_{12} = 1/(2n+2)!$$

$$m_{21} = \sum_{k=0}^{n} \frac{(2n+2)!}{(2k+2)!} \qquad m_{22} = \sum_{k=0}^{n} \frac{(2n+1)!}{(2k+1)!}$$

For each $n \in \mathbf{N}$, let det $\mathbf{M}_n$ denote the determinant of $\mathbf{M}_n$. Evaluate det $\mathbf{M}_n$ for $n = 1, 2, 3, 4, 5$. These five numbers are increasingly accurate approximations of a number N. Conjecture the value of N.

2.4. Write out the proof of Theorem 2.1.

2.5. Discuss the boundedness of each sequence $\{\mathbf{x}_n\}_{n=1}^{\infty}$:

(a) $x_n = (\sin n)/n$;

(b) $x_n = n^3 e^{-n/2}$;

(c) $x_n = (n^3 + 18)/(10n^2 + n - 10)$;

(d) $x_n = (n+1)/\ln(n+1)$;

(e) $\mathbf{x}_n = (1 - \sin^{-1}(1/n), \cos n)$.

2.6. This Exercise looks at $\{y_k\}_{k=1}^{\infty}$ from Example 2.5. Recall **Newton's Binomial Theorem** for $n \in \mathbf{N}$:

$$(a+b)^n = \sum_{j=0}^{n} \binom{n}{j} a^{n-j} b^j, \quad \binom{n}{j} = \frac{n!}{(n-j)!j!}.$$

(a) Write out the binomial expansion of $(1 + k^{-1})^k$, $k \in \mathbf{N}$.

(b) Prove that for $j = 1, 2, \ldots, k$, $\binom{k}{j} \frac{1}{k^j} \leq \frac{1}{2^{j-1}}$.

(c) Recall the formula for the sum of a finite geometric series:

$$\sum_{k=0}^{n-1} ar^k = a + ar + ar^2 + \cdots + ar^{n-1} = \frac{a(1-r^n)}{1-r}, \quad r \neq 1.$$

Show that for any $k \in \mathbf{N}$ the expansion in (a) does not exceed $3 - 2\left(\frac{1}{2}\right)^k$.

(d) By taking a simple limit, deduce that $\{y_k\}_{k=1}^{\infty}$ is bounded from above by 3.

2.7. Describe each sequence as increasing, decreasing, becomes increasing, becomes decreasing, or is never monotonic: $\{x_n\}_{n=1}^{\infty}$, where

(a) $x_n = (n+6)/[n - (1/2)]$;

(b) Exercise 2.5(b);

(c) $x_n = \mathrm{Tan}^{-1} n$;

(d) $x_n = 1 - (F_n/F_{n+1})^2$ (see Example 2.8);

(e) Exercise 2.5(d).

2.8. Refer to Example 2.8. Write a short program to compute F_{n+1}/F_n for $n = 1$ to 30. There are two obvious subsequences; describe them.

2.9. Follow the spirit of Example 2.7 and analyze the closely related sequence $\{x_n\}_{n=1}^{\infty}$, where $x_n = (1 + n^{-1})^{n+1}$. (Your result will be a particular case of the general theorem proved in Kang and Yi, 2007.)

2.10. If $c \in \mathbf{R}^1$ and $n \in \mathbf{N}$, then the **generalized binomial coefficient** $\binom{c}{n}$ is

$$\binom{c}{n} = \frac{c(c-1)(c-2)\cdots(c-n+1)}{n!}.$$

Let $b_n = (-1)^n\binom{-1/2}{n}$; prove that $\{b_n\}_{n=1}^{\infty}$ is a decreasing sequence.

2.11. (*For those with some knowledge of infinite series*) An important sequence in analysis is $\{x_n\}_{n=1}^{\infty}$, where $x_n = \sum_{k=1}^{n} \frac{1}{k} - \ln(n)$. Prove that this is a decreasing sequence.[11]

Section 2.2

2.12. How do you know that the following sequences converge?

(a) $\{x_n\}_{n=1}^{\infty}$, $x_n = [(4n+1)/(2n)]^{1/2}$;
(b) the sequence in Exercise 2.5(b);
(c) $\{x_n\}_{n=1}^{\infty}$, $x_n = \sqrt{n} - \sqrt{n+1}$;
(d) the sequence in Exercise 2.7(a);
(e) $\{x_n\}_{n=1}^{\infty}$, $x_n = (\ln n)^{1/n}$.

2.13. A sequence $\{x_n\}_{n=1}^{\infty}$ is defined as follows (Priestley, 1999):

$$x_n = \begin{cases} 4 & n = 1 \\ \frac{2x_{n-1}}{3} + \frac{4}{x_{n-1}^2} & n > 1. \end{cases}$$

(a) Compute x_2, x_4, x_6, x_8 to eight decimals.
(b) Show that the sequence is bounded and monotonic.
(c) Conjecture $\lim_{n\to\infty} x_n$.

2.14. Define a sequence $\{x_n\}_{n=1}^{\infty}$ by $x_n = \frac{\alpha^n - \beta^n}{\alpha - \beta}$, $\alpha = \frac{1+\sqrt{5}}{2}$, $\beta = \frac{1-\sqrt{5}}{2}$.

(a) Compute x_1 through x_6; conjecture a result.
(b) Show that $\{x_n\}_{n=1}^{\infty}$ is monotonic and diverges to ∞.
(c) Show that the sequence is equivalent to the recursive definition $x_1 = x_2 = 1$, $x_n = x_{n-1} + x_{n-2}$ for all $n > 2$.
(d) Write a short program to compute values of x_{n+1}/x_n. Conjecture $\lim_{n\to\infty} x_{n+1}/x_n$.

2.15. Consider the function $f(x) = (2^x - 1)/x$, $x > 0$.

(a) Prove that this is an increasing function.
(b) Now define $\{x_n\}_{n=2}^{\infty}$ by $x_n = n(\sqrt[n]{2} - 1)$. Use the result from part (a) to explain how you know that $\lim_{n\to\infty} x_n$ exists.
(c) Conjecture the value of $\lim_{n\to\infty} x_n$.

Section 2.3

2.16. Complete the proof of Theorem 2.3.

2.17. Show, first, that the sequence in Exercise 2.7(a) converges and then, second, show that it is bounded, in accordance with Theorem 2.4.
Discuss the sequence in Exercise 2.5(e) in connection with the converse of Theorem 2.4.

---

[11] Consideration of this sequence leads to the number known as **Euler's constant** (γ), about which there is an extensive body of literature that grows continually (De Temple, 2006).

2.18. Complete the proof of Theorem 2.5.

2.19. Write out all the details of Theorem 2.6; refer to Theorem 1.10, as suggested.

2.20. Draw a geometric figure that gives (in $\mathbf{R}^2$) an interpretation of the Quadrilateral Inequality referred to in footnote 3.

2.21. Complete the proofs of Lemma 2.3.2 and Theorem 2.7.

2.22. **Theorem 2.18 (Squeeze Theorem).** If $a_n \le b_n \le c_n$ for all natural numbers n larger than some $N \in \mathbf{N}$, and if $\lim_{n\to\infty} a_n = \lim_{n\to\infty} c_n = L$, then $\lim_{n\to\infty} b_n = L$.

Let $\varepsilon > 0$ be given. There is a natural number N such that for all $n > N$ we have $-\varepsilon < a_n - L < \varepsilon, -\varepsilon < c_n - L < \varepsilon, a_n \le b_n \le c_n$ simultaneously (why?). Go on to deduce $|b_n - L| < \varepsilon$. Complete the proof.

2.23. Refer to Exercise 2.10.

(a) Show that b_n is given by $b_n = \left[\frac{1\cdot3\cdot5\cdots(2n-1)}{2\cdot4\cdot6\cdots(2n)}\,\frac{1\cdot3\cdot5\cdots(2n-1)}{2\cdot4\cdot6\cdots(2n)}\right]^{1/2}$.

(b) Next, show that $b_n < (2n+1)^{-1/2}$.

(c) Use Exercise 2.22 to give the value of $\lim_{n\to\infty} b_n$.

2.24. Refer to parts (a) and (b) of Example 2.13.

(a) Prove those parts.

(b) Use all available limit theorems, including the two that you have just proved, to determine the following limits. These are not ε-proofs, but justify all steps.

i. $\{x_n\}_{n=1}^{\infty}$, $x_n = (n^3 + 5n^2 - 4)/(2n^3 - n + 1)$;

ii. $\{x_n\}_{n=1}^{\infty}$, $x_n = \left[2\sqrt{n^2+4}/(n+2)\right] - [(3-n)/(n+1)]$;

iii. $\{x_n\}_{n=1}^{\infty}$, $x_n = (n + \sin n)(n+1)^n/n^{n+1}$;

iv. $\{x_n\}_{n=1}^{\infty}$, $x_n = (1 + n^{-2})\cos(1/\sqrt{n})$.

2.25. Write out separately definitions for what it means for a sequence $\{x_n\}_{n=1}^{\infty}$ in $\mathbf{R}^1$ to diverge to ∞, and for a sequence $\{y_n\}_{n=1}^{\infty}$ in $\mathbf{R}^1$ to diverge to $-\infty$.

2.26. Let $\{x_n\}_{n=1}^{\infty}, \{y_n\}_{n=1}^{\infty}$ be two sequences in $\mathbf{R}^1$, and suppose that for all $n > N_1$ we have $x_n \le y_n$. Prove that if $\lim_{n\to\infty} x_n = \infty$, then $\lim_{n\to\infty} y_n = \infty$. Apply this variation of Theorem 2.7 to the sequence $\{y_n\}_{n=1}^{\infty}$, where $y_n = 2^n/[50n^3]$.

2.27. A sequence in $\mathbf{R}^2$ is defined geometrically by the diagram in Figure 2.5 (Vanden Eynden, 1994). All lines shown are horizontal, vertical, or have slope ± 1. The sequence is $\{\mathbf{p}_n\}_{n=1}^{\infty}$, where $\mathbf{p}_n = (x_n, y_n)$. Determine $\lim_{n\to\infty} \mathbf{p}_n$.

Sections 2.4 and 2.5

2.28. Conjecture the identity of any cluster points of these sequences $\{x_n\}_{n=1}^{\infty}$:

(a) $x_n = n^2[1 - \cos(1/n)]$;

(b) $x_n = n^{(-1)^n}/\sqrt{n^2+1}$;

(c) $x_n = \begin{cases} (\ln n)/n & n = 2k \\ 2n/(n+6) & n = 2k-1. \end{cases}$

2.29. **(a)** The following sequence of intervals is defined: $\{\mathbf{I}_n\}_{n=1}^{\infty}, \mathbf{I}_n = [2 - 2^{1-n}, 2)$. What is $\bigcap_{n=1}^{\infty} \mathbf{I}_n$? Is this a violation of the Nested Intervals Theorem?

(b) The following sequence of intervals is defined: $\{\mathbf{I}_n\}_{n=1}^{\infty}, \mathbf{I}_n = (-\infty, -n+1]$. What is $\bigcap_{n=1}^{\infty} \mathbf{I}_n$? Is this a violation of the Nested Intervals Theorem?

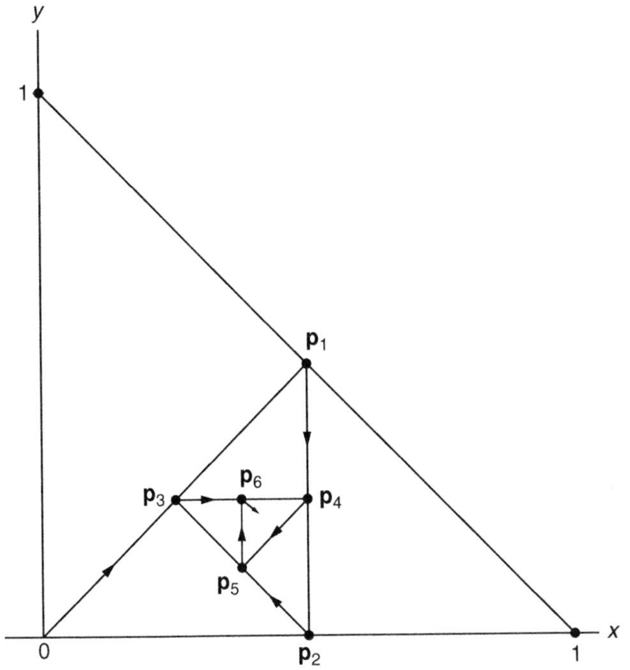

FIGURE 2.5
Diagram for Exercise 2.27.

2.30. Refer to Example 2.16; proceed as follows.

(a) Suppose n is odd. Then for intervals $\mathbf{I}_n = [a_n, b_n]$, $\mathbf{I}_{n+1} = [a_{n+1}, b_{n+1}]$, show that $b_n > \sqrt{2}$ implies $b_n > b_{n+1} > \sqrt{2}$, and that $a_n = a_{n+1}$. Hence, $\mathbf{I}_{2k-1} \supset \mathbf{I}_{2k}$ for any $k \in \mathbf{N}$.

(b) Now let n be even. Then for intervals $\mathbf{I}_n$, $\mathbf{I}_{n+1}$ show that $\sqrt{2} > a_n$ implies $\sqrt{2} > a_{n+1} > a_n$, and that $b_n = b_{n+1}$. Hence, $\mathbf{I}_{2k+1} \subset \mathbf{I}_{2k}$, so we have a nested sequence of intervals.

(c) How do you know that $\lim_{n\to\infty} a_n$, $\lim_{n\to\infty} b_n$ both exist?

(d) Determine these limits, thereby showing that the widths of the intervals approach 0. Hence, what is the unique point $p \in \bigcap_{n=1}^{\infty} \mathbf{I}_n$?

2.31. Here, we return to Example 2.18. Finding a monotonic subsequence of $\{x_n\}_{n=1}^{\infty}$, $x_n = \cos(n)$ would be facilitated if we knew where the cluster points are in $[-1, 1]$ (Ogilvy, 1969). Let $\varepsilon > 0$ be given, and *assume* that on the unit circle (Figure 2.6) there is an arc from θ_0 to $\theta_0 + \varepsilon$ such that no point $(1, \theta)$ on it has a natural number for its θ-coordinate.

(a) Why would the arcs from $\theta_0 + 1$ to $\theta_0 + 1 + \varepsilon$, $\theta_0 + 2$ to $\theta_0 + 2 + \varepsilon$, $\theta_0 + 3$ to $\theta_0 + 3 + \varepsilon$, and so on, then also not have any points with θ-coordinates that are natural numbers?

(b) Let $N > 2\pi/\varepsilon$ be a natural number. Why will N arcs like those in (a) cover the circle? By the way, how do you know that such an N as this will exist?

(c) We arrive at an absurdity. What is it, and what may we conclude?

(d) Consider the projection of the points of the circle onto the interval $[-1, 1]$ of the x-axis. Let $0 < \varepsilon_0 < 1$ be given and let $x_0 \in (-1, 1)$ be arbitrary. Also, let $\mathbf{I}_0 \subset (-1, 1)$ be an

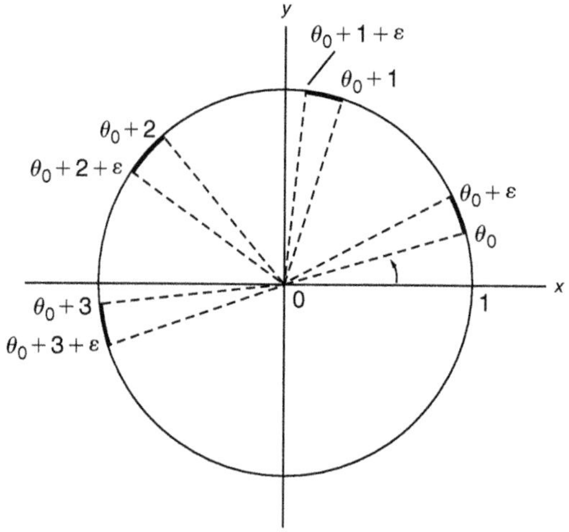

FIGURE 2.6
Arcs on the unit circle.

open interval of width ε_0 that contains x_0. Go on to show how we can construct a sequence $\{x_n\}_{n=1}^{\infty}$ such that for infinitely many n we have $\cos n \in \mathbf{B}(x_0;\varepsilon_0)$. What does this say?

2.32. For the sequence in Exercise 2.31, indicate in principle how we could construct an increasing convergent subsequence of $\{x_n\}_{n=1}^{\infty}$. Use part (d).

2.33. **(a)** Explain how the existence of the ceiling function in footnote 7 is a consequence of the Archimedean Property of $\mathbf{R}$ (Theorem 1.5) and the Well-Ordering Property (Theorem 1.8).
(b) Prove the other half of Theorem 2.11.

2.34. Consult the paper from Bell (1964) and digest the proof there that in any sequence in $\mathbf{R}^1$ we can find a monotonic subsequence. Write the proof in your own words.

Section 2.6

2.35. Suppose that $\{x_n\}_{n=1}^{\infty}$ in $\mathbf{R}^1$ is unbounded from above (below). Prove that there is a subsequence $\{x_{n_k}\}_{k=1}^{\infty}$ such that $\lim_{k\to\infty} x_{n_k} = \infty(-\infty)$.

2.36. Show, as implied (more generally) in Example 2.21, that if $\{x_n\}_{n=1}^{\infty}$ is a sequence in $\mathbf{R}^1$ with infinitely many values in its range and $\mathbf{S}$ is the set of these values, then

$$\inf \mathbf{S} \le \lim_{n\to\infty} \inf x_n \le \lim_{n\to\infty} \sup x_n \le \sup \mathbf{S}.$$

2.37. The analog of Theorem 2.14 for $\liminf$'s says: If $\{x_n\}_{n=1}^{\infty}$ is a bounded sequence in $\mathbf{R}^1$ and if $\liminf_{n\to\infty} x_n = w \in \mathbf{R}^1$, then (a) $L > w$ implies that for infinitely many $n \in \mathbf{N}$ we have $x_n < L$; (b) $l < w$ implies that there is an $N \in \mathbf{N}$ such that for all $n > N$ we have $x_n > l$. Prove this analog.

2.38. Reword the theorem given in Exercise 2.37 so as to produce statements (a′), (b′) analogous to the statements (a′), (b′) of Theorem 2.14.

2.39. Complete the little that remains in the proof of Theorem 2.15.

Section 2.7

2.40. Let $\{x_n\}_{n=1}^{\infty}$ in $\mathbf{R}^1$ be defined as follows: $x_n = \sum_{k=1}^{n} \frac{1}{k}$. A student argues as follows: "This is a Cauchy sequence because if $\varepsilon > 0$ is given, then choose $N = \lceil 1/\varepsilon \rceil$, so that now for any $n > N$,

$$|x_{n+1} - x_n| = \left| \sum_{k=1}^{n+1} \frac{1}{k} - \sum_{k=1}^{n} \frac{1}{k} \right| = \frac{1}{n+1} < \frac{1}{n} < \frac{1}{N} \le \varepsilon,$$

as required." Comment on this argument.

2.41. Demonstrate from the definition that the following $\{x_n\}_{n=1}^{\infty}$ are Cauchy sequences:

(a) $x_n = \ln n/n$;

(b) $x_n = (-1)^n/n^3$;

(c) $x_n = (n^2+1)/n^2$;

(d) $x_n = \sqrt{n} - \sqrt{n+1}$;

(e) $x_n = e^{-1/n}$;

(f) $x_n = n^{-1}\csc(n^{-1})$.

(Parts (e) and (f) require some knowledge of alternating series.)

2.42. Is the sequence $\{x_n\}_{n=1}^{\infty}$, $x_n = 1 + (-1)^n(1+n^{-1})$, a Cauchy sequence? Does it converge in $\mathbf{R}^1$? Does it have any cluster points in $\mathbf{R}^1$?

2.43. Let $\{x_n\}_{n=1}^{\infty}$, $\{y_n\}_{n=1}^{\infty}$ be Cauchy sequences in $\mathbf{Q}$. Is the sequence $\{x_n\}_{n=1}^{\infty} \oplus \{y_n\}_{n=1}^{\infty}$ also Cauchy in $\mathbf{Q}$? How about $\{x_n\}_{n=1}^{\infty} \otimes \{y_n\}_{n=1}^{\infty} = \{x_n y_n\}_{n=1}^{\infty}$?

2.44. Prove that the sequence $\{x_n\}_{n=1}^{\infty}$ given in the text, where $x_n =$ the first n digits in the base-10 expansion of π, is a Cauchy sequence in $\mathbf{Q}$.

2.45. Complete the proof of Theorem 2.16.

2.46. Complete the proof of Theorem 2.17.

2.47. The convergence of a Cauchy sequence $\{x_n\}_{n=1}^{\infty}$ in $\mathbf{R}^1$ can be considered a consequence of the Bolzano-Weierstrass Theorem. Proceed as follows:

(a) Prove in a manner differently from that in Theorem 2.16 that a Cauchy sequence in $\mathbf{R}^1$ is bounded.

(b) In view of part (a), apply the Bolzano-Weierstrass Theorem (Theorem 2.10, or a corollary thereof).

(c) Go on to show that $\lim_{n\to\infty} x_n$ exists in $\mathbf{R}^1$. Fill in the missing details and complete the proof.

Since the Bolzano-Weierstrass Theorem was proved from the Nested Intervals Theorem and this, in turn, was proved from the Axiom of Completeness, then the convergence of Cauchy sequences in $\mathbf{R}^m$ ultimately depends on this powerful axiom. Recall the italicized clause at the very end of Section 1.3.

SUPPLEMENTARY PROBLEMS

Additional problems on sequences for your enrichment and for challenge can be found in Appendix C. Good luck!

REFERENCES

Cited Literature

Bell, H.E., "Proof of a Fundamental Theorem on Sequences," *Amer. Math. Monthly*, **71**, 665–666 (1964). An elegant, though not the quickest, proof of the theorem that any sequence in $\mathbf{R}^1$ contains a monotonic subsequence.

De Temple, D.W., "A Geometric Look at Sequences that Converge to Euler's Constant," *Coll. Math. J.*, **37**, 128–131 (2006). Very accessible article that draws on the imagery of geometry; stimulating.

Grabiner, J.V., *The Origins of Cauchy's Rigorous Calculus*, MIT Press, Cambridge, 1981. A fine historical work, important for any serious student of mathematics; see especially pp. 9–12, 97–109. The concepts of lim sup, lim inf also date from the time of Bolzano (1817) and Cauchy (1821), albeit in primitive form (Grabiner, private communication).

Hoggatt, Jr., V.E., *Fibonacci and Lucas Numbers*, Fibonacci Association, Aurora, S.D., 1972. An early monograph by one of the pioneers in the field.

Kang, C.-X. and Yi, E., "The Convergence Behavior of $f_\alpha(x) = (1 + x^{-1})^{x+\alpha}$," *Coll. Math. J.*, **38**, 385–387 (2007). This perennially attractive problem is discussed nicely here; see also the papers below by Dence (1981) and De Temple (2005).

Ogilvy, C.S., "The Sequence $\{\sin n\}$," *Math. Mag.*, **42**, 94 (1969). The inspiration for our Exercise 2.31.

Priestley, W.M., "From Square Roots to nth Roots: Newton's Method in Disguise," *Coll. Math. J.*, **30**, 387–388 (1999). A nice Classroom Capsule that encourages algorithmic thinking; the source of our Exercise 2.13.

Shilov, G.E., *Elementary Real and Complex Analysis*, Dover Publications, NY, 1996, pp. 53–54, 73–76, 83–85. A well-written book that is highly recommended. The citations are to the Quadrilateral Inequality, limit (cluster) points of sequences in a metric space, the Cantor Intersection Theorem for sets in a complete metric space.

Simmons, G.F., *Topology and Modern Analysis*, Krieger Publishing Co., Malabar, FL, 1983, pp. 73–74. An alternative proof of the Cantor Intersection Theorem for complete metric spaces; it is surprisingly simple.

Sloane, N.J.A. (2005). www.research.att.com/~njas/sequences/. A scan of the index to the database shows the variety of topics covered. Browsing the database is claimed to be addictive.

Vajda, S., *Fibonacci & Lucas Numbers, and the Golden Section*, Ellis Horwood Ltd., Chichester, England, 1989. This book is packed with neat stuff.

Vanden Eynden, C., "Problem No. 1439," *Math. Mag.*, **67**, 66 (1994). The source of our Exercise 2.27.

Additional Literature

Ayoub, A.B., "Fibonacci-Like Sequences and Pell Equations," *Coll. Math. J.*, **38**, 49–53 (2007).

Biermann, K.-R., "Karl Theodor Wilhelm Weierstrass," in Gillispie, C.C. (ed.), *Dictionary of Scientific Biography*, Vol. XIV, Charles Scribner's Sons, NY, 1976, pp. 219–224.

Dence, T.P., "On the Monotonicity of a Class of Exponential Sequences," *Amer. Math. Monthly*, **88**, 341–344 (1981).

DeTemple, D.W., "An Elementary Proof of the Monotonicity of $(1 + n^{-1})^n$ and $(1 + n^{-1})^{n+1}$," *Coll. Math. J.*, **36**, 147–149 (2005).

Freudenthal, H., "Augustin-Louis Cauchy," in Gillispie, C.C. (ed.), *Dictionary of Scientific Biography*, Vol. III, Charles Scribner's Sons, NY, 1971, pp. 131–148.

Gardner, M., "Catalan Numbers: An Integer Sequence that Materializes in Unexpected Places," *Sci. Amer.*, **234**, 120–125 (1976).

Grattan-Guiness, I., "Bolzano, Cauchy and the 'New Analysis' of the Early Nineteenth Century," *Arch. Hist. Exact Sci.*, **6**, 372–400 (1970).

Guillera, J., "Easy Proofs of Some Borwein Algorithms for π," *Amer. Math. Monthly*, **115**, 850–854 (2008) (very efficient, creative uses of some sequences).

Kline, M., *Mathematical Thought from Ancient to Modern Times*, Oxford University Press, NY, 1972, pp. 947–978 (the installation of rigor in analysis).

Meyer, B., "On the Cauchy Convergence Criterion," *Amer. Math. Monthly*, **62**, 488 (1955) (use of Dedekind cuts to prove that $\mathbf{R}^1$ is a complete metric space).

Pulapaka, H., "On Generalized Alternating Galileo Sequences," *Missouri J. Math. Sci.*, **20**, 178–199 (2008).

Russ, S., "Bolzano's Analytic Programme," *Math. Intelligencer*, **14**, 45–53 (1992).

Russ, S., *The Mathematical Works of Bernard Bolzano*, Oxford University Press, Oxford, 2004.

Saliga, L.M., "Excitement from an Error," *Math. Mag.*, **80**, 304–305 (2007) (use of the Bolzano-Weierstrass Theorem).

Sprecher, D.A., *Elements of Real Analysis*, Dover Publications, NY, 1987, pp. 72–74 (convergence of Cauchy sequences in $\mathbf{R}^1$ implies the Axiom of Completeness).

van Rootselaar, B., "Bernard Bolzano," in Gillispie, C.C. (ed.), *Dictionary of Scientific Biography*, Vol. II, Charles Scribner's Sons, NY, 1970, pp. 273–279.

Wästlund, J., "An Elementary Proof of the Wallis Product Formula for Pi," *Amer. Math. Monthly*, **114**, 914–917 (2007) (uses elementary sequences).

Zimmerman, S. and Ho, C.-W., "On Infinitely Nested Radicals," *Math. Mag.*, **81**, 3–15 (2008).

CHAPTER 3

Infinite Series

"The sum of an infinite series of harmonic proportionals, $\frac{1}{1}+\frac{1}{2}+\frac{1}{3}+\frac{1}{4}+\frac{1}{5}+\cdots$ etc., is infinite."

Jakob Bernoulli

CONTENTS

Reviewed in this chapter	Elementary features of infinite series; ratio, root tests.
New in this chapter	Convergence of series in $\mathbf{R}^k$; proofs of convergence tests; absolute vs. conditional convergence; Cauchy multiplication; series of functions.

3.1 CONVERGENCE OF INFINITE SERIES

In this chapter we outline the theory of infinite series, primarily as an extension of the theory of sequences. Our focus will be on $\mathbf{R}^k$, principally on $\mathbf{R}^1$. If $\{\mathbf{a}_n\}_{n=1}^{\infty}$ is a sequence of elements in $\mathbf{R}^k$, there is automatically associated with it an allied sequence $\{\mathbf{s}_n\}_{n=1}^{\infty}$ of **partial sums**, where $\mathbf{s}_n = \mathbf{a}_1 \oplus \mathbf{a}_2 \oplus \cdots \oplus \mathbf{a}_n = \sum_{i=1}^{n} \mathbf{a}_i$. An **infinite series**, denoted $\sum_{n=1}^{\infty} \mathbf{a}_n$, is a multilayered mathematical object that incorporates the notions of both sequences. Front and center is the notion of convergence of an infinite series.

Definition. *An infinite series in $\mathbf{R}^k$ is said to* ***converge*** *to $\mathbf{S} \in \mathbf{R}^k$, called the* ***sum*** *of the series, iff, given any $\varepsilon > 0$, there is an $N \in \mathbf{N}$ such that for all $n > N$ we have $\mathbf{s}_n \in \mathbf{B}_k(\mathbf{S}; \varepsilon)$. A series that does not converge is said to* ***diverge*** *(or to be* ***divergent****).*

The sum $\mathbf{S}$ in the definition is also commonly denoted by $\sum_{n=1}^{\infty} \mathbf{a}_n$. Thus, this symbol does double duty by standing for both the series itself and its sum, *if it has one*. The context will imply the intended meaning.

In order to make series useful for mathematical operations, we endow them with certain elementary properties. The following definitions are made ($\cong$ means "left-hand side is defined by the right-hand side"):

Property 1. (Scalar Multiplication) If $c \in \mathbf{R}$ and a series is defined in $\mathbf{R}^k$, then

$$c \cdot \sum_{n=1}^{\infty} \mathbf{a}_n \cong \sum_{n=1}^{\infty} (c \cdot \mathbf{a}_n).$$

Property 2. (Series Addition) If two series are defined in $\mathbf{R}^k$, then

$$\sum_{n=1}^{\infty} \mathbf{a}_n \oplus \sum_{n=1}^{\infty} \mathbf{b}_n \cong \sum_{n=1}^{\infty} (\mathbf{a}_n \oplus \mathbf{b}_n).$$

Note that the properties have nothing to do with convergence. Fortunately, they lead to desirable practical consequences.

Theorem 3.1. *Suppose that two series in $\mathbf{R}^k$ have the following sums: $\sum_{n=1}^{\infty} \mathbf{a}_n = \mathbf{A}$, $\sum_{n=1}^{\infty} \mathbf{b}_n = \mathbf{B}$, and that $c \in \mathbf{R}$. Then we have*

(i) $c \cdot \sum_{n=1}^{\infty} \mathbf{a}_n = c \cdot \mathbf{A}$;

(ii) $\sum_{n=1}^{\infty} \mathbf{a}_n \oplus \sum_{n=1}^{\infty} \mathbf{b}_n = \mathbf{A} \oplus \mathbf{B}$.

Proof.

(i) For each $n \in \mathbf{N}$, let $\mathbf{s}_n = \sum_{i=1}^{n} \mathbf{a}_i$. Then by Property 1, $c \cdot \sum_{n=1}^{\infty} \mathbf{a}_n \cong \sum_{n=1}^{\infty} c \cdot \mathbf{a}_n \cong \lim_{n\to\infty} \sum_{i=1}^{n} c \cdot \mathbf{a}_i = c \cdot \lim_{n\to\infty} \sum_{i=1}^{n} \mathbf{a}_i = c \cdot \lim_{n\to\infty} \mathbf{s}_n = c \cdot \mathbf{A}$.

(ii) This part is left to you. ■

■ Example 3.1

The series $\sum_{n=1}^{\infty} a_n$ is defined for all $n \in \mathbf{N}$ by $a_n = c^n, 0 < |c| < 1$. Then $(1-c)s_n = c + c^2 + c^3 + \cdots + c^n - [c^2 + c^3 + c^4 + \cdots + c^{n+1}]$

$$s_n = (c - c^{n+1})/(1-c),$$

and since $\lim_{n\to\infty} c^{n+1} = 0$ because $|c| < 1$, then

$$S = \lim_{n\to\infty} s_n = c/(1-c).$$

The series is called a **geometric series** because for each $n \in \mathbf{N}$, the ratio a_{n+1}/a_n is a fixed constant c. Geometric series are useful throughout mathematics. ■

■ Example 3.2

Let the two series $\sum_{n=1}^{\infty} a_n, \sum_{n=1}^{\infty} b_n$ be defined for all $n \in \mathbf{N}$ by $a_n = [\sin(\pi/4)]^n$, $b_n = [-\cos(\pi/4)]^n$. Then by Theorem 3.1

$$\sum_{n=1}^{\infty} a_n \oplus \sum_{n=1}^{\infty} b_n = \frac{\sin(\pi/4)}{1-\sin(\pi/4)} - \frac{\cos(\pi/4)}{1+\cos(\pi/4)}$$
$$= (\sqrt{2}+1) - (\sqrt{2}-1) = 2.$$ ■

■ Example 3.3

Let r_1, c be given, $0 < c < 1$. Circles tangent to each other, centered at O, O′, and of radii $r_1, r_2 = cr_1$, are constructed (Figure 3.1). Isosceles triangle ABC is circumscribed about the two circles; this triangle is uniquely determined by r_1, c. Line segment AM passes through O, O′, and is perpendicular to BC; let $h = \overline{\text{AM}}$ and $b = \overline{\text{BC}}$. Triangles AMB, ADO, AD′O′ are similar, so we have

$$\frac{\overline{\text{AB}}}{\overline{\text{MB}}} = \frac{\overline{\text{AO}}}{\overline{\text{DO}}} = \frac{\overline{\text{AO}'}}{\overline{\text{D}'\text{O}'}}, \quad \text{or} \quad \frac{\sqrt{h^2 + (b^2/4)}}{b/2} = \frac{h - r_1}{r_1} = \frac{h - 2r_1 - cr_1}{cr_1}.$$

The two equalities yield $h = 2r_1/(1-c)$ and $b = 2r_1/\sqrt{c}$. The figure corresponds roughly to $c = 0.5$.

We notice now that if the ascending tower of inscribed circles were continued, then each $r_n = cr_{n-1} = c^{n-1}r_1$, and the height H of the tower would be

$$H = \sum_{n=1}^{\infty} 2r_n = \sum_{n=1}^{\infty} 2c^{n-1}r_1 = 2r_1 \sum_{k=1}^{\infty} c^k = 2r_1\left(\frac{1}{1-c}\right) = h.$$

So the diameters of the circles are the terms of an (infinite) geometric series whose sum is exactly the height of the circumscribing triangle ABC. ■

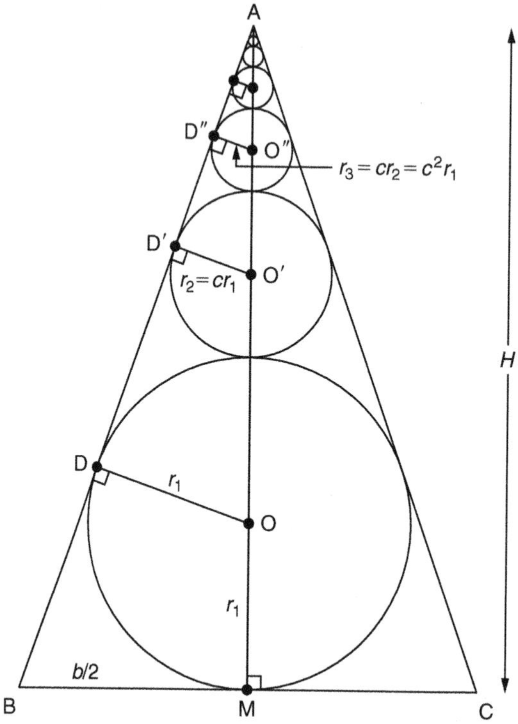

FIGURE 3.1

A geometric series in plane geometry.

Theorem 3.2. *For each $n \in \mathbf{N}$, let $\mathbf{x}_n = (x_{n1}, x_{n2}, \ldots, x_{nk})$ be a vector in $\mathbf{R}^k$. Then $\sum_{n=1}^{\infty} \mathbf{x}_n$ converges to $\mathbf{S} = (S_1, S_2, \ldots, S_k) \in \mathbf{R}^k$ iff each component series $\sum_{j=1}^{\infty} x_{ji}$ converges to $S_i, i = 1, 2, \ldots, k$.*

Proof. ($\rightarrow$) Suppose that $\sum_{n=1}^{\infty} \mathbf{x}_n$ converges to $\mathbf{S} = (S_1, S_2, \ldots, S_k) \in \mathbf{R}^k$. If $\varepsilon > 0$ is given, there is an $N \in \mathbf{N}$ such that for all $m > N$

$$\|\mathbf{s}_m - \mathbf{S}\| = \left[\sum_{i=1}^{k} (s_{mi} - S_i)^2\right]^{1/2} < \varepsilon,$$

where $\mathbf{s}_m = \sum_{i=1}^{m} \mathbf{x}_i = (s_{m1}, s_{m2}, \ldots, s_{mk})$ and each $s_{mi} = \sum_{j=1}^{m} x_{ji}$. But, $\max\limits_{1 \le i \le k} \{|s_{mi} - S_i|\} \le \left[\sum_{i=1}^{k} (s_{mi} - S_i)^2\right]^{1/2}$ holds, so each $|s_{mi} - S_i| < \varepsilon$. This says that each $\sum_{j=1}^{\infty} x_{ji}$ converges to $S_i \in \mathbf{R}^1, i = 1, 2, \ldots, k$.

($\leftarrow$) Completion of this part of the proof is left to you. ■

■ Example 3.4

Let $\sum_{n=1}^{\infty} \mathbf{x}_n$ be defined in $\mathbf{R}^2$ by $\mathbf{x}_n = \left([\sin(\pi/4)]^n, [-\cos(\pi/4)]^n\right)$.

From Example 3.2 we have

$$\sum_{n=1}^{\infty} [\sin(\pi/4)]^n = \sqrt{2} + 1, \quad \sum_{n=1}^{\infty} [-\cos(\pi/4)]^n = -\sqrt{2} + 1.$$

Hence, by Theorem 3.2, $\sum_{n=1}^{\infty} \mathbf{x}_n = (\sqrt{2} + 1, -\sqrt{2} + 1)$. ■

3.2 ELEMENTARY THEOREMS ON SUMS

Any method of attaching a sum to an infinite series may be called a **method of summability**. In this book we shall always understand sum to mean the limit of the sequence of partial sums, when this limit is in $\mathbf{R}^1$ (or, more generally, in $\mathbf{R}^k$). Otherwise, the symbol $\sum_{n=1}^{\infty} \mathbf{a}_n$, which always represents a series, will not represent a sum. There is, however, an extensive literature on methods of summability (Hardy, 1949), in which other definitions of sum are used (Exercises 3.11, 3.12). It is historically interesting that a fairly clear definition of (the usual) sum of an infinite series of real constants was given as early as 1742 by the Scottish mathematician **Colin MacLaurin** (1698–1746) in his book, *Treatise of Fluxions* (Grabiner, 1997).

It is easy in some cases to determine that a series does not have a (standard) sum. For this, the following theorem is useful.

Theorem 3.3. *If $\sum_{n=1}^{\infty} \mathbf{x}_n$ converges in $\mathbf{R}^k$, then* $\lim_{n\to\infty} \mathbf{x}_n = \mathbf{0}$.

Proof. By hypothesis, the sequence $\{\mathbf{s}_n\}_{n=1}^{\infty}$ of partial sums converges, so it must be a Cauchy sequence in $\mathbf{R}^k$ (Theorem 2.17). Hence, if $\varepsilon > 0$ is given, then there is an $N \in \mathbf{N}$ such that for all $n > N$ one has $||\mathbf{s}_{n+1} - \mathbf{s}_n|| = ||\mathbf{x}_{n+1} - \mathbf{0}|| < \varepsilon$. This says that $\lim_{n\to\infty} \mathbf{x}_n = \mathbf{0}$. ■

Notice how closely Theorem 3.3 depends upon the nature of real numbers, namely, upon the completeness of the metric spaces $\mathbf{R}^k$. This latter, in turn, is a consequence of the Completeness Axiom.

■ Example 3.5

Let the series $\sum_{n=1}^{\infty} a_n$ be defined for all $n \in \mathbf{N}$ by $a_n = (n^2 + 2)/n^2$. Then $\lim_{n\to\infty} a_n = \lim_{n\to\infty} (1 + 2n^{-2}) = 1$, from Example 2.13(b); hence, the series diverges according to the contrapositive of Theorem 3.3. ■

■ Example 3.6

The series in Example 3.4 converges, so $\lim_{n\to\infty} \mathbf{x}_n = \mathbf{0}$; indeed,

$$\lim_{n\to\infty} [\sin(\pi/4)]^n = \lim_{n\to\infty} [-\cos(\pi/4)]^n = 0.$$

■

Since Theorem 3.3 is a necessary condition for convergence, it cannot be used to show that a series converges. In Example 3.8, shortly, you will see a series that diverges, even though $\lim_{n\to\infty} a_n = 0$. In this regard, special interest attaches to **positive series** (each term is positive). The following important result is quite general.

Theorem 3.4 (Comparison Test). *Suppose two positive series, $\sum_{n=1}^{\infty} \|\mathbf{a}_n\|$ and $\sum_{n=1}^{\infty} \|\mathbf{b}_n\|$, defined in $\mathbf{R}^k$, are such that for all n larger than some $N \in \mathbf{N}$ we have $0 < \|\mathbf{a}_n\| \leq \|\mathbf{b}_n\|$. Then if the* **b***-series has a sum in $\mathbf{R}^1$, so does the* **a***-series.*

Proof. For each $n \in \mathbf{N}$, let $s_n = \sum_{k=N+1}^{N+n} \|\mathbf{a}_k\|$, $t_n = \sum_{k=N+1}^{N+n} \|\mathbf{b}_k\|$. By hypothesis, the sequence $\{t_n\}_{n=1}^{\infty}$ converges, so by Theorem 2.4 each t_n is bounded from above by some $M > 0$. Since, for $k \geq N+1$, $\|\mathbf{a}_k\| \leq \|\mathbf{b}_k\|$ holds, then each s_n is also bounded from above by M. But the sequence $\{s_n\}_{n=1}^{\infty}$ is increasing, so by Theorem 2.2 it converges to some $L > 0$. Hence, the **a**-series has the sum $\sum_{n=1}^{\infty} \|\mathbf{a}_n\| = L + \sum_{n=1}^{N} \|\mathbf{a}_n\|$. ■

The theorem is useful in the manner stated and also in the form of its contrapositive. The Comparison Test has been generalized somewhat very recently (Longo and Valori, 2006).

■ Example 3.7

Define $\sum_{n=1}^{\infty} \|\mathbf{a}_n\|$ and $\sum_{n=1}^{\infty} \|\mathbf{b}_n\|$ in $\mathbf{R}^2$ by $\mathbf{a}_n = (\sqrt{2}e^{-n^2/5}, \sqrt{2}e^{-n^2/5})$, and $\mathbf{b}_n = \left(\frac{3}{5}(2^{-n}), \frac{1}{5}(2^{2-n})\right)$. Then for all $n > N = 4$ we have $0 < \|\mathbf{a}_n\| < \|\mathbf{b}_n\|$ (verify!). Since the **b**-series is geometric and converges, then by Theorem 3.4 the **a**-series also converges. ■

■ Example 3.8

The **harmonic series** $\sum_{n=1}^{\infty} b_n$ is defined for each $n \in \mathbf{N}$ by $b_n = 1/n$.

An a-series is constructed as follows:

$$a_1 = a_2 = 1/2, \quad a_3 = a_4 = 1/4, \quad a_5 = a_6 = a_7 = a_8 = 1/8,$$

$$a_9 = a_{10} = \cdots = a_{16} = 1/16, \quad a_{2^{k-1}+1} = a_{2^{k-1}+2} = \cdots = a_{2^k} = 1/2^k.$$

By induction we have $s_{2^k} = \sum_{j=1}^{2^k} a_j = (k+1)/2$ (verify!). Thus, the subsequence of partial sums, $\{s_{2^k}\}_{k=1}^{\infty}$, diverges to $\lim_{k\to\infty} (k+1)/2 = \infty$. By Theorem 2.12, $\{s_n\}_{n=1}^{\infty}$ cannot converge and $\sum_{n=1}^{\infty} a_n$ must diverge to ∞.

Each term in the a-series satisfies $a_n \leq b_n$, equality holding only when $n = 2^k$, $k \in \mathbf{N}$. Hence, from Theorem 3.4, $\sum_{n=1}^{\infty} b_n = \sum_{n=1}^{\infty} (1/n)$ diverges to ∞.[1] ■

Even though it is divergent, the harmonic series is actually a very interesting series. For example, it is a very *slowly* diverging series. We begin to get an inkling of this when we compute some of the partial sums: $s_5 = 2.283333$, $s_{15} = 3.318229, s_{30} = 3.994987$. These sums are increasing fairly slowly. The divergence of the harmonic series has been studied in some detail (Boas, Jr. and Wrench, Jr., 1971). Let $n(A)$ denote the smallest integer n such that s_n just exceeds A. Table 3.1 gives some numerical data. The results are certainly impressive. Amazingly, there are series that diverge to ∞ even more slowly than the harmonic series (Agnew, 1947)!

Table 3.1 Selected Partial Sums of the Harmonic Series

A	$n(A)$	$s_{n(A)}$
3	11	3.01987734
6	227	6.00436670
10	12,367	10.00004301
13	248,397	13.00000123
18	36,865,412	18.000000004

What do you think would happen if the harmonic series were to be thinned out by the removal randomly of a billion terms? Or even by the selective removal of an infinite subset of terms? A **block** of some set of digits is a string $b_1b_2b_3 \ldots b_k$ of consecutively written digits. Thus, the integer 140561 contains the block 056 but not the block 450. Let $\{X_n(b_1b_2b_3 \ldots b_k)\}_{n=1}^{\infty}$ denote the increasing sequence of natural numbers that *do not* contain the block $b_1b_2b_3 \ldots b_k$. The associated series

$$\sum_{n=1}^{\infty} \frac{1}{X_n(b_1b_2b_3 \ldots b_k)}$$

may be called a **thinned-out harmonic series.**

It was proved in 1914 that when the block is the one-digit block 9, the thinned-out harmonic series converges. It is now known that convergence holds for any block (Hegyvári, 1993). Isn't that interesting! Of the nature of the sum in each

[1]The essence of this proof, albeit with much less rigor, was apparently first given by the French scholar **Nicole Oresme** (ca. 1323–1382) in a tract *Quaestiones super Geometriam Euclidis* (ca. 1360) (Kline, 1972). Oresme eventually became a Bishop of Lisieux, a town in the northwestern French department of Calvados. An excellent historical discussion of the harmonic series is given in Dunham (1987); this is a must read.

case, very little is known. It may be that in many (most? all?) cases the sum is irrational, but nobody knows.

However, it is known that these thinned-out harmonic series converge *very slowly*, too slowly to make direct estimation of their sums feasible. A very recent paper presents some algorithmic procedures for estimating the sums to high precision (Schmelzer and Baillie, 2008). For example, the thinned-out series of terms in which no 9 appears in the denominator has a sum (rounded to eight decimals) of 22.92067662.

We can, however, carry out the following highly simplified calculation. Partition all the terms of the harmonic series into (a) those with denominators in the interval $[1, 10)$, (b) those with denominators in $[10, 100)$, (c) those with denominators in $[100, 1000)$, and so on. The number of integers of the form $X_n(9)$ in $[10^{n-1}, 10^n)$ that have no 9 in their base-10 expansions is $8(9^{n-1})$ (why?). You can then establish that the thinned-out harmonic series

$$\sum_{n=1}^{\infty} \frac{1}{X_n(9)}$$

converges to a sum that does not exceed 74 ¾ (Exercise 3.14). This upper bound is, as seen, rather crude.

3.3 ADDITIONAL CONVERGENCE TESTS

Although the Comparison Test is powerful, it requires us to have on hand a sufficient arsenal of convergent and divergent positive series. So far, we have only geometric series and the harmonic series; further, use of the Comparison Test sometimes necessitates a certain amount of ingenuity. The present section introduces some alternative techniques that can be easier to apply. We refer to these as **convergence tests**; several are known, and without belaboring the point, we shall present only a few here and a few more (as optional items) in the Exercises.

Theorem 3.5 (Limit Comparison Test). *Suppose that* $\sum_{n=1}^{\infty} \|\mathbf{a}_n\|, \sum_{n=1}^{\infty} \|\mathbf{b}_n\|$ *are such that* $\lim_{n\to\infty} (\|\mathbf{a}_n\|/\|\mathbf{b}_n\|) = L \in \mathbf{R}^1, L > 0$. *Then the two series converge or diverge together. But if* $\lim_{n\to\infty} (\|\mathbf{a}_n\|/\|\mathbf{b}_n\|) = 0$, *then the only allowed conclusion is that convergence of the* **b***-series implies convergence of the* **a***-series.*

Proof. Suppose that $0 < L < \infty$; then there is an $N \in \mathbf{N}$ such that for all $n > N$, we have $(9/10)L < \|\mathbf{a}_n\|/\|\mathbf{b}_n\| < (10/9)L$, or equivalently,

$$\frac{9}{10}L\,\|\mathbf{b}_n\| < \|\mathbf{a}_n\| < \frac{10}{9}L\,\|\mathbf{b}_n\|.$$

If $\sum_{n=1}^{\infty} \|\mathbf{b}_n\|$ converges, then so does $\sum_{n=1}^{\infty} \frac{10}{9} L\|\mathbf{b}_n\|$ (Theorem 3.1) and, hence, also $\sum_{n=1}^{\infty} \|\mathbf{a}_n\|$ (Theorem 3.4). If $\sum_{n=1}^{\infty} \|\mathbf{a}_n\|$ converges, then so does $\sum_{n=1}^{\infty} \frac{9}{10} L\|\mathbf{b}_n\|$ and, hence, also $\sum_{n=1}^{\infty} \|\mathbf{b}_n\|$. Similarly, by reasoning contrapositively, it is seen that $\sum_{n=1}^{\infty} \|\mathbf{a}_n\|$ diverges to ∞ iff $\sum_{n=1}^{\infty} \|\mathbf{b}_n\|$ diverges to ∞.

Suppose that $L = 0$; as $\{\|\mathbf{a}_n\| / \|\mathbf{b}_n\|\}_{n=1}^{\infty}$ converges, then by Theorem 2.4 there is a number $M > 0$ such that for each $n \in \mathbf{N}$, we have $\|\mathbf{a}_n\| \,/\, \|\mathbf{b}_n\| < M$, or equivalently,

$$0 < \|\mathbf{a}_n\| < M\|\mathbf{b}_n\|.$$

From this we can only conclude that if $\sum_{n=1}^{\infty} \|\mathbf{b}_n\|$ converges, then so does $\sum_{n=1}^{\infty} M\|\mathbf{b}_n\|$ and, hence, also $\sum_{n=1}^{\infty} \|\mathbf{a}_n\|$. ■

The Limit Comparison Test has one advantage over the Comparison Test: It does not require us to be quite so clever in how to choose a series $\sum_{n=1}^{\infty} \|\mathbf{b}_n\|$ in order to establish the convergence (or not) of $\sum_{n=1}^{\infty} \|\mathbf{a}_n\|$.

■ Example 3.9

We establish for future reference the useful fact that $\sum_{n=1}^{\infty} \frac{1}{n^2}$ converges. First observe that for all $n \in \mathbf{N}$,

$$\frac{1}{n^2} \le \frac{2}{n(n+1)} = \frac{2}{n} - \frac{2}{n+1}.$$

$$\begin{aligned}
\text{Then } s_n &= \sum_{k=1}^{n} \left(\frac{2}{k} - \frac{2}{k+1}\right) \\
&= \left(\frac{2}{1} - \frac{2}{2}\right) + \left(\frac{2}{2} - \frac{2}{3}\right) + \left(\frac{2}{3} - \frac{2}{4}\right) + \cdots + \left(\frac{2}{n} - \frac{2}{n+1}\right) \\
&= 2 - \frac{2}{n+1} \quad \text{after extensive cancellation.}
\end{aligned}$$

Thus, $\lim_{n\to\infty} s_n = 2 = \sum_{n=1}^{\infty} \frac{2}{n(n+1)}$, so by Theorem 3.4 $\sum_{n=1}^{\infty} \frac{1}{n^2}$ has a sum; it is known that this sum is $\pi^2/6$. The series $\sum_{n=1}^{\infty} \frac{2}{n(n+1)}$ is termed a **telescoping series** because each partial sum is a shortening or condensation of several terms of the series. ■

■ Example 3.10

Let $\sum_{n=1}^{\infty} a_n$ be $\sum_{n=1}^{\infty} \frac{n}{2n^3-3n+4}$ and let $\sum_{n=1}^{\infty} b_n$ be $\sum_{n=1}^{\infty} \frac{1}{n^2}$. Then $\lim_{n\to\infty} (a_n/b_n) = \lim_{n\to\infty} \frac{n^3}{2n^3-3n+4} = \lim_{n\to\infty} \frac{1}{2-3n^{-2}+4n^{-3}} = 1/2$, after making use of Example 2.13(b). Hence, by Theorem 3.5 the series $\sum_{n=1}^{\infty} \frac{n}{2n^3-3n+4}$ converges. ■

Note that since Theorem 3.5 depends upon Theorem 3.4, both theorems ultimately are consequences of the Axiom of Completeness, since this was needed in the proof of Theorem 3.4 (where?).

Before reading the next two theorems, please review Section 2.6, especially Theorem 2.14(a) and Exercise 2.37. Note that the two theorems to follow are not restricted to positive series.

Theorem 3.6 (Ratio Test). *Let $\sum_{n=1}^{\infty} a_n$ be a series of nonzero real numbers, and denote*

$$v = \limsup_{n\to\infty} |a_{n+1}/a_n|, \quad w = \liminf_{n\to\infty} |a_{n+1}/a_n|.$$

Consequently,

- **(a)** *if $v < 1$, then $\sum_{n=1}^{\infty} |a_n|$ converges;*
- **(b)** *if $w > 1$, then $\sum_{n=1}^{\infty} a_n$ diverges;*
- **(c)** *if $w \le 1 \le v$, no conclusion can be reached.*

Proof.

- **(a)** Suppose $v < 1$; choose L such that $v < L < 1$. From Theorem 2.14(a), there is an $N \in \mathbf{N}$ such that for all $n > N$ we have $|a_{n+1}/a_n| < L < 1$; hence, by induction, $k \in \mathbf{N}$ implies that $|a_{N+1+k}| < L^k |a_{N+1}|$. But $\sum_{k=1}^{\infty} L^k |a_{N+1}|$ is a convergent geometric series (Example 3.1), so $\sum_{n=1}^{\infty} |a_n| = \sum_{n=1}^{N+1} |a_n| + \sum_{n=N+2}^{\infty} |a_n|$ converges by the Comparison Test.
- **(b)** Suppose $w > 1$; now select, arbitrarily, l such that $1 < l < w$. Then from Exercise 2.37(b) there is an $N \in \mathbf{N}$ such that for all $n > N$ we have $|a_{n+1}/a_n| > l > 1$. Thus, $|a_n| < |a_{n+1}| < |a_{n+2}| < \cdots$, so $\lim_{n\to\infty} |a_n| = 0$ is impossible. Consequently, $\lim_{n\to\infty} a_n = 0$ is also impossible, so by Theorem 3.3 the series $\sum_{n=1}^{\infty} a_n$ cannot converge.
- **(c)** Suppose $w \le 1 \le v$ and let $\varepsilon > 0$ be given. Then there is an $N \in \mathbf{N}$ such that for all $m > N$ we have $w - \varepsilon < |a_{n+1}/a_n| < v + \varepsilon$. But we cannot judge whether these ratios $|a_{n+1}/a_n|$ are larger or smaller than 1, so the Ratio Test is inconclusive in this case. ■

Theorem 3.7 (Root Test). *Let $\sum_{n=1}^{\infty} a_n$ be a series of real numbers and denote $v = \limsup_{n\to\infty} \sqrt[n]{|a_n|}$. Consequently,*

- **(a)** *if $v < 1$, then $\sum_{n=1}^{\infty} |a_n|$ converges;*
- **(b)** *if $v > 1$, then $\sum_{n=1}^{\infty} a_n$ diverges;*
- **(c)** *if $v = 1$, the test is inconclusive.*

Proof. Reasoning proceeds along lines similar to those used in the proof of the Ratio Test. The completion of the proof is left to you. ■

You will see shortly from Theorem 3.8 that the conclusions in Theorems 3.6(a) and 3.7(a) can be strengthened to read "then $\sum_{n=1}^{\infty} a_n$ converges."

■ Example 3.11

Consider $\sum_{n=1}^{\infty}(-1)^{n+1}/F_n$, where F_n is the nth Fibonacci number (Section 2.1). Let $a_{n+1} = (-1)^{n+2}/F_{n+1}$ and $|a_{n+1}/a_n| = F_n/F_{n+1}$. It is known that (see Exercises 2.8 and 2.14)

$$\lim_{n\to\infty}\frac{F_n}{F_{n+1}} = \lim_{n\to\infty}\left(\frac{F_{n+1}}{F_n}\right)^{-1} = \left(\frac{1+\sqrt{5}}{2}\right)^{-1} = \frac{\sqrt{5}-1}{2} < 1.$$

For this special case, $w = v = (\sqrt{5}-1)/2$, so from Theorem 3.6(a) we see that $\sum_{n=1}^{\infty}|(-1)^{n+1}/F_n|$ converges. This sum has been proved to be irrational. ■

■ Example 3.12

Consider $\sum_{n=1}^{\infty} a_n$, where $a_n = (3/2)^{\lfloor n/2\rfloor}\cdot 2^{1-n}$, and $\lfloor x\rfloor$ is the **floor function** of x: $\lfloor x\rfloor$ is the largest integer that does not exceed x. The partial sum s_7, for example, is

$$s_7 = 1 + \frac{3}{4} + \frac{3}{8} + \frac{9}{32} + \frac{9}{64} + \frac{27}{256} + \frac{27}{512}.$$

We observe that for even $n(n = 2k)$, we have $a_{2k} = (3/2)^k 2^{1-2k} = (3/8)^k \cdot 2$, and for odd $n(n = 2k-1)$, we have $a_{2k-1} = (3/2)^{k-1}2^{1-(2k-1)} = (3/8)^{k-1}$. Hence, for each $k \in \mathbf{N}$, we have $|a_{2k}/a_{2k-1}| = 3/4$ and $|a_{2k+1}/a_{2k}| = 1/2$, and so $v = \limsup_{n\to\infty}|a_{n+1}/a_n| = 3/4 < 1$ and $w = \liminf_{n\to\infty}|a_{n+1}/a_n| = 1/2$. By Theorem 3.6(a) we conclude that $\sum_{n=1}^{\infty} a_n$ converges. The sum, in fact, is 14/5 (Exercise 3.18(a)). ■

■ Example 3.13

Consider $\sum_{n=2}^{\infty} a_n$, where $a_n = [(\ln n)/n]^n$. Since the general term involves an nth power, use of the Root Test is suggested. The sequence $\{\sqrt[n]{|a_n|}\}_{n=2}^{\infty} = \{(\ln n)/n\}_{n=2}^{\infty}$ becomes decreasing for $n \geq 3$, and is bounded below by 0; hence, it converges. We observe that $\ln n < \sqrt{n}$ is true for $n = 4$, and that the function $f(x) = \sqrt{x} - \ln x$ is an increasing function for $x > 4$ (Exercise 3.18(b)). Thus, $0 < \ln n < \sqrt{n}$ is true for all integers $n \geq 4$, so we have

$$0 \leq \frac{\ln n}{n} \leq \frac{\sqrt{n}}{n}.$$

By the Squeeze Theorem (Exercise 2.22), we deduce that $\lim_{n\to\infty}[(\ln n)/n] = 0$, and so for $\{(\ln n)/n\}_{n=2}^{\infty}$ we have $w = v = 0$. Thus, by Theorem 3.7(a) it follows that $\sum_{n=2}^{\infty}[(\ln n)/n]^n$ converges. ■

We make two comments about the two previous theorems. First, versions of the Ratio Test have been known for more than 225 years (Kline, 1972). In spite of this, the last word on this marvelous result has not yet been uttered. A very recent, exciting paper presents a so-called Second Ratio Test, along with several interesting, worked examples (Ali, 2008).

Second, an especially interesting and instructive application of Cauchy's Root Test is to an efficient proof of this important limit (Wiener, 1987):

$$\lim_{n\to\infty}\frac{\sqrt[n]{n!}}{n} = \frac{1}{e}.$$

Suppose that for a positive series $\sum_{n=1}^{\infty} a_n$ we have $\lim_{n\to\infty}(a_{n+1}/a_n) = A \in \mathbf{R}^1$. Hence, if $\varepsilon > 0$ is given, then there is an $N \in \mathbf{N}$ such that for all $k > N$ we have $A - \varepsilon < a_{k+1}/a_k < A + \varepsilon$. Write this pair of inequalities for $k = N+1, N+2, \ldots, n-1$ and multiply corresponding sides throughout. Extensive cancellation leads to

$$(A-\varepsilon)^{n-N} < \frac{a_n}{a_N} < (A+\varepsilon)^{n-N}.$$

Multiply through by a_N and take the nth root throughout:

$$a_N^{1/n}(A-\varepsilon)^{(n-N)/n} < a_n^{1/n} < a_N^{1/n}(A+\varepsilon)^{(n-N)/n}.$$

The Comparison Theorem for sequences in $\mathbf{R}^1$ (Theorem 2.8) then gives

$$A - \varepsilon < \lim_{n\to\infty} a_n^{1/n} < A + \varepsilon,$$

which says that $\lim_{n\to\infty} a_n^{1/n} = A$.

Hence, if a positive series is predicted by the Ratio Test to converge because $\lim_{n\to\infty}(a_{n+1}/a_n) = A < 1$, then the Root Test will make the same prediction, and $\lim_{n\to\infty}\sqrt[n]{a_n} = \lim_{n\to\infty}(a_{n+1}/a_n) = A$. For the series whose general term is $a_n = n!/n^n$, we then obtain

$$\lim_{n\to\infty}\frac{\sqrt[n]{n!}}{n} = \lim_{n\to\infty}\frac{(n+1)!/(n+1)^{n+1}}{n!/n^n} = \lim_{n\to\infty}\left[1+\frac{1}{n}\right]^{-n} = \frac{1}{e},$$

from Example 2.9. This limit leads to one form of the useful Stirling's Approximation for factorials (Exercise 3.19).

3.4 ABSOLUTE AND CONDITIONAL CONVERGENCE; ALTERNATING SERIES

Series in $\mathbf{R}^1$ may contain some negative terms. Convergence can be considered, therefore, in two contexts. Although "positive" and "negative" are meaningless in $\mathbf{R}^k$, $k > 1$, the duality of convergence contexts can still be extended to the series in $\mathbf{R}^k$.

Definition. *Let $\sum_{n=1}^{\infty} \mathbf{a}_n$ be defined in $\mathbf{R}^k$, $k \geq 1$. The series is* ***absolutely convergent*** *iff the related series $\sum_{n=1}^{\infty} \|\mathbf{a}_n\|$ converges in $\mathbf{R}^1$. The original series is* ***conditionally convergent*** *iff it converges but not absolutely.*

Intuition suggests that absolute convergence should be a stronger condition than convergence. The meaning of this is as follows:

Theorem 3.8. *An absolutely convergent series of terms defined in $\mathbf{R}^k$ is automatically a convergent series.*

Proof. This follows readily from Cauchy's Convergence Criterion for Series (Exercises 3.6, 3.24). The proof is left to you. ■

■ Example 3.14

The series $\sum_{n=1}^{\infty} a_n$, where $a_n = (-1)^{n+1}/(2n-1)!$, converges absolutely (use Comparison Test with the series $\sum_{n=1}^{\infty} b_n$, $b_n = 2^{2-n}$). In fact,

$$\sum_{n=1}^{\infty} \frac{1}{(2n-1)!} = \sinh(1).$$

The series $\sum_{n=1}^{\infty} (-1)^{n+1}/(2n-1)!$ then also converges (to $\sin(1)$). On the other hand, the series $\sum_{n=1}^{\infty} (-1)^{n+1}/n$ is only conditionally convergent (to $\ln 2$). ■

■ Example 3.15

The series $\sum_{n=1}^{\infty} \mathbf{a}_n$, where $\mathbf{a}_n = \left(\frac{\sin n}{n^2}, \frac{\cos n}{n^2}\right) \in \mathbf{R}^2$, converges absolutely, since $\sum_{n=1}^{\infty} \|\mathbf{a}_n\| = \sum_{n=1}^{\infty} \frac{1}{n^2}$, and Example 3.9 now applies. ■

Of special interest are those series in $\boldsymbol{R}^k$, such as $\sum_{n=1}^{\infty} a_n$ in Example 3.14, in which there is strict alternation of algebraic signs. These alternating series were studied by the German father of calculus, **Gottfried Wilhelm von Leibniz** (1646–1716), and later by Cauchy.

Theorem 3.9 (Alternating Series Test). *Let $\{b_n\}_{n=1}^{\infty}$ be a decreasing sequence of positive numbers that converges to 0. Then the series $\sum_{n=1}^{\infty} a_n \cong \sum_{n=1}^{\infty} (-1)^{n+1} b_n$ converges.*

Proof. We have for the partial sums and for any $n \in \mathbf{N}$,

$$s_{2n+1} = s_{2n-1} - (b_{2n} - b_{2n+1}) \leq s_{2n-1}$$
$$s_{2n+2} = s_{2n} - (b_{2n+1} - b_{2n+2}) \geq s_{2n},$$

and, therefore, $\{s_{2n}\}_{n=1}^{\infty}$ is an increasing sequence and $\{s_{2n-1}\}_{n=1}^{\infty}$ is a decreasing sequence. Further, we observe from

$$s_2 \leq s_4 \leq \cdots \leq s_{2n-2} \leq s_{2n} \leq (s_{2n} + b_{2n+1}) = s_{2n+1} \leq s_{2n-1} \leq \cdots \leq s_3 \leq s_1,$$

valid for any $n \in \mathbf{N}$, that the s_{2n}'s are bounded from above by each s_{2n-1} and the s_{2n-1}'s are bounded from below by each s_{2n}. It follows that

$$\begin{cases} v = \lim_{n\to\infty} \sup s_n = \lim_{n\to\infty} s_{2n-1} \\ w = \lim_{n\to\infty} \inf s_n = \lim_{n\to\infty} s_{2n}, \end{cases}$$

and, finally,

$$0 \leq v - w \leq s_{2n+1} - s_{2n} = b_{2n+1}.$$

Since $\lim_{n\to\infty} b_{2n+1} = 0$, then $w = v$ and we conclude $\sum_{n=1}^{\infty} a_n = \lim_{n\to\infty} s_n = v$. ■

Note that in order to use Theorem 3.9 as a test for convergence, three conditions must hold for all terms of the series beyond a certain point ($n > N$):

1. The terms must alternate in sign.
2. The terms must be decreasing in magnitude.
3. The nth term must approach 0 as $n \to \infty$.

■ Example 3.16

Consider the series $\sum_{n=2}^{\infty} (-1)^n \ln n/\sqrt{n}$. We observe that the function $f(x) = (\ln x)/\sqrt{x}$ becomes a decreasing function, since

$$\begin{aligned} f'(x) &= \left[1 - \frac{1}{2}\ln x\right] / x^{3/2} \\ &< 0 \quad \text{if } x > e^2 \approx 7.38. \end{aligned}$$

Hence, the terms of the series decrease in magnitude for $n \geq 8$. Next, we define $h(x) = \sqrt[3]{x} - \ln x$ and then obtain $h'(x) = \frac{1}{3}x^{-2/3} - x^{-1}$; this is positive for all

$x > 27$. Further, $\ln x < \sqrt[3]{x}$ for $x = 94$, and as $h(x)$ is an increasing function for $x > 27$, the inequality $\ln x < \sqrt[3]{x}$ holds for all $x \geq 94$. Borrowing the strategy shown in Example 3.13, we now write

$$0 \leq \frac{\ln n}{\sqrt{n}} \leq \frac{\sqrt[3]{n}}{\sqrt{n}} = \frac{1}{\sqrt[6]{n}} \quad (n \geq 94).$$

Since $\lim_{n\to\infty} 1/\sqrt[6]{n} = 0$, then by the Squeeze Theorem (Exercise 2.22) we must have $\lim_{n\to\infty} [(\ln n)/\sqrt{n}] = 0$. The three conditions for use of Theorem 3.9 now hold for $n \geq 8$, so $\sum_{n=2}^{\infty} (-1)^n \ln n/\sqrt{n}$ converges. ■

■ Example 3.17

Consider the series $\sum_{n=1}^{\infty} (-1)^{n+1}|\sin n|/n^2$. Requirement 2 for the use of Theorem 3.9 is not met. This does not mean that the series diverges, but only that we cannot apply Theorem 3.9. In fact, the series converges absolutely because $|a_n| \leq 1/n^2$, and Theorem 3.4 and Example 3.9 are now pertinent. Finally, Theorem 3.8 furnishes the conclusion that $\sum_{n=1}^{\infty} (-1)^{n+1}|\sin n|/n^2$ converges. ■

An important use of convergent infinite series is to the representation and estimation of various irrational numbers. Table 3.2 gives some examples of representations of irrational numbers by alternating series. Theorem 3.10 shows that some alternating series, including all of those in Table 3.2, have a useful approximation feature.

Table 3.2 Representations of Certain Irrational Numbers x by Alternating Series

x	Series	x_{app}^{a}	s_6^{b}
$\sin 1$	$\sum_{n=1}^{\infty} (-1)^{n+1}/(2n-1)!$	0.84147098	0.84147098
$\tan^{-1}(1/2)$	$\sum_{n=1}^{\infty} (-1)^{n+1}/[(2n-1)(2)^{2n-1}]$	0.46364761	0.46363989
$\pi^3/4$	$8\sum_{n=1}^{\infty} (-1)^{n+1}/(2n-1)^3$	7.75156917	7.74934351
$\ln(3/2)$	$\sum_{n=1}^{\infty} (-1)^{n+1}/[n(2)^n]$	0.40546511	0.40468750

[a] *The "exact value," rounded to eight decimals.*
[b] *The sixth partial sum of each indicated series.*

Theorem 3.10. *Let $\{b_n\}_{n=1}^{\infty}$ be a sequence in $\mathbf{R}^+$ such that $b_{n+1} \leq b_n$ for all $n > N$ and $\lim_{n\to\infty} b_n = 0$. Then the series $\sum_{n=1}^{\infty} (-1)^{n+1} b_n$ has partial sums s_n, $n > N$, that satisfy $0 < |S - s_n| < b_{n+1}$, where $S = \lim_{n\to\infty} s_n$.*

Proof. The proof of Theorem 3.9 has shown that for all integers $2k$, $2k+1 > N$ we have

$$s_{2k+1} - b_{2k+2} = s_{2k+2} < S < s_{2k+1},$$

or equivalently,

$$0 < s_{2k+1} - S < b_{2k+2}, \tag{*}$$

and also

$$s_{2k} - S < s_{2k+1} = s_{2k} + b_{2k+1},$$

or equivalently,

$$0 < S - s_{2k} < b_{2k+1}. \tag{**}$$

Equations (*) and (**) are equivalent to the single expression

$$0 < |S - s_n| < b_{n+1},$$

for all $n > N$. ■

In words, the theorem says that the magnitude of the error in approximating the sum of a convergent alternating series by its nth partial sum (subject to the condition in the hypothesis) is less than the magnitude of the first term not retained (Exercise 3.27).

■ Example 3.18

How many terms in the series expansion of $\pi^3/4$ in Table 3.2 shall be taken in order to give an error of magnitude less than 10^{-3}?

We desire $|\pi^3/4 - 8\sum_{k=1}^{n}(-1)^{k+1}/(2k-1)^3| < \frac{8}{(2n+1)^3} < 10^{-3}$, or equivalently, $(2n+1)^3 > 8 \times 10^3$; hence, $n > 19/2$, so 10 terms will suffice. Indeed, we find $s_{10} \approx 7.75107643$ and $|\pi^3/4 - 7.75107643| \approx 0.00049274 < 10^{-3}$. ■

It sometimes happens, as is true in the case of the series for $\pi^3/4$, that the differences $\Delta b_n = b_n - b_{n+1}$ themselves monotonically decrease beyond a certain point ($n > N$). Then it is known that the bounds given in the conclusion of Theorem 3.10 can be rewritten as follows (Young, 1985):

$$\frac{b_{n+1}}{2} < |S - s_n| < \frac{b_n}{2},$$

for all $n > N$. The change may or may not lead to modest improvement (Exercise 3.30(b)).

■ Example 3.19

From Table 3.2 we have $|\ln(3/2) - s_6| = 0.000778$. We compute $b_7/2 = 1/[7(2^8)] = 0.000558$ and $b_6/2 = 1/[6(2^7)] = 0.001302$; Young's result is upheld for $|\ln(3/2) - s_n|$ in the case of $n = 6$. More work is needed to show that Young's theorem applies to all n larger than some $N \in \mathbf{N}$. ■

3.5 SOME CONSEQUENCES OF ABSOLUTE CONVERGENCE

Absolute convergence is a strong condition, and series that possess it inherit special properties. For example, the following hold:

1. Absolute convergence of series is preserved in linear combinations (Exercise 3.33). That is, if $\sum_{n=1}^{\infty} \mathbf{x}_n$ and $\sum_{n=1}^{\infty} \mathbf{y}_n$ in $\mathbf{R}^k$ are absolutely convergent, then so are $\sum_{n=1}^{\infty} c \cdot \mathbf{x}_n$ and $\sum_{n=1}^{\infty} (\mathbf{x}_n \oplus \mathbf{y}_n)$, where $c \in \mathbf{R}$ is nonzero.

2. Sums of absolutely convergent series in $\mathbf{R}^1$ are invariant upon arbitrary rearrangements of the terms. Thus, the series $\sum_{n=1}^{\infty} (-1)^{n+1}/n^2$, which contains infinitely many terms of each sign, has the same sum (namely, $\pi^2/12$) if the series is written in any way whatsoever by a resequencing of terms.

   Absolute convergence is a *necessary* condition for invariance of the sum of a convergent series, for **Georg F. B. Riemann** (1826–1866) proved this gem:

Theorem 3.11 (Riemann's Rearrangement Theorem). *A given conditionally convergent series of real numbers can always be rearranged so as to converge to any real number, or even to diverge to $\pm\infty$.*

Proofs of this can be found in (Galanor, 1987) and (Rudin, 1976). Not surprisingly, the alternating harmonic series has been much studied; one result appears in (Dence, 2008).

In the remainder of this section we consider some questions regarding multiplication of infinite series, in which absolute convergence assumes importance. In Section 3.1 we defined the sum of two infinite series of real numbers, but we gave no definition for their product. One way is motivated by the way we multiply polynomials:

$$\left(a_0 + a_1 x + a_2 x^2 + \cdots + a_n x^n\right)\left(b_0 + b_1 x + b_2 x^2 + \cdots + b_m x^m\right) =$$
$$a_0 b_0 + (a_0 b_1 + a_1 b_0)x + (a_0 b_2 + a_1 b_1 + a_2 b_0)x^2 + \cdots + a_n b_m x^{n+m}.$$

Setting $x = 1$, we see that we have expressed the product of two sums of numbers as a sum of partial products:

$$\left(\sum_{k=0}^{n} a_k\right)\left(\sum_{k=0}^{m} b_k\right) = \sum_{k=0}^{n+m} c_k$$

where $c_0 = a_0 b_0$, $c_1 = a_0 b_1 + a_1 b_0 = \sum_{k=0}^{1} a_k b_{1-k}$, $c_2 = a_0 b_2 + a_1 b_1 + a_2 b_0 = \sum_{k=0}^{2} a_k b_{2-k}$, and so on. For convenience with the coefficients in the polynomials, we have started the indexing in the series at $k = 0$. Continuing the list of properties in Section 3.1, we have

Property 3. (Cauchy Multiplication) Multiplication of two series defined in $\mathbf{R}^1$ is defined by

$$\sum_{n=0}^{\infty} a_n \otimes \sum_{n=0}^{\infty} b_n \cong \sum_{n=0}^{\infty} c_n, \quad c_n = \sum_{k=0}^{n} a_k b_{n-k}, \quad n \in \mathbf{N} \cup \{0\}.$$

The right-hand side is called the **Cauchy product** of the two series on the left-hand side. For Property 3 to be useful to us, we would surely like to know:

1. If two series converge, does their Cauchy product converge?
2. If the sums of two convergent series are S_1, S_2 and if the Cauchy product converges, is its sum S_1S_2?
3. What conditions on two component series will guarantee that their Cauchy product will converge?
4. When a Cauchy product does converge, is this convergence necessarily absolute convergence?

The answer to Question 1 is "not necessarily." Suppose we multiply the series whose general term is $(-1)^n/\sqrt{n+1}$, $n \geq 0$, by itself.

a_n's	1	$\frac{-1}{\sqrt{2}}$	$\frac{1}{\sqrt{3}}$	$\frac{-1}{\sqrt{4}}$	...
b_n's	1	$\frac{-1}{\sqrt{2}}$	$\frac{1}{\sqrt{3}}$	$\frac{-1}{\sqrt{4}}$	...
c_n's	1	$-\left(\frac{1}{\sqrt{2}}+\frac{1}{\sqrt{2}}\right)$	$\left(\frac{1}{\sqrt{3}}+\frac{1}{\sqrt{4}}+\frac{1}{\sqrt{3}}\right)$	$-\left(\frac{1}{\sqrt{4}}+\frac{1}{\sqrt{6}}+\frac{1}{\sqrt{6}}+\frac{1}{\sqrt{4}}\right)$	...
	(c_0)	(c_1)	(c_2)	(c_3)	

We see (by induction) that when $n = 2k$, the term c_n contains $n+1$ addends, and the smallest addend is $1/(\frac{n+2}{2})$. Hence, $c_{2k} > (2k+1)/(k+1)$ for all k, so $\lim_{k\to\infty} c_{2k} \geq \lim_{k\to\infty} (2k+1)/(k+1) = 2$. Thus, we have found a subsequence of the sequence $\{c_n\}_{n=0}^{\infty}$ that does not converge to 0. By Theorem 3.3, the series $\sum_{n=0}^{\infty} c_n$ then cannot converge, even though the two component alternating series do converge (Theorem 3.9).

The answer to Question 2, in brief, is "yes." For Question 3, a sufficient condition on the two component series is that both series converge and at least one of them converges absolutely.[2] The following algebraic lemma is needed:

[2]The answer to Question 2 is **Abel's Convergence Theorem** (after **Niels Henrik Abel** (1802–1829)). A proof different from Abel's is sketched in footnote 4. The answer to Question 3 was given by the Austrian **Franz Mertens** (1840–1927) (see Theorem 3.12).

Lemma 3.5.1. *For each* $n \in \mathbf{N} \cup \{0\}$ *define* $A_n = \sum_{k=0}^{n} a_k$, $B_n = \sum_{k=0}^{n} b_k$, *and* $C_n = \sum_{k=0}^{n} c_k$, *where* $c_k = \sum_{j=0}^{k} a_j b_{k-j}$. *Then* $A_n B_n - C_n$ *is given by* $\sum_{k=1}^{n} b_k \left[\sum_{j=0}^{k-1} a_{n-j}\right]$.

Proof.

$$
\begin{aligned}
A_nB_n - C_n &= \sum_{k=0}^{n} a_k \sum_{k=0}^{n} b_k - \sum_{k=0}^{n} c_k \\
&= \sum_{k=0}^{n} a_k \left[b_0 + b_1 + b_2 + \cdots + b_n\right] - [a_0b_0 + (a_0b_1 + a_1b_0) + (a_0b_2 + \\
&\quad a_1b_1 + a_2b_0) + \cdots + (a_0b_n + a_1b_{n-1} + a_2b_{n-2} + \cdots + a_nb_0)] \\
&= \sum_{k=0}^{n} a_k \left[b_0 + b_1 + b_2 + \cdots + b_n\right] - \left[a_0 \sum_{k=0}^{n} b_k + a_1 \sum_{k=0}^{n-1} b_k + \right. \\
&\quad \left. a_2 \sum_{k=0}^{n-2} b_k + \cdots + a_n \sum_{k=0}^{0} b_k \right] \\
&= a_1b_n + a_2(b_{n-1} + b_n) + a_3(b_{n-2} + b_{n-1} + b_n) + \cdots + \\
&\quad a_n(b_1 + b_2 + b_3 + \cdots + b_n) \\
&= b_n(a_1 + a_2 + a_3 + \cdots + a_n) + b_{n-1}(a_2 + a_3 + \cdots + a_n) \\
&\quad + b_{n-2}(a_3 + a_4 + \cdots + a_n) + \cdots + b_1a_n \\
&= \sum_{k=1}^{n} b_k \left[\sum_{j=0}^{k-1} a_{n-j}\right].
\end{aligned}
$$

■

Theorem 3.12 (Mertens' Theorem). *If* $\sum_{n=0}^{\infty} a_n$, $\sum_{n=0}^{\infty} b_n$ *are convergent, with sums A and B, respectively, and if the b-series converges absolutely, then the series obtained by Cauchy multiplication is convergent, with sum AB.*

Proof. Let $\{A_n\}_{n=0}^{\infty}$, $\{B_n\}_{n=0}^{\infty}$ be the sequences of partial sums of the a-series and b-series, and let $\sum_{n=0}^{\infty} |b_n| = L$, $0 < L < \infty$. Also let $\{C_n\}_{n=0}^{\infty}$ be the sequence of partial sums of the Cauchy product,

$$
C_n = \sum_{k=0}^{n} c_k, \quad c_k = \sum_{j=0}^{k} a_j b_{k-j}.
$$

The idea is to look at $A_nB_n - C_n$; if this can be shown to be arbitrarily small in magnitude for all sufficiently large n, then we are done.

The following are immediate:

1. $\{A_n\}_{n=0}^{\infty}$ is bounded, so $M = \sup_{n \geq 0} |A_n|$ is real, $M > 0$.

2. For any $a, b \in \mathbf{N}$, $|A_a - A_b| \leq |A_a| + |A_b| \leq 2M$.

3. Given $\varepsilon > 0$, there is an $N_1 \in \mathbf{N}$ such that $n > m - 1 > N_1$ implies

$$|A_n - A_{m-1}| = \left|\sum_{k=m}^{n} a_k\right| < \varepsilon/(2L).$$

4. Given $\varepsilon > 0$, there is an $N_2 \in \mathbf{N}$ such that $n > N_2$ implies

$$\sum_{k=n}^{\infty} |b_k| < \varepsilon/(4M).$$

Now let $N = \max\{N_1, N_2\}$; from Lemma 3.5.1 and for all n that satisfy $n - N + 1 > N + 1$, we have

$$\begin{aligned}
|A_nB_n - C_n| &= \left|\sum_{k=1}^{N} b_k \left[\sum_{j=0}^{k-1} a_{n-j}\right] + \sum_{k=N+1}^{n} b_k \left[\sum_{j=0}^{k-1} a_{n-j}\right]\right| \\
&\le \left|\sum_{k=1}^{N} b_k \left[\sum_{i=n-k+1}^{n} a_i\right]\right| + \left|\sum_{k=N+1}^{n} b_k(A_n - A_{n-k})\right| \qquad (i = n - j) \\
&\le \sum_{k=1}^{N} |b_k| \left(\frac{\varepsilon}{2L}\right) + \sum_{k=N+1}^{n} \left(\frac{\varepsilon}{4M}\right) |A_n - A_{n-k}| \\
&\le L\left(\frac{\varepsilon}{2L}\right) + \left(\frac{\varepsilon}{4M}\right)(2M) \\
&= \varepsilon.
\end{aligned}$$

Hence, $\lim_{n\to\infty} C_n = \lim_{n\to\infty} A_nB_n = (\lim_{n\to\infty} A_n)(\lim_{n\to\infty} B_n) = AB.$ ■

■ Example 3.20

Let $\sum_{n=0}^{\infty} a_n$ have general term $a_n = (-1)^n/\sqrt{n+1}$ and $\sum_{n=0}^{\infty} b_n$ have general term $b_n = (-1)^n/2^n$. The first series converges and the second series converges absolutely. By Mertens' Theorem the Cauchy product, which is

$$\sum_{n=0}^{\infty} a_n \otimes \sum_{n=0}^{\infty} b_n = \left[\frac{1}{\sqrt{1}} - \frac{1}{\sqrt{2}} + \frac{1}{\sqrt{3}} - \frac{1}{\sqrt{4}} + \sum_{n=4}^{\infty} (-1)^n/\sqrt{n+1}\right]$$
$$\otimes \left[1 - \frac{1}{2} + \frac{1}{3} - \frac{1}{4} + \sum_{n=4}^{\infty} (-1)^n/2^n\right]$$

$$= 1 + \left(\frac{-1}{2} - \frac{1}{\sqrt{2}}\right) + \left(\frac{1}{4} + \frac{1}{2\sqrt{2}} + \frac{1}{\sqrt{3}}\right)$$
$$+ \left(\frac{-1}{8} - \frac{1}{4\sqrt{2}} - \frac{1}{2\sqrt{3}} - \frac{1}{2}\right) + \cdots$$
$$\approx 1 - (1.207) + (1.181) - (1.091) + \cdots,$$

has the value AB, where $A = \lim_{n\to\infty} \sum_{k=0}^{n} (-1)^k/\sqrt{k+1} \approx 0.6049$ (Bromwich, 1926) and $B = \lim_{n\to\infty} \sum_{k=0}^{n} (-1)^k/2^k = 2/3$. Observe that the pattern of the first few terms is not encouraging. ■

The answer to Question 4, posed earlier, is "not necessarily." We see (by induction) that for each $n \geq 1$ the term $|c_n|$ in the Cauchy product in Example 3.20 exceeds $1/\sqrt{n+1}$. Use of the p-Series Test (Exercise 3.23(b)) shows that the series $\sum_{n=0}^{\infty} 1/\sqrt{n+1}$ diverges to ∞. However, absolute convergence is a strong enough property to be preserved upon multiplication, *if* both of the component series are absolutely convergent.

Theorem 3.13. *The Cauchy product obtained from two absolutely convergent series is itself absolutely convergent.*

Proof. Let $\sum_{n=0}^{\infty} a_n$, $\sum_{n=0}^{\infty} b_n$ be absolutely convergent, and let $\{c_n\}_{n=0}^{\infty}$ be the sequence of terms in their Cauchy product. By Theorem 3.12, the Cauchy product converges to AB, where $A = \sum_{n=0}^{\infty} a_n$ and $B = \sum_{n=0}^{\infty} b_n$.

$$\sum_{n=0}^{\infty} c_n = \sum_{n=0}^{\infty}\left[\sum_{k=0}^{n} a_k b_{n-k}\right]$$

But as these two series are absolutely convergent, then the series $\sum_{n=0}^{\infty} |a_n|$ and $\sum_{n=0}^{\infty} |b_n|$ also converge (absolutely) to, say, A' and B', respectively, so their Cauchy product converges to $A'B'$, by Theorem 3.12:

$$\sum_{n=0}^{\infty} |a_n| \otimes \sum_{n=0}^{\infty} |b_n| = \sum_{n=0}^{\infty} c_n' = \sum_{n=0}^{\infty}\left[\sum_{k=0}^{n} |a_k b_{n-k}|\right]$$
$$= A'B'.$$

But from the Triangle Inequality we have

$$|c_n| = \left|\sum_{k=0}^{n} a_k b_{n-k}\right| \leq \sum_{k=0}^{n} |a_k b_{n-k}| = c_n',$$

so by the Comparison Test the series $\sum_{n=0}^{\infty} |c_n|$ must also converge. ■

■ Example 3.21

The following two geometric series have the indicated sums:

$$\sum_{n=0}^{\infty} a_n = \sum_{n=0}^{\infty} (-1)^n \left(\frac{1}{2}\right)^n = \frac{2}{3}$$

$$\sum_{n=0}^{\infty} b_n = \sum_{n=0}^{\infty} (-1)^n \left(\frac{2}{3}\right)^n = \frac{3}{5},$$

and both series are absolutely convergent. By Theorem 3.12 the Cauchy product of $\sum_{n=0}^{\infty} a_n, \sum_{n=0}^{\infty} b_n$ has the sum

$$\sum_{n=0}^{\infty} c_n = 1 - \frac{7}{6} + \frac{37}{36} - \frac{175}{216} + \frac{781}{1296} - \cdots = \frac{2}{3}\left(\frac{3}{5}\right) = \frac{2}{5}.$$

Additionally, Theorem 3.13 guarantees that the following sum exists:

$$\sum_{n=0}^{\infty} |c_n| = 1 + \frac{7}{6} + \frac{37}{36} + \frac{175}{216} + \frac{781}{1296} + \cdots.$$

What do you think is the value of this sum? ■

3.6 A FIRST LOOK AT SERIES OF FUNCTIONS

The theory of infinite series is a continuation of the theory of sequences. Let $\{f_n\}_{n=0}^{\infty}$ be a sequence of functions defined on a common domain $\mathbf{D} \subseteq \mathbf{R}^1$ (or, possibly, $\mathbf{R}^k$) and with values in $\mathbf{R}^1$. Then the series $\sum_{n=0}^{\infty} f_n(x)$ is termed a **series of functions**. A common such type of series is one where each $f_n(x) = c_n(x-a)^n$, $c_n, a \in \mathbf{R}^1$, called a **power series about the point a**.

Any series of functions $\sum_{n=0}^{\infty} f_n(x)$ is said to **converge pointwise** at $x = x_0 \in \mathbf{D}$ iff $\sum_{n=0}^{\infty} f_n(x_0)$ converges to some $s_0 \in \mathbf{R}^1$. That is, if $\varepsilon > 0$ is given, then an $N \in \mathbf{N}$ exists such that $n > N$ implies that

$$\sum_{i=0}^{n} f_i(x_0) \in \mathbf{B}(s_0; \varepsilon).$$

There are two interesting issues that lurk beneath the surface here.

1. The series may converge at each point in $\mathbf{D}$; if so, prove this. If not, determine that maximal subset $\mathbf{D}' \subset \mathbf{D}$ on which $\sum_{n=0}^{\infty} f_n(x)$ does converge for each $x \in \mathbf{D}'$. In either case, the series converges to some function f with domain $\mathbf{D}'$. What are some of the properties of f? For example,

if each $f_n(x)$ is continuous (differentiable) on $\mathbf{D}'$, is f itself continuous (differentiable) on $\mathbf{D}'$?

■ Example 3.22

The series $\sum_{n=0}^{\infty} \frac{x^{2n}}{(2n)!}$ converges for all real x, according to the Ratio Test, to the function $f(x) = \cosh x$, which is everywhere differentiable.

The series $\sum_{n=0}^{\infty} \frac{(-1)^n}{n+1}(x-1)^{n+1}$ converges only if $0 < x \leq 2$ (again, use the Ratio Test); it converges to $\ln x$, which is differentiable on $(0, 2]$.

The series $\sum_{n=0}^{\infty} \frac{e^{nx}}{\sqrt{x}}$ converges (in $\mathbf{R}^1$) for no real x. ■

2. Suppose that a series $\sum_{n=0}^{\infty} f_n(x_0)$ converges to s_0. When $\varepsilon > 0$ is given, the N needed to ensure that $\sum_{i=0}^{n} f_i(x_0) \in \mathbf{B}(s_0; \varepsilon)$ for all $n > N$ depends, in general, upon ε and upon the x_0. Change x_0 to $x_0' \in \mathbf{D}'$, then a different N may be needed. For some series of functions, however, the same N may work for a given ε and for all $x \in \mathbf{D}'$. When this happens, the series is said to **converge uniformly on D′**. How can we determine if a series is a uniformly convergent series? We anticipate that this question will be important because a uniformly convergent series will probably possess some nice properties. Does point (1) suggest any possibilities to you?

The ideas touched upon already, plus other related notions, will be treated substantially in Chapter 7. However, there is a common situation in which the convergence of a series of functions, especially a power series, is not of overriding importance. Suppose that $\{K_n\}_{n=0}^{\infty}$ is a given sequence of real numbers and that a function $f(x)$ can be written in closed form but also has associated with it the expansion $\sum_{n=0}^{\infty} K_n x^n$.[3] The function $f(x)$ is then called a **generating function** for the sequence $\{K_n\}_{n=0}^{\infty}$.

It would be nice to be able to state more, namely, that

$$f(x) = \sum_{n=0}^{\infty} c_n K_n x^n, \quad c_n \in \mathbf{R}^1. \tag{*}$$

We may not know precisely in what interval of x-values the series in (*) converges to $f(x)$. But for *some* purposes we do not need this information. Commonly, though, the equality in (*) will hold for some interval of nonzero

[3]There are two issues here. First, a theorem (which we do not prove in this book) says that every power series is a Taylor series. That is, to any series $\sum_{n=0}^{\infty} c_n x^n$ there is a function f such that $f(0) = c_0$ and for all $n \in \mathbf{N}$ we have $f^{(n)}(0)/n! = c_n$. Second, it may be very hard to actually represent f in terms of standard elementary functions. We are interested presently only in cases where f can be found in closed form in order that we can proceed with certain practical objectives.

length. Consequently, generating functions are more common than you might expect (Watkins, 1987) (Exercises 3.42, 3.44).

■ Example 3.23

If $r \in \mathbf{N}$, then $(1+x)^r$ is the generating function for the finite sequence $\left\{\binom{r}{n}\right\}_{n=0}^{r}$.

The function $\frac{1}{1+x}$ is the generating function for the simple sequence $\left\{(-1)^n\right\}_{n=0}^{\infty}$.

The function $\frac{x}{1-x-x^2}$ is the generating function for the sequence of Fibonacci numbers, $\left\{F_n\right\}_{n=1}^{\infty}$ (see later). ■

In accordance with Properties 1, 2, and 3 for series, we can multiply generating functions and the associated series by scalars, add and subtract generating functions, and multiply any two generating functions. These operations may be denoted as **formal operations**, since no attention is paid to issues of convergence of the series (Niven, 1969). The objective in formal operations is usually to obtain relations among the members of the sequence of numbers of interest or to derive certain other properties of them.

■ Example 3.24

For the last example in Example 3.23, let us write

$$\begin{aligned}
\frac{x}{1-x-x^2} &= \frac{x}{\left(1-x+\frac{x^2}{4}\right)-\frac{5x^2}{4}} \\
&= \frac{x}{\left[\left(1-\frac{x}{2}\right)-\frac{\sqrt{5}x}{2}\right]\left[\left(1-\frac{x}{2}\right)+\frac{\sqrt{5}x}{2}\right]} \\
&= \frac{1}{\sqrt{5}}\left[\frac{1}{1-\frac{1+\sqrt{5}}{2}x}-\frac{1}{1-\frac{1-\sqrt{5}}{2}x}\right].
\end{aligned}$$

But the function $\frac{1}{1-ax}$, $a \neq 0$, is the generating function for $\{a^n\}_{n=0}^{\infty}$, so subtraction of the two generating functions above gives, from Property 2,

$$\begin{aligned}
\frac{x}{1-x-x^2} = \sum_{n=1}^{\infty} F_n x^n &= \frac{1}{\sqrt{5}}\left[\sum_{n=0}^{\infty}\left(\frac{1+\sqrt{5}}{2}\right)^n x^n - \sum_{n=0}^{\infty}\left(\frac{1-\sqrt{5}}{2}\right)^n x^n\right] \\
&= \frac{1}{\sqrt{5}}\sum_{n=1}^{\infty}\left[\left(\frac{1+\sqrt{5}}{2}\right)^n - \left(\frac{1-\sqrt{5}}{2}\right)^n\right]x^n.
\end{aligned}$$

Hence, for each $n \in \mathbf{N}$ we have

$$F_n = \frac{1}{\sqrt{5}}\left[\left(\frac{1+\sqrt{5}}{2}\right)^n - \left(\frac{1-\sqrt{5}}{2}\right)^n\right],$$

which agrees with Exercise 2.14. ■

An especially interesting example of a formal operation occurs in connection with the famous **Bernoulli numbers**, $\{B_n\}_{n=0}^{\infty}$. These fascinating numbers, which arose in connection with Jakob Bernoulli's work on the summation of the pth powers of m consecutive natural numbers (Exercise 3.48(c)), are now known to occur in many places in mathematics. Euler, a generation after Bernoulli, discovered a generating function for $\{B_n\}_{n=0}^{\infty}$:

$$\frac{x}{e^x - 1} = \sum_{n=0}^{\infty} \frac{B_n}{n!} x^n. \qquad (**)$$

Note that the equality sign is not to be interpreted too rigorously, since no mention of convergence has been made.

It is tedious to work out the values of even the first few B_n's from the generating function because l'Hôpital's Rule must be used repeatedly. The formal operations shown next circumvent this difficulty.

■ Example 3.25

Recalling from elementary calculus (or looking ahead in Section 5.7) that the MacLaurin series for e^x is $\sum_{k=0}^{\infty} \frac{x^k}{k!}$, we rewrite equation $(**)$ as

$$\begin{aligned} x &= (e^x - 1)\sum_{n=0}^{\infty} \frac{B_n}{n!} x^n \\ &= \left(\sum_{k=1}^{\infty} \frac{x^k}{k!}\right) \otimes \left(\sum_{n=0}^{\infty} \frac{B_n}{n!} x^n\right) \\ &= \frac{B_0}{0!\,1!} x + \sum_{m=2}^{\infty} c_m x^m, \end{aligned} \qquad (***)$$

where $B_0 = 1$ and each $c_m = \sum_{k=1}^{m} \frac{1}{k!} \frac{B_{m-k}}{(m-k)!}$, from Property 3 (Section 3.5). As there are no powers of x higher than the first on the left-hand side of equation $(***)$, it follows that for all $m \geq 2$ we have

$$0 = \sum_{k=1}^{m} \frac{1}{k!} \frac{B_{m-k}}{(m-k)!} = \sum_{k=1}^{m} \binom{m}{k} B_{m-k} = \sum_{r=0}^{n} \binom{n+1}{r} B_r,$$

Table 3.3 Some Bernoulli Numbers[a]

n	B_n	n	B_n
0	1	5	0
1	−1/2	6	1/42
2	1/6	7	0
3	0	8	−1/30
4	−1/30	9	0

[a] *A larger compilation of B_n's, as well as a listing of* ***Bernoulli polynomials****, can be found in Abramowitz, M. and Stegun, I. A. (eds.), Handbook of Mathematical Functions, Dover Publications, NY, 1965, pp. 809, 810.*

where $n = m - 1 \geq 1$ and $m - k = r$. The result is a useful recursion relationship for obtaining early Bernoulli numbers. ■

Infinite series, especially those in $\mathbf{R}^1$, offer a fertile ground for the implementation or testing of many numerical procedures. Bernoulli numbers (and the associated Bernoulli polynomials) are a meeting ground of the practical and the theoretical sides of infinite series. If you would like to dip into some of this literature, you will find (Apostol, 2008; Dence and Dence, 1999; Dilcher, Skula and Slavutskii, 1991; Lehmer, 1998) useful and interesting.

EXERCISES

Section 3.1

3.1. Translate into the symbolism of inequalities the definition of an infinite series in $\mathbf{R}^k$ that converges to $S \in \mathbf{R}^k$.

3.2. Series (of nonvariable terms) can be considered in spaces other than $\mathbf{R}^1$.

(a) How would Property 1 be implemented if the terms a_n were 2×2 matrices of real numbers?

(b) How would Property 2 then be implemented?

3.3. Prove the second part of Theorem 3.1.

3.4. An equilateral triangle ABC has sides that are 6 units in length. The midpoints of the sides are connected to form the inner equilateral triangle A′B′C′. The midpoints of the sides of A′B′C′ are connected to form A″B″C″ (Figure 3.2). We imagine that this bisection process is continued indefinitely. If a_n is the general term of an infinite series in which a_n is the area of the nth constructed triangle (a_1 = area of ΔABC), obtain a formula for the nth partial sum s_n, and then determine $\lim_{n\to\infty} s_n$.

3.5. Write out the proof of the sufficiency part of Theorem 3.2.

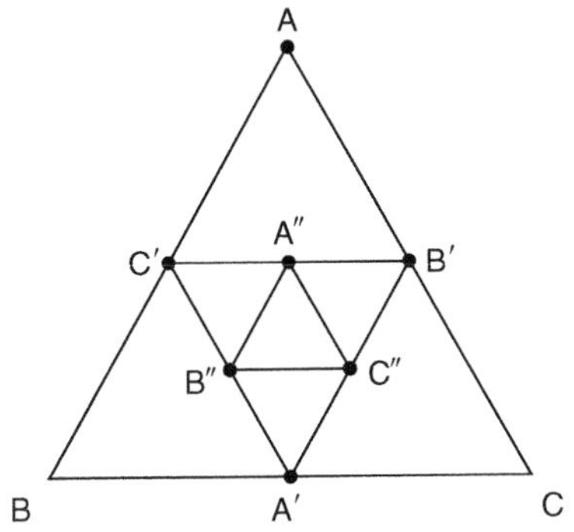

FIGURE 3.2
Diagram for Exercise 3.4.

3.6. Let $\sum_{n=1}^{\infty} \mathbf{x}_n$ be a series in $\mathbf{R}^k$ and $\{\mathbf{s}_n\}_{n=1}^{\infty}$ the sequence of partial sums in $\mathbf{R}^k$.
(a) Express in the language of k-balls what it means for the sequence to be Cauchy.
(b) The following general theorem is very useful; prove this Criterion.

Cauchy's Convergence Criterion for Series

> The series $\sum_{n=1}^{\infty} \mathbf{x}_n$ converges in $\mathbf{R}^k$ iff, given any $\varepsilon > 0$, there is an $N \in \mathbf{N}$ such that for any $n > m > N$ we have $\|\mathbf{x}_{m+1} \oplus \mathbf{x}_{m+2} \oplus \ldots \oplus \mathbf{x}_n\| < \varepsilon$.

(c) Let $\sum_{n=1}^{\infty} \mathbf{x}_n$ be a series in $\mathbf{R}^k$. Prove that if the series $\sum_{n=1}^{\infty} \|\mathbf{x}_n\|$ converges in $\mathbf{R}^1$, then $\sum_{n=1}^{\infty} \mathbf{x}_n$ converges in $\mathbf{R}^k$.

Section 3.2

3.7. The **Leibniz series** is the series whose terms are $a_n = \begin{cases} 1 & n = 2k-1 \\ -1 & n = 2k. \end{cases}$ Prove that $\sum_{n=1}^{\infty} a_n$ diverges.

3.8. Determine, using only material in Sections 3.1 and 3.2, whether each series converges or diverges:
(a) $\sum_{n=1}^{\infty} \left(1 - \frac{1}{2^n}\right)$;
(b) $\sum_{n=1}^{\infty} n^{-n/10}$;
(c) $\sum_{n=1}^{\infty} (n!)^{-1/2}$;
(d) $\sum_{n=1}^{\infty} \cos(1/n)$;
(e) $\sum_{n=1}^{\infty} a_n$, where $a_n = \begin{cases} 2^{-k} & n = 2k-1 \\ 3^{-k} & n = 2k. \end{cases}$

3.9. Verify, in Example 3.7, that for all $n > 4$ we have $0 < \|\mathbf{a}_n\| < \|\mathbf{b}_n\|$. What upper bound do you then have for $\sum_{n=1}^{\infty} \|\mathbf{a}_n\|$?

3.10. In $\mathbf{R}^1$ the two series $\sum_{n=1}^{\infty} a_n$ and $\sum_{n=1}^{\infty} b_n$ are such that for all $n \in \mathbf{N}$ we have $a_n = b_{n+1} - b_n$.

(a) Prove that convergence of the b-series implies convergence of the a-series.

(b) Prove that convergence of the a-series implies convergence of $\{b_n\}_{n=1}^{\infty}$, but does not imply convergence of the b-series.

3.11. A series $\sum_{n=1}^{\infty} a_n$, with partial sums $\{s_n\}_{n=1}^{\infty}$, is said to be **summable** Y iff $\lim_{n\to\infty} y_n = y \in \mathbf{R}^1$, where

$$y_n = \begin{cases} s_1/2 \quad (= a_1/2) & n = 1 \\ (s_{n-1} + s_n)/2 & n \geq 2. \end{cases}$$

We then write $(Y)\sum_{n=1}^{\infty} a_n = y$.

(a) Prove that if a series is summable Y, then $\lim_{n\to\infty} (a_{n-1} + a_n) = 0$.

(b) Prove that the Leibniz series (Exercise 3.7) is summable Y, and determine the value of $(Y)\sum_{n=1}^{\infty} a_n$.

3.12. A series $\sum_{n=1}^{\infty} a_n$, with partial sums $\{s_n\}_{n=1}^{\infty}$, is said to be summable by the method of arithmetic means (**summable** M) iff $\lim_{n\to\infty} \sigma_n = \sigma \in \mathbf{R}^1$, where

$$\sigma_n = \begin{cases} s_1 \quad (= a_1) & n = 1 \\ \frac{s_1+s_2+\cdots+s_n}{n} & n \geq 2. \end{cases}$$

We then write $(M)\sum_{n=1}^{\infty} a_n = \sigma$.

(a) Consider the series $\sum_{n=1}^{\infty} b_n$, where $b_n = \begin{cases} 1 & n = 4k+1 \\ 0 & n = 2k \\ -1 & n = 4k+3. \end{cases}$ Show that this does not converge, that it is not summable Y, but that it is summable M.

(b) Suppose a series $\sum_{n=1}^{\infty} a_n$ converges to $L \in \mathbf{R}^1$; let $\varepsilon > 0$ be given. Show that there is an $N \in \mathbf{N}$ such that for all $n > N$ we have

$$(n-N)\left(L - \frac{\varepsilon}{2}\right) < s_{N+1} + s_{N+2} + \cdots + s_n < (n-N)\left(L + \frac{\varepsilon}{2}\right).$$

(c) From this obtain

$$\frac{-N}{n}\left(L - \sigma_N - \frac{\varepsilon}{2}\right) + L - \frac{\varepsilon}{2} < \sigma_n < L + \frac{\varepsilon}{2} + \frac{N}{n}\left(\sigma_N - L - \frac{\varepsilon}{2}\right).$$

(d) Choose N_0 so large that $N_0 > N$ and $\frac{1}{2}N_0\varepsilon > N(|L| + \frac{\varepsilon}{2} + |\sigma_N|)$. Show that this implies $L - \varepsilon < \sigma_N < L + \varepsilon$ for $n > N_0$ and, hence, convergence of $\sum_{n=1}^{\infty} a_n$to L implies that the series is summable M also to L. Such a method of summability is said to be **regular** (Hardy, 1949).[4]

3.13. Write a short program and construct a table analogous to Table 3.1, but for the series $\sum_{n=2}^{\infty}[n\ln n]^{-1}$. Let the A-column consist of just five entries: $1\frac{1}{2}, 2, 2\frac{1}{2}, 3, 3\frac{1}{2}$. Make a conjecture.

3.14. Verify the convergence of the particular series described at the end of Section 3.2, and confirm the upper bound of 74.75 for the sum.

Section 3.3

3.15. Determine whether each series converges or diverges:

(a) $\sum_{n=7}^{\infty}\frac{1}{(n-6)^{4/3}}$;

(b) $\sum_{n=2}^{\infty}\frac{3}{4}\frac{9^{n-1}-1}{(2n-1)!}$;

(c) $\sum_{n=1}^{\infty}\left(1+\frac{1}{n}\right)^{n^2}$;

(d) $\sum_{n=1}^{\infty}\frac{\sqrt{n}}{2n^3+n+1}$;

(e) $\sum_{n=1}^{\infty}\frac{n!}{1\cdot 3\cdot 5\cdots(2n-1)}$;

(f) $\sum_{n=1}^{\infty}a_n,\quad a_n=\begin{cases}(-2/3)^n & n=2k\\ 2(-2/3)^n & n=2k-1;\end{cases}$

(g) $\sum_{n=1}^{\infty}a_n,\quad a_n=\begin{cases}1/n^2 & n\neq k^2\\ 1/n^{2/3} & n=k^2.\end{cases}$ (Bromwich, 1926).

3.16. Complete the proof of the Root Test (Theorem 3.7).

3.17. Refer to the last line of Example 3.11. An approximate value for the sum $\sum_{n=1}^{\infty}\frac{1}{F_n}$ is 3.35988566. Prove (on your own) that the indicated sum is less than 27/8.

3.18. Verify that (a) the sum of the series in Example 3.12 is 14/5, and (b) the function $f(x)$ in Example 3.13 is an increasing function for $x > 4$.

3.19. Show in the application of Theorem 3.7 to the limit result

$$\lim_{n\to\infty}\frac{\sqrt[n]{n!}}{n}=\frac{1}{e},$$

that this result implies $\ln(n!)\approx n\ln n-n$. This is the simplest form of what is known in applied mathematics as **Stirling's Approximation**.[5] More accurate approximations abound.

---

[4] It was proved (as a special case of a wider theorem) in 1890 by the Italian mathematician **Ernesto Cesàro** (1859–1906) that if $\sum_{n=1}^{\infty}a_n$ and $\sum_{n=1}^{\infty}b_n$ converge to A and B, respectively, then the Cauchy product is summable M to the value AB. Hence, in view of part (d) above, if it is known independently that the Cauchy product is convergent, then it necessarily converges to AB (Hardy, *ibid.*, 228–229).

[5] After the Scottish mathematician **James Stirling** (1692–1770), who presented the result in his 1730 book, *Methodus Differentialis* (Kline, 1972). MacLaurin derived Stirling's formula from (what we now call) the Euler-MacLaurin Summation Formula. This appeared as early as 1737 in circulating copies of the *Treatise of Fluxions* (Grabiner, 1997). Stirling, in 1738, communicated this fact to Euler.

(a) Compute the percent errors incurred by using the Approximation to estimate $\ln(n!)$ for $n = 10, 15, 20, 30, 60$.

(b) The next level of approximation is given by

$$\ln(n!) \approx \frac{1}{2}\ln(2\pi n) + n\ln(n) - n.$$

What percent errors are incurred now for $n = 10, 15, 20, 30, 60$?

3.20. **(Differentiation Test)** Let $\sum_{n=1}^{\infty} a_n$ be a positive series, and let f be any real-valued function on $[0, 1/N]$ such that $f(x) = a_n$ when $x = 1/n, n \geq N$, and f'' exists at $x = 0$. Then the series converges if $f(0) = f'(0) = 0$, and diverges otherwise (Abu-Mostafa, 1984).

(a) Apply the test to the series $\sum_{n=1}^{\infty} \sin(n^{-1})$.

(b) Apply the test to the series $\sum_{n=1}^{\infty} \sinh\left(\tanh \frac{1}{n}\right)$.

3.21. A student applies the Ratio Test as follows: "The series whose general term is $a_n = n/(n^2 + 1)$ must converge because

$$\frac{a_{n+1}}{a_n} = \frac{n+1}{(n+1)^2 + 1} \cdot \frac{n^2 + 1}{n} = 1 - \frac{n^2 + n - 1}{n^3 + 2n^2 + 2n},$$

and this is less than 1 for all n." How would you respond?

3.22. If, upon attempted application of the Ratio Test to $\sum_{n=1}^{\infty} a_n$, we find that $\lim_{n\to\infty}(a_{n+1}/a_n) = 1$, then a more delicate test is required. One such test is that due to the Swiss mathematician **Josef Ludwig Raabe** (1801–1859):

Raabe's Test

Let a_n be the general term of a positive series. Then $\sum_{n=1}^{\infty} a_n$ converges if $\lim_{n\to\infty} \inf \left[n\left(\frac{a_n}{a_{n+1}} - 1\right)\right] > 1$, and diverges if there is an $N \in \mathbf{N}$ such that for all $n > N$ we have $n\left(\frac{a_n}{a_{n+1}} - 1\right) \leq 1$.

After convincing yourself that the Ratio Test is useless for the series whose general term is $a_n = \frac{1\cdot 3\cdot 5\cdots(2n-1)}{4\cdot 6\cdot 8\cdots(2n+2)}$, apply Raabe's Test.

3.23. A good alternative to the well-known Cauchy Integral Test (Theorem 6.23) is the much less well-known Cauchy Condensation Test (Porter, 1972):

Cauchy's Condensation Test

If a_n is the general term of a positive series that is decreasing for all $n \geq N$, then this series converges or diverges accordingly as the sum

$$\sum_{n=1}^{\infty} 2^n a_{2^n}$$

does or does not exist.

(a) Develop a proof of the Condensation Test using the technique illustrated in Oresme's treatment of the harmonic series (see Example 3.8). To get started, establish that for any $k \in \mathbf{N}$ we have

$$2^{k-1}a_{2^k} \leq a_{2^{k-1}+1} + a_{2^{k-1}+2} + \cdots + a_{2^k} \leq 2^{k-1}a_{2^{k-1}}.$$

(b) Use Cauchy's Condensation Test to prove the useful **p-Series Test**: The series whose general term is $a_n = 1/n^p, p \in \mathbf{R}$, converges iff $p > 1$.

(c) If $\sum_{n=1}^{\infty} a_n, \sum_{n=1}^{\infty} b_n$ are two positive, divergent series, then the second series is said to diverge **more slowly** than the first if $\lim_{n\to\infty}(b_n/a_n) = 0$ (Agnew, 1947). Use Cauchy's Condensation Test to prove that the series in Exercise 3.13 diverges, and does so more slowly than the harmonic series.

Section 3.4

3.24. Prove Theorem 3.8. Then give an example of a series in $\mathbf{R}^2$ that is conditionally convergent.

3.25. Give an example of a series in $\mathbf{R}^1$ that has infinitely many each of positive and negative terms and $\lim_{n\to\infty} x_n = 0$, but which nevertheless diverges.

3.26. Consider $\sum_{n=1}^{\infty} a_n$, where $a_n = \begin{cases} 1 & n = 1 \\ (-1)^{n-1}\dfrac{2^{2n-3}(\pi/4)^{2n-2}}{(2n-2)!} & n \geq 2. \end{cases}$

(a) Show that $\{|a_n|\}_{n=1}^{\infty}$ is decreasing for all $n \in \mathbf{N}$.

(b) Complete the analysis needed to show that $\sum_{n=1}^{\infty} a_n$ converges to some $S \in \mathbf{R}^1$.

(c) What is the upper bound on $|S - s_6|$ if Theorem 3.10 is used?

3.27. A series expansion of the number $(\pi - 3)/4$ is given by:

$$\sum_{n=1}^{\infty} (-1)^{n+1}/[2n(2n+1)(2n+2)].$$

How many terms of this shall be taken in order to give an error of magnitude less than 10^{-6}? Determine by computer if fewer terms would suffice.

3.28. The series in Exercise 3.27 would be a poor one in order to estimate π to high accuracy. A better series is this one (Chan, 2006):

$$\pi = \sqrt{3}\sum_{n=0}^{\infty}\frac{1}{64^n}\left(\frac{1}{6n+1} + \frac{3}{2(6n+2)} + \frac{1}{4(6n+3)} - \frac{1}{8(6n+5)}\right).$$

What is the magnitude of the error if π is approximated by:

$$\sqrt{3}s_5 = \sqrt{3}\sum_{k=0}^{5}\frac{1}{64^k}\left(\frac{1}{6k+1} + \frac{3}{2(6k+2)} + \frac{1}{4(6k+3)} - \frac{1}{8(6k+5)}\right)?$$

For a much more amazing procedure, see the paper by J. Guillera at the end of Chapter 2.

3.29. Look up the paper by Young (1985). Digest the author's proof of his theorem, and then write this up in your own words.

3.30. **(a)** Refer to the series for $\tan^{-1}(1/2)$ in Table 3.2. If the sum S is approximated by the partial sum s_{11}, show that the error conforms to Young's result.

(b) Refer to Exercise 3.26. Show that the sequence $\{\Delta b_n\}_{n=1}^{\infty}$, $\Delta b_n = |a_n| - |a_{n-1}|$, is a decreasing sequence for all $n > 1$. Then what is the estimated upper bound on $|S - s_6|$ according to Young (1985)? How does this estimate compare with that in Exercise 3.26(c), and how does it compare with the actual value of $|S - s_6|$, given that $S = 1/2$?

3.31. Determine whether each series $\sum_{n=1}^{\infty} a_n$ converges absolutely, converges conditionally, or diverges.

(a) $a_n = (-1)^{n-1}\frac{(n+1)}{3n}, n \geq 1;$

(b) $a_n = (-1)^n \frac{1}{2n-1}, n \geq 1;$

(c) $a_n = (-1)^{n-1}\frac{n^{1/2}}{(\ln(n))^3}, n \geq 2;$

(d) $a_n = (-1)^n \frac{1+2n}{(7n^2-1)^2}, n \geq 1;$

(e) $a_n = (-1)^n \left(\frac{2n+5}{3n+1}\right)^n, n \geq 1;$

(f) $a_n = (-1)^{n-1}(\cos n)[1 - \cos(n^{-1})], n \geq 1;$

(g) $a_n = (-1)^n \frac{1}{n[2+(-1)^n]}, n \geq 1;$

(h) $a_n = (-1)^{n-1}\frac{\sqrt{n}}{2n+1}, n \geq 1;$

(i) $a_n = (-1)^{n-1}\frac{(n+2)!}{2[3\cdot 6\cdot 9\cdots(3n)]}, n \geq 1;$

(j) $a_n = (-1)^n \frac{n!}{10^{n+1}}, n \geq 1;$

(k) $a_n = (-1)^{n-1}\frac{3^n}{n^2\cdot 2^{n+1}}, n \geq 1;$

(l) $a_n = (-1)^n \frac{2}{(-1)^n+\sqrt{n}}, n \geq 2.$

3.32. Series representations of irrational numbers are not necessarily alternating series, despite the appearance of Table 3.2. There is an enormous literature (Nunemacher, 1992) on **Euler's Constant**, γ, first mentioned here in Exercise 2.11. The following *positive* series for γ appeared in Gerst (1969):[6]

$$1 - \gamma = \lim_{N\to\infty} \sum_{n=1}^{N} \left[\sum_{k=2^{n-1}+1}^{2^n} \frac{n}{(2k-1)(2k)} \right].$$

The value of γ is known to be 0.57721566, to eight decimals. Write a short program and compute estimates of γ corresponding to $N = 5, 10, 15, 25$.

Section 3.5

3.33. Prove, as stated in the text, that absolute convergence of series is preserved in linear combinations.

---

[6]Actually, it has *never* been proved if γ is irrational or not!

3.34. The article (Galanor, 1987) is quite accessible. Read through it and then write up in your own words the proof of Riemann's Rearrangement Theorem (Part 2).

3.35. Perform Cauchy multiplications of the following pairs of series: simplify, where possible.

(a) $\left(\sum_{n=0}^{\infty} \frac{n+1}{n!}\right) \otimes \left(\sum_{n=0}^{\infty} \frac{2}{n!}\right)$;

(b) $\left(\sum_{n=0}^{\infty} \frac{1}{n!}\left(\frac{1}{2}\right)^n\right) \otimes \left(\sum_{n=1}^{\infty} (-1)^{n+1} \frac{1}{n!}\left(\frac{1}{2}\right)^n\right)$;

(c) $\left(\sum_{n=0}^{\infty} \frac{(2x)^n}{n!}\right) \otimes \left(\sum_{n=0}^{\infty} (-1)^{n+1} \frac{x^n}{n!}\right)$;

(d) $\left(\sum_{n=1}^{\infty} \frac{x^{2n-1}}{(2n-1)!}\right) \otimes \left(\sum_{n=1}^{\infty} \frac{x^{2n-2}}{(2n-2)!}\right)$.

3.36. The series $\sum_{n=1}^{\infty} (-1)^{n+1}/n$ is known as the **Alternating Harmonic Series**; it is clearly conditionally convergent. With respect to Question 4 on Cauchy products, show how a series could, in principle, be constructed such that its Cauchy product with itself is the Alternating Harmonic Series.

3.37. Supply reasons for the "immediate" steps (1) through (4) in the proof of Theorem 3.12.

3.38. **(a)** Two series, $\sum_{n=1}^{\infty} a_n$ and $\sum_{n=1}^{\infty} b_n$, are such that the a-series is absolutely convergent and for all $n \in \mathbf{N}$ we have $|b_n| \leq |a_n|$. Prove that the b-series is convergent.

(b) Suppose that $\sum_{n=1}^{\infty} x_n$ is absolutely convergent. Prove that the allied series $\sum_{n=1}^{\infty} (-1)^{n+1} \frac{x_n^2}{x_n^2+(1/2)}$ is convergent.

(c) Suppose that $\sum_{n=1}^{\infty} x_n$ is absolutely convergent and $\{y_n\}_{n=1}^{\infty}$ is a bounded sequence. Prove that $\sum_{n=1}^{\infty} x_n y_n$ is also absolutely convergent.

3.39. In Example 3.21 compute c_5, c_6, c_7. If $s_n = \sum_{k=0}^{n} c_k$, how much error is made by approximating $\sum_{n=0}^{\infty} a_n \otimes \sum_{n=0}^{\infty} b_n$ by s_6? What is the value of the sum $\sum_{n=0}^{\infty} |c_n|$, and how much error is made by approximating it by $\sum_{n=0}^{7} |c_n|$?

Section 3.6

3.40. Consider the sequence of functions $\{f_n\}_{n=1}^{\infty}$, where for each $n \in \mathbf{N}, f_n(x) = 1/(1+x^n), \mathbf{D} = \{x : x \geq 0\}$.

(a) Determine the maximal subset $\mathbf{D}' \subseteq \mathbf{D}$ such that $\sum_{n=1}^{\infty} f_n(x)$ converges for each $x \in \mathbf{D}'$.

(b) Let f denote the function to which $\sum_{n=1}^{\infty} f_n(x)$ converges on $\mathbf{D}'$. Show that $\frac{3}{4} < f(2) < 1$.

3.41. In Example 3.22, how do you know that $\sum_{n=0}^{\infty} \frac{e^{nx}}{\sqrt{x}}$ fails to converge (in $\mathbf{R}^1$) for any real x? Also, show that $\sum_{n=0}^{\infty} \frac{(-1)^n}{n+1}(x-1)^{n+1}$ converges iff $0 < x \leq 2$.

3.42. What function $f(x)$ could be used as a generating function for the sequence $\left\{\frac{1}{2^n n!}\right\}_{n=0}^{\infty}$? What function $g(x)$ could be used as a generating function for the sequence $\{a_n\}_{n=0}^{\infty}$, where

$$a_n = \begin{cases} 1 & n = 0 \\ 1/2^2 & n = 1 \\ \frac{-(2n-3)}{2^2} a_{n-1} & n > 1? \end{cases}$$

A convenient form for this is $g(x) = \sum_{n=0}^{\infty} \frac{a_n}{n!} x^n$.

3.43. In the third example in Example 3.23, initiate a long division of $1 - x - x^2$ into x. Look at the coefficients of the first six terms in the quotient. Are they as you would expect?

3.44. Refer to Section 1.7, where the **Euler ϕ-function** was mentioned. A recent survey (Gould and Shonhiwa, 2008) gives a generating function for the sequence $\{\phi(n)\}_{n=1}^{\infty}$:

$$\frac{\zeta(s-1)}{\zeta(s)} = \sum_{n=1}^{\infty} \frac{\phi(n)}{n^s}, \tag{‡}$$

where $\zeta(s) = \sum_{k=1}^{\infty} \frac{1}{k^s}, s > 1$, is the **Riemann zeta function**. Choose $s = 3$, arbitrarily. Rearrange equation (‡) and perform a partial Cauchy multiplication on the right. Show that the first six terms on both sides agree with each other.[7]

3.45. Use the generating function for the Bernoulli numbers (equation (**) in the text) to show that $B_{2n+1} = 0, n \in \mathbb{N}$. To get started, transpose in equation (**) the term containing $B_1 = -1/2$ to the left-hand side and then show that the new left-hand side is an even function of x.

3.46. Show that in some region of convergence we have

$$\cot x = \frac{1}{x} + \sum_{n=1}^{\infty} \frac{(-4)^n}{(2n)!} B_{2n} x^{2n-1}.$$

To get started, begin with the definition of coth x, then reexpress this in terms of Bernoulli numbers, make use of Exercise 3.45, and make use of $\coth(ix) = -i \cot x$ (verify!).

3.47. Further analysis is needed to determine the region of convergence of the series in Exercise 3.46. Let us assume that $x = \pi/4$ falls in that region. Estimate $\cot(x)$.

3.48. The sequence of **Bernoulli polynomials** can be defined by

$$B_n(x) \cong \sum_{k=0}^{n} \binom{n}{k} B_k x^{n-k}, \quad n \geq 0.$$

(a) Work out polynomials $B_0(x)$ through $B_5(x)$.

(b) A way to remember the earlier definition is by the simpler, **symbolic equation** $B_n(x) = (B + x)^n$. Explain this.

(c) **Bernoulli's Identity** involves both Bernoulli numbers and Bernoulli polynomials (Williams, 1997):

$$\sum_{k=1}^{m} k^p = \frac{B_{p+1}(m+1) - B_{p+1}}{p+1}, \quad p \geq 1, m \geq 1.$$

Compute (from this formula) the sum of the fifth powers of the first 100 natural numbers.[8]

---

[7] The right-hand side of equation (‡) is termed a **Dirichlet series**, an important class of series distinct from power series.

[8] At a time when there were no pocket calculators, no computers, nor even any batteries (!), Bernoulli computed the sum of the tenth powers of the first 1000 natural numbers in about seven or eight minutes (or so he said!).

3.49. Euler discovered the following generating function for the sequence of Bernoulli polynomials, $\{B_n(x)\}_{n=0}^{\infty}$:

$$\frac{ze^{xz}}{e^z - 1} = \sum_{n=0}^{\infty} \frac{B_n(x)}{n!} z^n.$$

(a) Obtain, by partial expansion of the generating function, the polynomials $B_0(x)$ through $B_4(x)$.

(b) Show that for each $n \in \mathbf{N}$ we have $B_n(1) = B_n$.

(c) If we assume that term-by-term differentiation of the previous series is permitted, show that for each $n \in \mathbf{N}$ we have

$$\frac{1}{n}\frac{dB_n(x)}{dx} = B_{n-1}(x).$$

(d) Hence, prove that for each $n \in \mathbf{N}$

$$B_n(x) = B_n + n\int_1^x B_{n-1}(t)\, dt.$$

Obtain, then, from part (a), polynomials $B_5(x)$, $B_6(x)$.[9]

3.50. Refer to Exercise 3.49. Use Cauchy multiplication to prove the following recursion relationship:

$$x^{n-1} = \frac{1}{n}\sum_{k=1}^{n}\binom{n}{k} B_{n-k}(x).$$

What happens if x is set equal to 0 in this equation?

SUPPLEMENTARY PROBLEMS

Additional problems on infinite series for your enrichment and for challenge can be found in Appendix C. Good luck!

REFERENCES

Cited Literature

Abu-Mostafa, Y.S., "A Differentiation Test for Absolute Convergence," *Math. Mag.*, 57, 228–231 (1984). This test deserves to be better known, as it is easy to apply. Cited in our Exercise 3.20.

Agnew, R.P., "A Slowly Divergent Series," *Amer. Math. Monthly*, 54, 273–274 (1947). This brief note will dispel the idea that despite what the p-Series Test might lead us to believe, the harmonic series is not the most slowly divergent series. Cited in our Exercise 3.23(c).

---

[9]A sequence of functions with the differential property shown in part (c) is called an **Appell sequence**, after the French mathematician **Paul Appell** (1855–1930).

Ali, S.A., "The mth Ratio Test: New Convergence Tests for Series," *Amer. Math. Monthly*, **115**, 514–524 (2008). Two convincing uses of this new test are to the convergence of the hypergeometric series and to a proof of Raabe's Test.

Apostol, T.M., "A Primer on Bernoulli Numbers and Polynomials," *Math. Mag.*, **81**, 178–190 (2008). The reader will certainly sense that a good deal more about these fascinating mathematical objects is being held up the author's proverbial sleeve.

Boas, Jr., R.P. and Wrench, Jr., J.W., "Partial Sums of the Harmonic Series," *Amer. Math. Monthly*, **78**, 864–870 (1971). Very interesting paper; the theoretical work in the paper makes use of the famous Euler-MacLaurin Summation Formula.

Chan, H.-C., "More Formulas for π,"*Amer. Math. Monthly*, **113**, 452–455 (2006). Interesting paper on a perennially interesting topic. Cited in our Exercise 3.28.

Dence, J.B. and Dence, T.P., *Elements of the Theory of Numbers*, Harcourt/Academic Press, San Diego, 1999, pp. 431–442. These few pages give you an overview of the Bernoulli numbers.

Dence, T.P., "Sums of Simple Rearrangements of the Alternating Harmonic Series," *Math. Gaz.*, **92**, 511–514 (2008). Let $P, N \in \mathbf{N}$ and let the alternating harmonic series be rearranged so that P consecutive positive terms are followed repeatedly by N negative terms. Then the series sums to $\ln\left[2\sqrt{P/N}\right]$.

Dilcher, K., Skula, L. and Slavutskii, I.S., (eds.), *Bernoulli Numbers: Bibliography (1713–1990)*, Queen's Papers in Pure and Applied Mathematics, No. 87, Queen's University, Kingston, Ontario, 1991. A very useful and very comprehensive bibliography (up to 1990) that could be a good place to mine for ideas.

Dunham, W., "The Bernoullis and the Harmonic Series," *Coll. Math. J.*, **18**, 18–23 (1987). A fine historical article that sketches the contributions of the brothers Johann and Jakob Bernoulli to the study of the harmonic series.

Galanor, S., "Riemann's Rearrangement Theorem," *Math. Teacher*, **80**, 675–681 (1987). Very readable article, which the author actually considered to be accessible to an Advanced Placement high school class. Cited in our Exercise 3.34.

Gerst, I., "Some Series for Euler's Constant," *Amer. Math. Monthly*, **76**, 273–275 (1969). Interesting paper; the source of our Exercise 3.32. There are still other series representations for γ in the paper besides the one cited in the exercise. There also exist many integral representations for γ, but these are not given in this paper.

Gould, H.W. and Shonhiwa, T., "A Catalog of Interesting Dirichlet Series," *Missouri J. Math. Sci.*, **20**, 2–18 (2008). There are many Dirichlet series that can serve as interesting generating functions. The author's bibliography looks quite useful. Cited in our Exercise 3.44.

Grabiner, J.V., "Was Newton's Calculus a Dead End? The Continental Influence of MacLaurin's Treatise of Fluxions," *Amer. Math. Monthly*, **104**, 393–410 (1997). It is clear that MacLaurin was an important figure in the historical evolution of the calculus. A synopsis of the man appears in Tweedie, C., "A Study of the Life and Writings of Colin MacLaurin," *Math. Gaz.*, **8**, 132–151 (1915).

Hardy, G.H., *Divergent Series*, Oxford University Press, Oxford, 1949, pp. 1–22. Read these few pages in order to get the flavor of summability techniques for series (especially, divergent series). Cited in our Exercise 3.12.

Hegyvári, N., "On Some Irrational Decimal Fractions," *Amer. Math. Monthly*, **100**, 779–780 (1993). Did you know that the number $\alpha = 0.235711131719\ldots$, where the sequence of digits is formed by the primes in ascending order, is irrational? Read about it in this interesting paper.

Kline, M., *Mathematical Thought from Ancient to Modern Times*, Oxford University Press, New York, 1972, pp. 437, 453, 963, 1096–1121. The citations are, respectively, to Oresme's discovery of the divergence of the harmonic series, to Stirling's presentation of the series that bears his name, to Cauchy's presentation of the Root Test, and to a nice summary chapter on divergent series. The Ratio Test was given as early as 1776 by **Edward Waring** (Cambridge University), some 13 years before Cauchy was even born.

Lehmer, D.H., "A New Approach to Bernoulli Polynomials," *Amer. Math. Monthy*, **95**, 905–911 (1988). This is a novel entry into these interesting polynomials by a master expositor.

Longo, M. and Valori, V., "The Comparison Test—Not Just for Nonnegative Series," *Math. Mag.*, **79**, 205–210 (2006). The generalization of the Comparison Test is as follows: If $\sum_{n=1}^{\infty} a_n, \sum_{n=1}^{\infty} b_n, \sum_{n=1}^{\infty} c_n$ are defined in $\mathbf{R}^1$ and $a_n \le b_n \le c_n$ for all sufficiently large $n \in \mathbf{N}$, then $\sum_{n=1}^{\infty} b_n$ converges if $\sum_{n=1}^{\infty} a_n, \sum_{n=1}^{\infty} c_n$ converge.

Niven, I., "Formal Power Series," *Amer. Math. Monthly*, **76**, 871–889 (1969). Very useful exposition.

Nunemacher, J., "On Computing Euler's Constant," *Math. Mag.*, **65**, 313–322 (1992). This is a stimulating but accessible lead paper into the literature on Euler's constant. It would make a good outside reading assignment. Cited in our Exercise 3.32.

Porter, G.J., "An Alternative to the Integral Test for Infinite Series," *Amer. Math. Monthly*, **79**, 634–635 (1972). The author recommends a more frequent use of Cauchy's Condensation Test (Exercise 3.23), as it may be used for all the series that are usually studied by the Integral Test.

Rudin, W., *Principles of Mathematical Analysis*, 3rd ed., McGraw-Hill, NY, 1976, pp. 76–77. A proof of the Riemann Rearrangement Theorem.

Schmelzer, T. and Baillie, R., "Summing a Curious, Slowly Convergent Series," *Amer. Math. Monthly*, **115**, 525–540 (2008). The authors state that "this paper provides many opportunities for further exploration."

Watkins, W., "Generating Functions," *Coll. Math. J.*, **18**, 195–211 (1987). A good introduction to this topic for the reader who has not yet had much exposure to generating functions.

Wiener, J., "A Shorter, More Efficient Proof of lim $\sqrt[n]{n!}/n = 1/e$," *Coll. Math. J.*, **18**, 319 (1987). The crux of this proof depends on the standard result $\lim_{n\to\infty} (1 + n^{-1})^n = e$.

Williams, K.S., "Bernoulli's Identity Without Calculus," *Math. Mag.*, **70**, 47–50 (1997). An elementary proof of the famous Identity. Cited in our Exercise 3.48(c).

Young, R.M., "The Error in Alternating Series," *Math. Gaz.*, **69**, 120–121 (1985). This result deserves to be better known to students of the calculus. It would be worth your while to write out the proof of Young's result. Cited in our Exercises 3.29 and 3.30.

Additional Literature

Berndt, B.C., "Elementary Evaluation of $\zeta(2n)$," *Math. Mag.*, **48**, 148–154 (1975) (connection with the Bernoulli numbers).

Bromwich, T.J.I'a., *An Introduction to the Theory of Infinite Series*, 2nd ed., Macmillan & Co., Ltd., London, 1926 (a bible).

Chand, H., "On Some Generalizations of Cauchy's Condensation and Integral Test," *Amer. Math. Monthly*, **46**, 338–341 (1939).

Cowen, C.C., Davidson, K.R. and Kaufman, R.P., "Rearranging the Alternating Harmonic Series," *Amer. Math. Monthly*, **87**, 817–819 (1980).

Dence, J.B., "A Development of Euler Numbers," *Missouri J. Math. Sci.*, **9**, 148–155 (1997) (material parallel to that of the Bernoulli numbers).

Grabiner, J.V., "Newton, Maclaurin, and the Authority of Mathematics," *Amer. Math. Monthly*, **111**, 841–852 (2004).

Kalman, D., "Six Ways to Sum a Series," *Coll. Math. J.*, **24**, 402–421 (1993) (the perennial favorite $\sum (1/n^2)$).

Leeming, D.J., "An Asymptotic Estimate for the Bernoulli and Euler Numbers," *Can. Math. Bull.*, **20**, 109–111 (1977).

Littlewood, J.E., "Note on the Convergence of Series of Positive Terms," *Messenger Math.*, **39**, 191–192 (1910), as reproduced in Baker, A. et al (eds.), *Collected Papers of J. E. Littlewood*, Vol. 1, Oxford University Press, 1982, pp. 755–756 (a new test).

Osler, T.J. and Stugard, N., "A Collection of Numbers Whose Proof of Irrationality is Like That of the Number *e*," *Math. Comp. Educ.*, **40** (2), 103–107 (2006).

Roy, R., "The Discovery of the Series Formula for π by Leibniz, Gregory and Nilakantha," *Math. Mag.*, **63**, 291–306 (1990).

Szasz, O., *Introduction to the Theory of Divergent Series*, Hafner Publ. Co., NY, 1948.

Vanden Eynden, C.L., "Proofs that $\sum (1/p)$ Diverges," *Amer. Math. Monthly*, **87**, 394–397 (1980).

Widder, D.V., *Advanced Calculus*, Prentice-Hall, NY, 1947, pp. 261–265 (methods of summability).

CHAPTER 4

Continuity

"... every continuous function of x which is positive for one value of x, and negative for another, must be zero for some intermediate value of x."

Bernhard Bolzano

CONTENTS

Reviewed in this chapter	Quantifiers and negations; discontinuities; arithmetic of continuity.

New in this chapter	Continuity in terms of balls, sequences; open and closed sets; compactness and connectedness; Heine-Borel, Intermediate-Value Theorems; uniform continuity; Contraction Mapping Theorem.

4.1 FIRST DEFINITION OF CONTINUITY

In elementary discussions a function f with a domain $\mathbf{D}(f)$ that is an open interval is said to be continuous at $x_0 \in \mathbf{D}(f)$ iff $\lim_{x \to x_0} f(x) = f(x_0)$. We proceed to generalize this by framing a definition in the language of balls.

Definition. *Let* $\mathbf{f} : \mathbf{D}(\mathbf{f}) \to \mathbf{R}^m, \mathbf{D}(\mathbf{f}) \subseteq \mathbf{R}^n$*, be a function, and let* $\mathbf{a} \in \mathbf{D}(\mathbf{f})$*. Then* $\mathbf{f}$ *is* ***continuous at*** $\mathbf{a}$ *iff given any* $\varepsilon > 0$*, there is a* $\delta > 0$ *such that if* $\mathbf{x} \in \mathbf{B}_n(\mathbf{a}; \delta) \cap \mathbf{D}(\mathbf{f})$ *and* $(\mathbf{x}, \mathbf{y}) \in \mathbf{f}$*, then* $\mathbf{y} \in \mathbf{B}_m(\mathbf{f}(\mathbf{a}); \varepsilon)$*. A function that is not continuous at* $\mathbf{a}$ *is termed* ***discontinuous at*** $\mathbf{a}$.

The following comments provide additional elaboration:

1. In contrast to the definition of the limit of $\mathbf{f}$ at $\mathbf{a}$ (see Section 1.9), in the preceding definition, the n-ball about $\mathbf{a}$ is not a deleted n-ball.
2. The point $\mathbf{a}$, which must belong to $\mathbf{D}(\mathbf{f})$, can be either a cluster point of $\mathbf{D}(\mathbf{f})$ or an isolated point.
3. If $\mathbf{a}$ is a cluster point that is an interior point of $\mathbf{D}(\mathbf{f})$ (Section 1.8), then the definition is equivalent to the statement that $\mathbf{f}$ is continuous

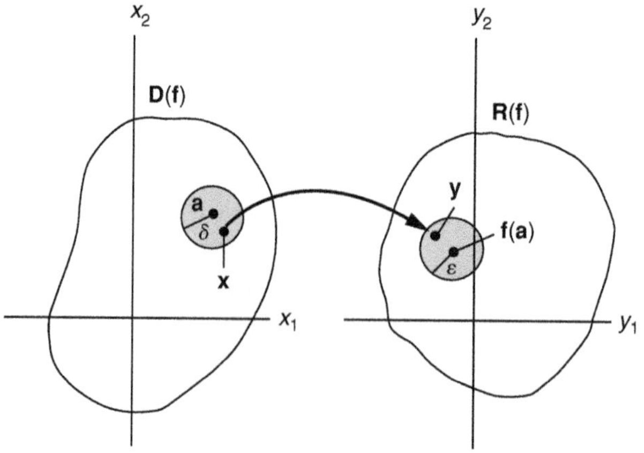

FIGURE 4.1
Continuity at $\mathbf{a}$ *of a function in* $\mathbf{R}^2$ *when* $\mathbf{a}$ *is an interior point.*

at a iff $\lim_{\mathbf{x}\to\mathbf{a}} \mathbf{f}(\mathbf{X}) = \mathbf{f}(\mathbf{a}) = \mathbf{f}\left(\lim_{\mathbf{x}\to\mathbf{a}} \mathbf{x}\right)$ (Figure 4.1). The definition is also equivalent to this: **f** is continuous at **a** iff, given any $\varepsilon > 0$, there is a $\delta > 0$ such that $\mathbf{f}[\mathbf{B}_n(\mathbf{a};\delta] \subseteq \mathbf{B}_m(\mathbf{f}(\mathbf{a});\varepsilon)$.

4. If **a** is a boundary point that belongs to **D(f)**, then the equalities in comment 3 still hold, but the limits are "sided" limits (Figure 4.2).

5. If **a** is an isolated point, it is not possible even to discuss $\lim_{\mathbf{x}\to\mathbf{a}} \mathbf{f}(\mathbf{x})$. The definition, however, implies that **f** is continuous at all of its isolated points (Figure 4.3). There is no alarm here since isolated points are usually a minor issue.

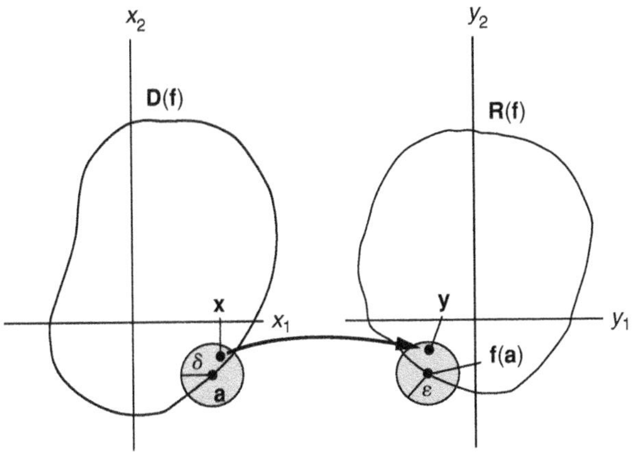

FIGURE 4.2
Continuity at **a** *of a function in* $\mathbf{R}^2$ *when* **a** *is a boundary point.*

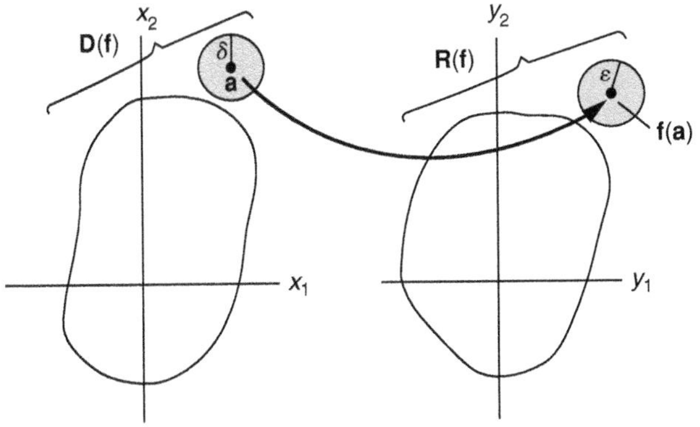

FIGURE 4.3
Continuity at **a** *of a function in* $\mathbf{R}^2$ *when* **a** *is an isolated point.*

■ Example 4.1

Let $\mathbf{f} : \mathbf{D(f)} \to \mathbf{R}^2, \mathbf{D(f)} \subset \mathbf{R}^2$ be defined by $\mathbf{y} = \mathbf{f(x)} = \mathbf{f}[(x_1, x_2)] = (y_1, y_2) = (3 - x_2, \frac{1}{2}x_1^2 - 1)$, and $\mathbf{D(f)} = \mathbf{S} \cup \{(1, 1)\}$, where $\mathbf{S}$ is the set of points on and in the square $\{(x_1, x_2)\colon 2 \le x_1 \le 4, 1 \le x_2 \le 3\}$ (Figure 4.4). Let $\mathbf{a} = (4, 1)$ and let $\varepsilon > 0$ be given; we have $\mathbf{f(a)} = (2, 7)$. For any δ satisfying $0 < \delta \le 2$, if $\mathbf{x} \in \mathbf{B}_2(\mathbf{a}; \delta) \cap \mathbf{D(f)}$, then $(4 - x_1)^2 + (x_2 - 1)^2 \le \delta^2$. Then for $\mathbf{y} = \mathbf{f(x)}$, we desire $(7 - y_2)^2 + (2 - y_1)^2 \le \varepsilon^2$, or equivalently,

$$\left(8 - \frac{1}{2}x_1^2\right)^2 + (x_2 - 1)^2 \le \varepsilon^2.$$

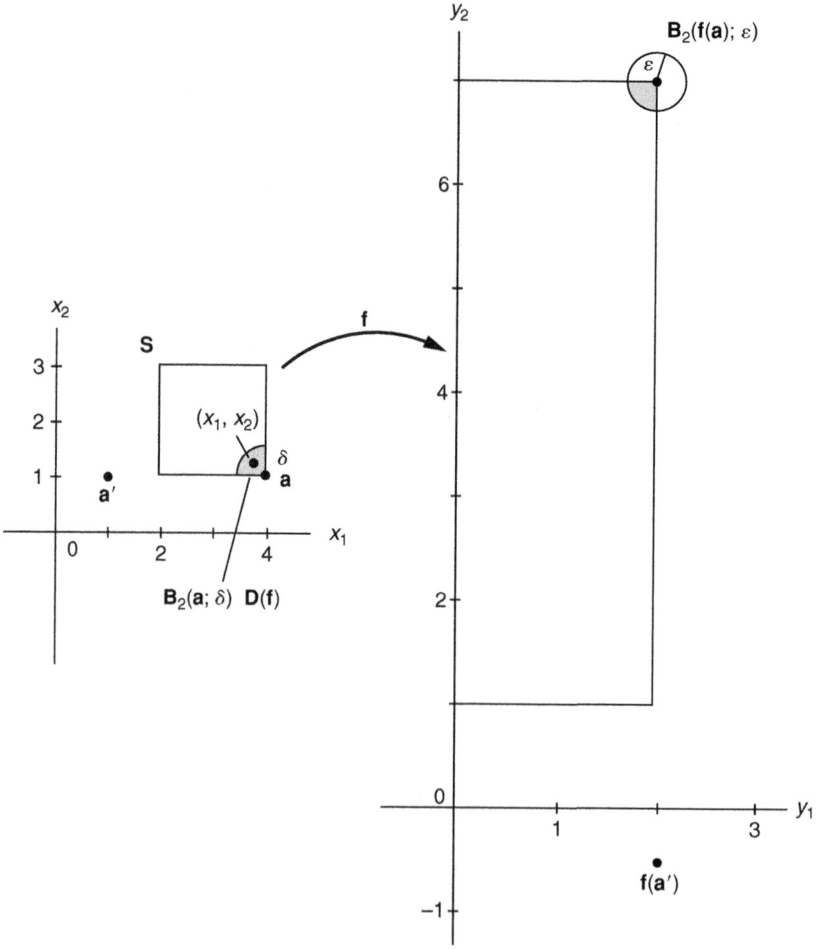

FIGURE 4.4
A function $\mathbf{D(f)} \to \mathbf{R}^2, \mathbf{D(f)} \subset \mathbf{R}^2$.

From the geometry of set $\mathbf{S}$, we have

$$8 - \frac{1}{2}x_1^2 = \frac{1}{2}\left(16 - x_1^2\right) = \frac{1}{2}(4 - x_1)(4 + x_1) \leq 4(4 - x_1),$$

and hence,

$$\left(8 - \frac{1}{2}x_1^2\right)^2 + (x_2 - 1)^2 \leq 16(4 - x_1)^2 + (x_2 - 1)^2$$
$$\leq 16\left[(4 - x_1)^2 + (x_2 - 1)^2\right] \leq 16\delta^2 \leq \varepsilon^2.$$

It follows that if we take $\delta = \varepsilon/4$, then corresponding to any $\mathbf{x} \in \mathbf{B}_2(\mathbf{a}; \varepsilon/4) \cap \mathbf{D}(\mathbf{f})$, the point $\mathbf{y} = \mathbf{f}(\mathbf{x})$ will lie in $\mathbf{B}_2(\mathbf{f}(\mathbf{a}); \varepsilon)$. Hence, $\mathbf{f}$ is continuous at $\mathbf{a}$. ■

■ Example 4.2

In Figure 4.4, let $\mathbf{a}' = (1, 1)$; we have $\mathbf{f}(\mathbf{a}') = \mathbf{y}' = (2, -\frac{1}{2})$. If $\varepsilon > 0$ is given, then choose $\delta = 0.5$. The only point in $\mathbf{B}_2(\mathbf{a}'; 0.5) \cap \mathbf{D}(\mathbf{f})$ is $\mathbf{a}'$, and corresponding to this, $\mathbf{y}' = (2, -\frac{1}{2}) \in \mathbf{B}_2(\mathbf{f}(\mathbf{a}'); \varepsilon)$. Hence, $\mathbf{f}$ is continuous at $\mathbf{a}'$. ■

In Section 4.3 we shall establish an equivalent, sequential definition of continuity at a point. The second half of the proof there involves the negation of a δ, ε-statement. We pause at this point in order to supply some needed background.

4.2 QUANTIFIERS AND NEGATIONS

A declarative sentence whose truth or falsity depends upon the "values" of one or more variables is called a **predicate** or a **propositional function** (Copi, 1986; Wolf, 2005). In contrast to a proposition such as

"sin x is continuous at x = 3,"

two examples of predicates are

"Every natural number can be expressed as the sum of no more than n squares of nonnegative integers;"
"The function f, which is continuous at $x = 1$, is differentiable there."

The variables are, respectively, n and f. Some predicates may be true for all choices of the variable(s) in some specified universe of discourse, whereas other predicates are true only for particular choices. The following terminology is useful.[1]

[1]The logical calculus of propositional functions containing quantifiers was worked out by the German mathematician **Friedrich Ludwig Gottlob Frege** (1848–1925) and the Italian mathematician **Giuseppe Peano** (1858–1932).

Definition. *The symbol ∀ is called the* **universal quantifier**, *and for the one-variable predicate* P(x) *the statement* (∀x)P(x) *is read "for all* x, P(x)." *This statement is true only when* P(x) *is true for every* x *in its specified universe of discourse.*

The symbol ∃ is called the **existential quantifier**, *and the statement* (∃x)P(x) *is read "there exists an* x *such that* P(x)." *This statement is true only when the set of* x*'s that make* P(x) *true is nonempty.*

■ Example 4.3

"(∀x)[{($x > 2$) AND (x is prime)} → x is odd]" expresses (the fact) that all primes larger than 2 are odd.

"(∃x) : ($x \in \mathbf{R}$ AND $x = x^{-1}$)" expresses (the fact) that there is a real number that is equal to its multiplicative inverse. ■

Note that the predicate following the universal quantifier is written as an implication; this is because we have to specify clearly the universe of discourse for the objects x. The predicate following the existential quantifier is written as a conjunction because it has to guarantee the existence of some x's.

It is frequently necessary to formulate the negation of a statement that contains a quantifier. Figure 4.5(a) shows that all A(x) are B(x). The negation of this, shown in Figure 4.5(b), is that some A(x) are not-B(x). Maybe all A(x) have this property, and maybe not, but anyway some A(x) have this property. Figure 4.5(c) shows the statement that some A(x) are B(x). "Some" could conceivably mean "all" in a particular case. The negation of "some A(x) are B(x)" is that "not-some A(x) are B(x)," that is, "no A(x) are B(x)" for any choice of x in the universe of discourse (Figure 4.5(d)).

Thus, the connection between negation (~) and the two quantifiers is as follows.

Definition. *If* P(x) *is a predicate with variable* x, *then*

(a) ~[(∀x)P(x)] *is equivalent to* (∃x) : [~P(x)];

(b) ~[(∃x) : P(x)] *is equivalent to* (∀x)[~P(x)].

In order to implement this definition, we need to know how to form negations of different forms for **P**(x). These can be worked out from truth tables (Exercises 1.1–1.3). We present the results in Table 4.1. It may be necessary to apply the definition and Table 4.1 repeatedly in order to construct the negation of some statements. As shown in the following example, the basic maneuver is to move the negation symbol (~) step-by-step from left to right.

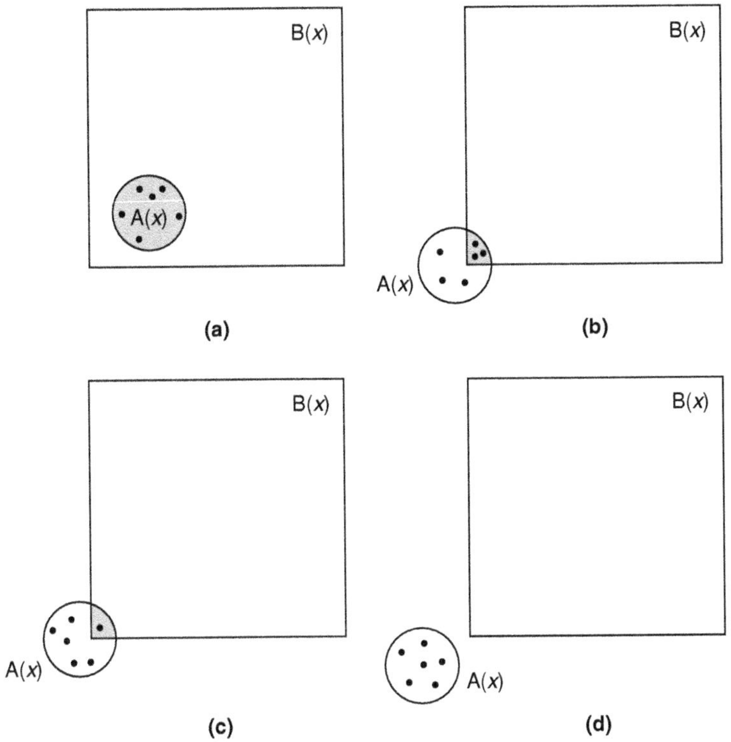

FIGURE 4.5
Negations of statements with one quantifier.

Table 4.1 Elementary Negations

Form of P(x)	Negation, ~P(x)
1. A(x) AND B(x)	A(x) → ~B(x)[a]
2. A(x) OR B(x)	~A(x) AND ~B(x)
3. A(x) → B(x)	A(x) AND ~B(x)

[a] *This negation is logically equivalent to ~A(x) OR ~B(x).*

■ Example 4.4

"$(\forall n)(n$ is prime $\rightarrow n$ is odd)"

The negation of this is:

"$\sim (\forall n)(n$ is prime $\rightarrow n$ is odd)"

iff "$(\exists n) : [\sim (n$ is prime $\rightarrow n$ is odd)]" iff "$(\exists n)$: [n is prime AND $\sim$ (n is odd)]"

iff "$(\exists n)$: [n is prime AND n is even]."

The negation is true, so the original sentence is false. At this point, please work Exercises 4.6 through 4.9. ■

4.3 THE SEQUENCE DEFINITION; DISCONTINUITIES

We return now to the main line of discussion that was interrupted at the end of Section 4.1.

Theorem 4.1. *Let* $\mathbf{f} : D(\mathbf{f}) \to \mathbf{R}^m, D(\mathbf{f}) \subseteq \mathbf{R}^n$, *be a function and suppose that* $\mathbf{p}_0 \in D(\mathbf{f})$ *is not an isolated point. Then* $\mathbf{f}$ *is continuous at* $\mathbf{p}_0$ *iff for each sequence* $\{\mathbf{p}_k\}_{k=1}^{\infty}$ *in* $D(\mathbf{f})$ *that converges to* $\mathbf{p}_0$, *the sequence* $\{\mathbf{f}(\mathbf{p}_k)\}_{k=1}^{\infty}$ *converges to* $\mathbf{f}(\mathbf{p}_0)\}$.

Proof. ($\to$) Suppose, first, that $\mathbf{f}$ is continuous at $\mathbf{p}_0$, and let $\{\mathbf{p}_k\}_{k=1}^{\infty}$ be any sequence in $D(\mathbf{f})$ for which $\lim_{k\to\infty} \mathbf{p}_k = \mathbf{p}_0$. Let $\varepsilon > 0$ be given. Then the continuity of $\mathbf{f}$ at $\mathbf{p}_0$ implies that there is a $\delta > 0$ such that if $\mathbf{p}_k \in \mathbf{B}_n(\mathbf{p}_0; \delta) \cap D(\mathbf{f})$, then $f(\mathbf{p}_k) \in \mathbf{B}_m(\mathbf{f}(\mathbf{p}_0); \varepsilon)$. But the convergence of $\{\mathbf{p}_k\}_{k=1}^{\infty}$ to $\mathbf{p}_0$ guarantees that for some $N \in \mathbf{N}$, $||\mathbf{p}_0 - \mathbf{p}_k|| < \delta$ for all $k > N$. Hence, for all such k we have $\mathbf{f}(\mathbf{p}_k) \in \mathbf{B}_m(\mathbf{f}(\mathbf{p}_0); \varepsilon)$. This says that $\lim_{k\to\infty} \mathbf{f}(\mathbf{p}_k) = \mathbf{f}(\mathbf{p}_0)$.

($\leftarrow$) We prove this direction (the converse of the above) in the form of its contrapositive. Suppose, then, that $\mathbf{f}$ is not continuous at $\mathbf{p}_0$. This statement is, from Exercise 4.9(e),

$$(\exists\varepsilon) : [\varepsilon > 0 \text{ AND } (\forall\delta)\{(\delta > 0) \to (\exists\mathbf{p}) : [(\mathbf{p} \in \mathbf{B}_n(\mathbf{p}_0; \delta) \cap D(\mathbf{f}))$$

$$\text{AND } \mathbf{f}(\mathbf{p}) \notin \mathbf{B}_m(\mathbf{f}(\mathbf{p}_0); \varepsilon)]\}]. \qquad (*)$$

Thus, there is some aberrant $\varepsilon' > 0$ that satisfies (*). If we define the sequence $\{\delta_k\}_{k=1}^{\infty}$ by $\delta_k = k^{-1}$, then (from (*)) for each $k \in \mathbf{N}$ there is a point $\mathbf{p}_k \in \mathbf{B}_n(\mathbf{p}_0; \delta_k) \cap D(\mathbf{f})$ for which $\mathbf{f}(\mathbf{p}_k) \notin \mathbf{B}_m(\mathbf{f}(\mathbf{p}_0); \varepsilon')$. That is, we have a sequence $\{\delta_k\}_{k=1}^{\infty}$ that converges to $\mathbf{p}_0$ (because $\lim_{k\to\infty} \delta_k = 0$) but for which $\{\mathbf{f}(\mathbf{p}_k)\}_{k=1}^{\infty}$ does not converge to $\mathbf{f}(\mathbf{p}_0)$ (because, always,

$$||\mathbf{f}(\mathbf{p}_0) - \mathbf{f}(\mathbf{p}_k)|| > \varepsilon').$$

■

Theorem 4.1 is useful in a practical sense for the investigation of discontinuities of functions.

■ Example 4.5

Let $f : \mathbf{R}^2 \to \mathbf{R}^1$ be defined by

$$f(\mathbf{x}) = \begin{cases} \frac{x_1 x_2}{x_1^2 + x_2^2} & \mathbf{x} \neq \mathbf{0} \\ 0 & \mathbf{x} = \mathbf{0}, \end{cases}$$

where $\mathbf{x} = (x_1, x_2) \in \mathbf{R}^2$. Choose the sequence $\{\mathbf{p}_n\}_{n=1}^{\infty}$, where $\mathbf{p}_n = (n^{-1}, n^{-1})$. Then $\lim_{n\to\infty} \mathbf{p}_n = \mathbf{0}$; further, for each $n \in \mathbf{N}, f(\mathbf{p}_n) = n^{-1}n^{-1}/\left[\left(n^{-1}\right)^2 + \left(n^{-1}\right)^2\right] = 1/2$. Hence, $\lim_{n\to\infty} f(\mathbf{p}_n) = 1/2 \neq f\left(\lim_{n\to\infty} \mathbf{p}_n\right)$, so by Theorem 4.1 f is discontinuous at $\mathbf{x} = \mathbf{0}$. ■

■ Example 4.6

Let $f : [0, 1] \to \mathbf{R}^1$ be defined by[2]

$$f(x) = \begin{cases} 1 & x \in \mathbf{Q} \\ 0 & x \notin \mathbf{Q}. \end{cases}$$

Fix $x \in [0, 1]$ arbitrarily. Then if x is rational, choose $\{x_n\}_{n=1}^{\infty}$ to be any sequence of irrationals in $(0, 1)$ that converges to x. For example, the following sequence $\{x_n\}_{n=1}^{\infty}$ converges to x:

$$x_n = x + \frac{\pi}{n}.$$

This follows because each x_n is irrational, and $\lim_{n\to\infty} (\pi/n) = 0$.

Then f is discontinuous at rational x because $\lim_{n\to\infty} x_n = x$, but $\lim_{n\to\infty} f(x_n) = \lim_{n\to\infty} 0 = 0 \neq f\left(\lim_{n\to\infty} x_n\right) = f(x) = 1$. ■

Discontinuities are classified as either essential or removable. A function $\mathbf{f} : D(\mathbf{f}) \to \mathbf{R}^m, D(\mathbf{f}) \subseteq \mathbf{R}^n$, has an **essential discontinuity** at $\mathbf{a} \in \mathbf{R}^n$ if $\mathbf{f}$ is discontinuous at $\mathbf{a}$, no matter what value in $\mathbf{R}^m$ has been assigned to $\mathbf{f}(\mathbf{a})$. The discontinuity in Example 4.5 is of this type. Two types of essential discontinuities are those commonly known as **infinite discontinuities** (Figure 4.6(a)) and **jump discontinuities** (Figure 4.6(b)).

On the other hand, a function $\mathbf{f}$ has a **removable discontinuity** at $\mathbf{a}$ if it is possible to define (or change) the value of $\mathbf{f}$ at $\mathbf{a}$ so as to make $\mathbf{f}$ continuous there. The function in Figure 4.7 has a removable discontinuity at $x = 2$, since the function becomes continuous there if we define $f(2) = 3$.

---

[2]Sometimes called the **Dirichlet function**, because it was introduced into analysis by the German mathematician **Peter G. L. Dirichlet** (1805–1859).

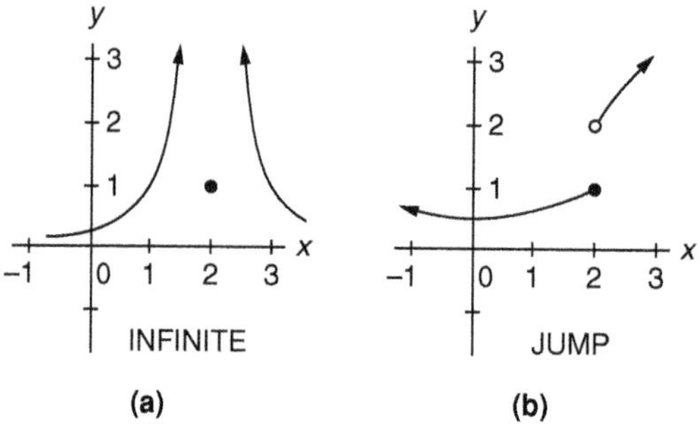

FIGURE 4.6
An essential discontinuity exists at $x = 2$ *for (a) the function* $f(x) = \begin{cases} (x-2)^{-2} & x \neq 2 \\ 1 & x = 2 \end{cases}$, *and (b) the function* $f(x) = \begin{cases} (x^2+4)/8 & x \leq 2 \\ \sqrt{5x-6} & x > 2 \end{cases}$.

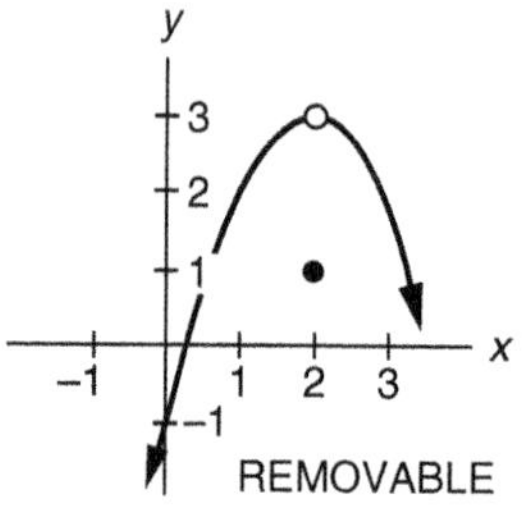

FIGURE 4.7
A removable discontinuity exists at $x = 2$ *for the function* $f(x) = \begin{cases} -x^2+4x-1 & x \neq 2 \\ 1 & x = 2 \end{cases}$.

4.4 ELEMENTARY CONSEQUENCES OF CONTINUITY

The most obvious consequence of a real-valued function f being continuous at a point $\mathbf{a}$ is that if $f(\mathbf{a}) \neq 0$, then there is a small region about $\mathbf{a}$ in which $f(\mathbf{x})$ is of one algebraic sign (Figure 4.8).

We make this precise in the following theorem, stated without loss of generality for the case $f(\mathbf{a}) > 0$.

Theorem 4.2. *Let* $f : \mathbf{D}(f) \to \mathbf{R}^1, \mathbf{D}(f) \subseteq \mathbf{R}^n$, *be a function that is continuous at* $\mathbf{a}$. *If* $f(\mathbf{a}) > 0$, *then there are numbers* $\delta, \varepsilon > 0$ *such that* $\mathbf{x} \in \mathbf{B}_n(\mathbf{a}; \delta) \cap \mathbf{D}(f)$ *implies* $f(\mathbf{x}) > \varepsilon$.

Proof. The continuity of f at a means that there is a $\delta > 0$ such that $\mathbf{x} \in \mathbf{B}_n(\mathbf{a}; \delta) \cap \mathbf{D}(f)$ *implies* $|f(\mathbf{x}) - f(\mathbf{a})| < \frac{1}{2}f(\mathbf{a})$. The completion of the proof is left to you. ■

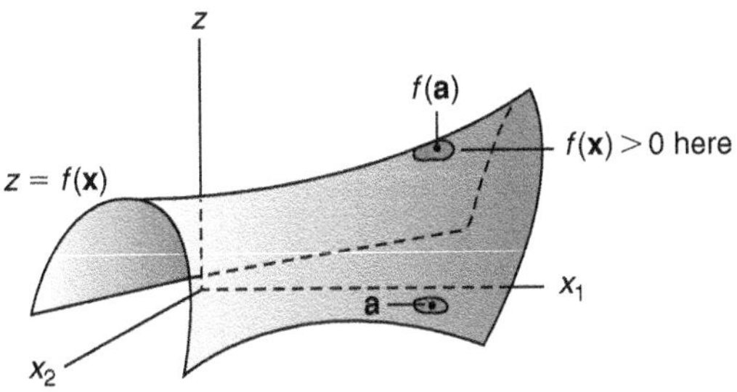

FIGURE 4.8
A localized region of values $f(\mathbf{x})$ of one algebraic sign.

The elementary arithmetic of the continuity of functions holds no surprises. All parts of the following theorem can be proved by use of Theorems 2.6, 4.1. We leave parts (a) and (b) for the exercises, but for variety we prove part (c) differently.

Theorem 4.3. *Let* $\mathbf{f}: \mathbf{D} \to \mathbf{R}^m, \mathbf{g} : \mathbf{D} \to \mathbf{R}^m, \mathbf{D} \subseteq \mathbf{R}^n$, *be functions defined on a common domain* $\mathbf{D}$; *let* $\mathbf{a} \in \mathbf{D}$ *and* $k \in \mathbf{R}$. *Then*

(a) *if* $\mathbf{f}, \mathbf{g}$ *are continuous at* $\mathbf{a}$, *then so are* $\mathbf{f} \oplus \mathbf{g}, k{\cdot}\mathbf{f}$.

(b) *Further, if* $m = 1$, *then* fg *is also continuous at* $\mathbf{a}$.

(c) *Additionally* ($m = 1$), *if* $g(\mathbf{a}) \neq 0$, *then* f/g *is continuous at* $\mathbf{a}$.

Proof. (c) Assume f, g are continuous at $\mathbf{a}$ and $g(\mathbf{a}) \neq 0$; let $\varepsilon > 0$ be given. Then there is a $\delta_1 > 0$ such that $\mathbf{x} \in \mathbf{B}_n(\mathbf{a}; \delta_1) \cap \mathbf{D}$ implies $|g(\mathbf{x}) - g(\mathbf{a})| < \frac{\varepsilon}{2}[g(\mathbf{a})]^2$. From Theorem 4.2, generalized slightly, there is a number $\delta_2 > 0$ such that $\mathbf{x} \in \mathbf{B}_n(\mathbf{a}; \delta_2) \cap \mathbf{D}$ implies $|g(\mathbf{x})| > \frac{1}{2}|g(\mathbf{a})| > 0$ (Exercise 4.16), or equivalently

$$0 < \frac{1}{|g(\mathbf{x})|} < \frac{2}{|g(\mathbf{a})|}.$$

Hence, if $\delta = \min\{\delta_1, \delta_2\}$, one has for all $\mathbf{x} \in \mathbf{B}_n(\mathbf{a}; \delta) \cap \mathbf{D}$

$$\left| \frac{1}{g(\mathbf{a})} - \frac{1}{g(\mathbf{x})} \right| = \frac{|g(\mathbf{x}) - g(\mathbf{a})|}{|g(\mathbf{a})|} \frac{1}{|g(\mathbf{x})|} < \frac{(\varepsilon/2)\,[g(\mathbf{a})]^2}{|g(\mathbf{a})|} \frac{2}{|g(\mathbf{a})|}$$
$$= \varepsilon.$$

Thus, $1/g$ is continuous at $\mathbf{a}$, and (c) follows from application of part (b). ■

The property of continuity (at a point) is preserved upon function composition, provided a simple condition is met. The proof is straightforward, and you should have no trouble with it; again, use Theorem 4.1.

Theorem 4.4. *Let* $\mathbf{f}: \mathrm{D}(\mathbf{f}) \to \mathbf{R}^m, \mathbf{g}: \mathrm{D}(\mathbf{g}) \to \mathbf{R}^n, \mathrm{R}(\mathbf{g}) \subseteq \mathrm{D}(\mathbf{f}) \subseteq \mathbf{R}^n, \mathrm{D}(\mathbf{g}) \subseteq \mathbf{R}^k$*, be functions. If* $\mathbf{g}$ *is continuous at* $\mathbf{a} \in \mathbf{R}^k$ *and* $\mathbf{f}$ *is continuous at* $\mathbf{g}(\mathbf{a})$*, then* $\mathbf{f}[\mathbf{g}]$ *is continuous at* $\mathbf{a}$.

Proof. The proof is left to you. ■

■ Example 4.7

Let f be defined by $f(x) = \frac{x^2+x+1}{x^3+1}, x \in \mathbf{R}^1\backslash\{-1\}$. Write this as

$$f(x) = x\left(\frac{x}{g(x)}\right) + \frac{x}{g(x)} + \frac{1}{g(x)},$$

where $g(x) = x^3 + 1$. Then $f(x)$ is continuous at $x = 2$ because $g(x)$ is continuous there, and so are all three of the indicated terms, upon making use of all parts of Theorem 4.3. A δ, ε-proof, as in Example 4.1, is completely unnecessary. ■

■ Example 4.8

Let f: $\mathrm{D}(f) \to \mathbf{R}^1, g : \mathrm{D}(g) \to \mathbf{R}^1$ be functions defined by $f(x) = \ln(x), x > 0$ and $g(x) = \sqrt{x}, x \geq 0$. Then g is continuous at $x = 1/4$, and f is continuous at $g(1/4) = 1/2$. By Theorem 4.4, $\ln(\sqrt{x})$ is continuous at $x = 1/4$. ■

A comment about Example 4.8 is in order. It is easy enough to show that g is continuous at any $x \geq 0$, but less easy to show that f is continuous at $1/2$. How to do this depends upon how $\ln(x)$ has been defined; several approaches are feasible.

APPROACH 1: The function e^x may already have been introduced and found to be increasing and continuous everywhere on its domain. A theorem then says that the inverse function, notated $\ln x$, exists and is also increasing and continuous everywhere on its domain.

APPROACH 2: The function $\ln x$ may have been defined by $\int_1^x t^{-1}dt, x > 0$. Since t^{-1} is continuous on the interval of integration, then a theorem says that the integral is continuous at any point in that interval.

APPROACH 3: The function $\ln x$ may have been defined by the series $\sum_{k=1}^{\infty}(-1)^{k+1}(x-1)^k/k$ and the interval of convergence found to be $0 < x \leq 2$. A theorem then says that a power series can be differentiated term-by-term at any point in its interval of convergence. A second theorem then says that $\ln x$ is continuous wherever it is differentiable. Finally, if $x > 2$, then from the identity $\ln(x) = -\ln(1/x)$, the continuity of $\ln(1/x)$ implies the continuity of $\ln(x)$.

The theorems just alluded to will appear later in the book. For the present, we take for granted the continuity of the standard transcendental functions everywhere on their domains.

4.5 MORE ON SETS—OPEN SETS

Continuity, as discussed in Section 4.1, is a property that holds locally at some point $\mathbf{p}_0 \in D(\mathbf{f})$; that is, it is a property that is true at $\mathbf{p}_0$ and at all points $\mathbf{p}$ sufficiently close to $\mathbf{p}_0$. In contrast, a property that holds at all points $\mathbf{p}$ of a given set $\mathbf{S}$ is said to hold **globally** on $\mathbf{S}$.

Definition. *If* $\mathbf{f} : D(\mathbf{f}) \to \mathbf{R}^m, D(\mathbf{f}) \subseteq \mathbf{R}^n$, *is a function that is continuous at each point of a given set* $\mathbf{S} \subseteq D(\mathbf{f})$, *then* $\mathbf{f}$ *is termed* ***continuous on*** $\mathbf{S}$.

Theorem 4.1 may be summarized as follows: the local property of continuity of a function preserves sequence convergence. It is of interest to extend this line of thought to the global version of continuity and examine some of the types of sets $\mathbf{S}$ whose nature is preserved by continuity in their image sets $\mathbf{f}(\mathbf{S})$. Our discussion here and in the next three sections will center, therefore, on sets (open and closed). This will extend the material on sets scattered throughout Chapter 1, and it will also give you a glance (only a glance, unfortunately!) at an entire branch of mathematics—**topology**—that is indispensable in many other areas of mathematics. For additional introductory coverage of topology, you may consult Baker (1997), Gamelin and Greene (1999), Lipschutz (1965), Messer and Straffin (2006), and Simmons (1983).

We will present a loosely-sequenced string of set-theoretic lemmas and theorems, some of which will be proved and others of which will be left for the Exercises. The general setting is that of a metric space $< \mathbf{M}, d >$ (Section 1.5), where the set $\mathbf{M}$ may be $\mathbf{R}^n$ or it may be left general and unspecified. We begin by generalizing the definition of open set that was given in Section 1.8.

Definition. *A subset* $\mathbf{S} \subseteq \mathbf{M}$ *in the metric space* $< \mathbf{M}, d >$ *is an* ***open set*** *(in* $\mathbf{M}$*) iff given any point* $\mathbf{p}_0 \in \mathbf{S}$ *there is a* $\delta > 0$ *such that the ball* $\mathbf{B}(\mathbf{p}_0; \delta)$ *is contained in* $\mathbf{S}$, *where* $\mathbf{B}(\mathbf{p}_0; \delta) = \{\mathbf{p} : \mathbf{p} \in \mathbf{M}, d(\mathbf{p}, \mathbf{p}_0) < \delta\}$.

Lemma 4.5.1. *In any metric space* $< \mathbf{M}, d >$, *the empty set* $\emptyset$ *and the entire set* $\mathbf{M}$ *are open sets. A singleton set (e.g.,* $\{2\} \subset \mathbf{R}^1$*), however, is not an open set.*

Lemma 4.5.2. *A ball* $\mathbf{B}(\mathbf{p}_0; \delta)$ *in any metric space* $< \mathbf{M}, d >$ *is itself an open set.*

Proof. Use the Triangle Inequality; the proof is left to you. ■

The next result, an important theorem, shows one way that new open sets can be constructed from other open sets.

Theorem 4.5. *In $< \mathbf{M}, d >$ the union of an arbitrary collection $\mathfrak{C}$ of open subsets of* $\mathbf{M}$ *is itself an open set.*

Proof. Let $\mathbf{T}$ denote generically a member of $\mathfrak{C}$; this latter could be finite, countably infinite, or uncountably infinite. Choose, arbitrarily, $\mathbf{p}_0 \in \bigcup_{\mathbf{T}\in\mathfrak{C}} \mathbf{T}$. Then for some $\mathbf{T}' \in \mathfrak{C}$ we have $\mathbf{p}_0 \in \mathbf{T}'$. Since $\mathbf{T}'$ is open, then by the earlier definition there is a $\delta > 0$ and a ball $\mathbf{B}(\mathbf{p}_0; \delta)$ such that

$$\mathbf{B}(\mathbf{p}_0; \delta) \subset \mathbf{T}' \subset \bigcup_{\mathbf{T}\in\mathfrak{C}} \mathbf{T}.$$

As $\mathbf{p}_0$ was arbitrary, the set inclusions say that $\bigcup_{\mathbf{T}\in\mathfrak{C}} \mathbf{T}$ is open. ■

The theorem that follows is the first real fruit of our topological labors. It sometimes is used as an equivalent definition of continuity of a function on a set.

Theorem 4.6. *Let $f : \mathbf{M} \to \mathbf{M}'$ be a function from the metric space $< \mathbf{M}, d >$ into the metric space $< \mathbf{M}', d' >$. Then f is continuous on $\mathbf{M}$ iff for every open set $\mathbf{Y} \subseteq \mathbf{M}'$, the inverse image $f^{-1}(\mathbf{Y})$ is an open subset in $\mathbf{M}$.*

Proof. $(\rightarrow)$ Assume f is continuous on $\mathbf{M}$ and let $\mathbf{Y}$ be an open set in $\mathbf{M}'$. If $f^{-1}(\mathbf{Y}) = \emptyset$, then by Lemma 4.5.1 $f^{-1}(\mathbf{Y})$ is open in $\mathbf{M}$. If $f^{-1}(\mathbf{Y}) \neq \emptyset$, then choose any $\mathbf{p} \in f^{-1}(\mathbf{Y})$. Then $f(\mathbf{p}) \in \mathbf{Y}$, and since $\mathbf{Y}$ is open in $\mathbf{M}'$, then there is an $\varepsilon > 0$ and a ball $\mathbf{B}'(f(\mathbf{p}); \varepsilon) \subseteq \mathbf{Y}$. However, f is continuous on $\mathbf{M}$, so from remark (3) in Section 4.1 there is a $\delta > 0$ and a ball $\mathbf{B}(\mathbf{p}; \delta)$ such that

$$f(\mathbf{B}(\mathbf{p}; \delta)) \subseteq \mathbf{B}'(f(\mathbf{p}); \varepsilon).$$

Combination of the two set inclusions yields

$$f(\mathbf{B}(\mathbf{p}; \delta)) \subseteq \mathbf{Y},$$

or $\mathbf{B}(\mathbf{p}; \delta) \subseteq f^{-1}[f(\mathbf{B}(\mathbf{p}; \delta))] \subseteq f^{-1}(\mathbf{Y})$ (Exercise 4.30(a)), so $f^{-1}(\mathbf{Y})$ is open in $\mathbf{M}$.

$(\leftarrow)$ Conversely, assume $f^{-1}(\mathbf{Y})$ is open in $\mathbf{M}$ whenever $\mathbf{Y}$ is open in $\mathbf{M}'$. Let $\mathbf{p} \in \mathbf{M}$ be chosen arbitrarily. By Lemma 4.5.2, for each $\varepsilon > 0$, $\mathbf{B}'(f(\mathbf{p}); \varepsilon)$ is open in $\mathbf{M}'$. But by assumption, $f^{-1}(\mathbf{B}'(f(\mathbf{p}); \varepsilon))$ is open in $\mathbf{M}$. Hence, there is a $\delta > 0$ and a ball $\mathbf{B}(\mathbf{p}; \delta)$ such that

$$\mathbf{B}(\mathbf{p}; \delta) \subseteq f^{-1}(\mathbf{B}'(f(\mathbf{p}); \varepsilon)),$$

or equivalently (Exercise 4.30(b)),

$$f(\mathbf{B}(\mathbf{p}; \delta)) \subseteq f[f^{-1}(\mathbf{B}'(f(\mathbf{p}); \varepsilon))] \subseteq \mathbf{B}'(f(\mathbf{p}); \varepsilon).$$

This shows that f is continuous at $\mathbf{p}$; since $\mathbf{p} \in \mathbf{M}$ was arbitrary, then f is continuous on $\mathbf{M}$. ■

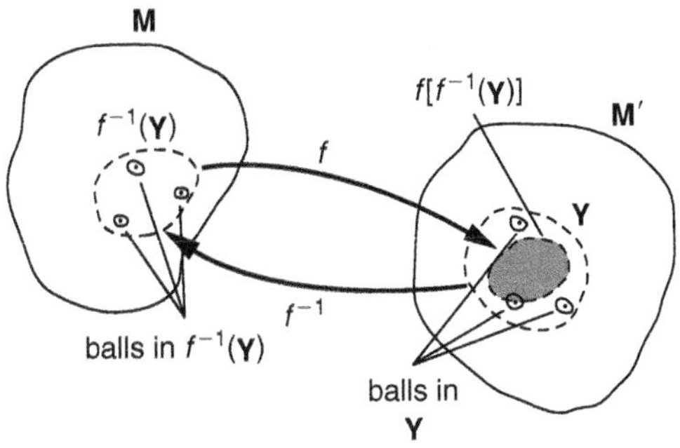

FIGURE 4.9
Illustration of the inverse image definition of a continuous function.

We recall from Exercise 1.39 that the inverse image of a set $\mathbf{Y} \subseteq \mathbf{M}'$ under f is the set $f^{-1}(\mathbf{Y}) = \{x : (x, y) \in f, y \in \mathbf{Y}\}$. Theorem 4.6 says, and Figure 4.9 indicates, that f is continuous on $\mathbf{M}$ iff when $\mathbf{Y}$ consists only of interior points (i.e., $\mathbf{Y}$ is open in $\mathbf{M}'$), then $f^{-1}(\mathbf{Y})$ is open in $\mathbf{M}$.

■ Example 4.9

Let $\mathbf{M} = [-1, 1]$, $\mathbf{M}' = [-1, 2]$, and define $f : \mathbf{M} \to \mathbf{M}'$ by

$$f(x) = \begin{cases} x^2 - x & x \neq 0 \\ 1 & x = 0. \end{cases}$$

Choose $\mathbf{Y} = \left(\frac{1}{2}, \frac{3}{2}\right)$; this is open in $\mathbf{M}'$. Solving $\frac{1}{2} < f(x) < \frac{3}{2}$ (by examination of two quadratic equations), we obtain (verify!)

$$f^{-1}(\mathbf{Y}) = \left(\frac{1 - \sqrt{7}}{2}, \frac{1 - \sqrt{3}}{2}\right) \cup \{0\}.$$

This is not open in $\mathbf{M}$, according to Lemma 4.5.1. Hence, by Theorem 4.6, f is not continuous on $\mathbf{M}$. ■

■ Example 4.10

With $\mathbf{M}, \mathbf{M}', f$ as in Example 4.9, a student chooses $\mathbf{Y} = \left(\frac{1}{2}, 1\right)$.

It is then determined that

$$f^{-1}(\mathbf{Y}) = \left(\frac{1 - \sqrt{5}}{2}, \frac{1 - \sqrt{3}}{2}\right),$$

which *is* open in M. It is concluded that f is continuous on M after all. What's happening here? Were either of the two examples a violation of Theorem 4.6? ■

4.6 CLOSED SETS

Definition. *A subset* $F \subseteq M$ *in the metric space* $< M, d >$ *is a* **closed set** *(in* M*) iff* $M \backslash F$ *is an open set.*[3]

Lemma 4.6.1. *In any metric space* $< M, d >$, *the empty set* $\emptyset$ *and the entire set* M *are closed sets.*

Lemma 4.6.2. *In any metric space* $< M, d >$, *the intersection of an arbitrary collection* $\mathfrak{C}$ *of closed subsets of* M *is itself a closed set.*

Proof. For each $F_\alpha \in \mathfrak{C}$, let $F_\alpha^c = M \backslash F_\alpha$. The idea is then to show that (a) $\bigcup_\alpha F_\alpha^c$ is open, and that (b) $\bigcup_\alpha F_\alpha^c = M \backslash \bigcap_\alpha F_\alpha$. The completion of the proof is left to you. ■

The definition of a closed set has the important consequence that if a set $F \subset M$ is closed, then it contains all of its boundary points, for $M \backslash F$ is then open and each point $p \in M \backslash F$ lies in some ball that is contained within $M \backslash F$. Hence, $M \backslash F$ contains none of its boundary points. Conversely, if $S \subset M$ contains all of its boundary points, then S is a closed set.

We recall from Section 1.8 that the **closure** of a set $S \subset M$ is the set $\overline{S} = S \cup Bd(S)$, where Bd(S) is the set of all boundary points of S. The previous discussion then says that $S \subset M$ is closed iff $S = \overline{S}$ (Exercise 4.35(a)).

■ Example 4.11

Let $< S, d' >$ be the **metric subspace** obtained from $< R^1, d >$ by defining $S = [0, 1] \cap Q$ and restricting the usual metric d (on R^1) to $S \times S$. Let H be the set of rationals in the open interval (1/5, 1/2). If H is viewed as a subset of R^1, then it is neither open nor closed because it consists only of boundary points, but it does not contain *all* of them (verify!).

However, if H is viewed as a subset of S in $< S, d' >$, then H is open because it contains none of its boundary points. ■

Theorem 4.7. *Let* $< M, d >$ *be a metric space and let* F *be an infinite subset of* M. *Then* F *is closed iff every sequence of points in* F *that converges in* M *converges to a point in* F.

[3]The choice of letter for a closed set derives from the French *fermé* (closed), and is rather conventional.

Proof. ($\leftarrow$) Suppose that all cluster points of $\mathbf{F}$ are contained in $\mathbf{F}$. Let $\mathbf{p}_0$ be any point in $\mathbf{M}\backslash\mathbf{F}$; then $\mathbf{p}_0$ cannot be a cluster point. This implies that there is a sufficiently small ball $\mathbf{B}(\mathbf{p}_0;\delta)$ such that $\mathbf{B}(\mathbf{p}_0;\delta)\cap\mathbf{F}=\emptyset$. Hence, $\mathbf{B}(\mathbf{p}_0;\delta)\subseteq\mathbf{M}\backslash\mathbf{F}$; since $\mathbf{p}_0$ was arbitrary, then $\mathbf{M}\backslash\mathbf{F}$ is open, so $\mathbf{F}$ is closed.

($\rightarrow$) Suppose that $\mathbf{F}$ is closed. Let $\mathbf{p}_0$ be a cluster point of $\mathbf{F}$ and assume, to the contrary, that $\mathbf{p}_0\notin\mathbf{F}$. Then $\mathbf{p}_0\in\mathbf{M}\backslash\mathbf{F}$; since $\mathbf{F}$ is closed, then $\mathbf{M}\backslash\mathbf{F}$ is open. Hence, there is a sufficiently small ball $\mathbf{B}(\mathbf{p}_0;\delta)$ such that $\mathbf{B}(\mathbf{p}_0;\delta)\cap\mathbf{F}=\emptyset$. This implies that $\mathbf{p}_0$ cannot be a cluster point, a contradiction. Hence, $\mathbf{p}_0\in\mathbf{F}$ and since $\mathbf{p}_0$ was arbitrary, then all cluster points of $\mathbf{F}$ lie in $\mathbf{F}$. ■

■ Example 4.12

In $<\mathbf{R}^1,d>$ let $\mathbf{S}\subset\mathbf{R}^1$ be defined by $\mathbf{S}=[0,1]\backslash\mathbf{Q}$. By Theorem 4.7, $\mathbf{S}$ is not closed because there are sequences in $\mathbf{S}$ that converge (in $\mathbf{R}^1$) to rational numbers; see Example 4.6. ■

Corollary 4.7.1. *Let $\mathbf{F}$ be an infinite subset in the complete metric space $<\mathbf{M},d>$. Then if $\mathbf{F}$ is closed, the metric subspace $<\mathbf{F},d'>$ is itself a complete metric space, where d' is the restriction of d to $\mathbf{F}\times\mathbf{F}$.*

■ Example 4.13

The metric subspace $<\mathbf{F},d'>$, where $\mathbf{F}=[0,1]\subset\mathbf{R}^1$ and d' is the restriction to $\mathbf{F}\times\mathbf{F}$ of the usual metric d on $\mathbf{R}^1$, is complete. ■

Lemma 4.6.3. *The closure of a set $\mathbf{S}\subseteq\mathbf{M}$ in a metric space $<\mathbf{M},d>$ is the intersection of all closed sets $\mathbf{F}$ in $\mathbf{M}$ that contain $\mathbf{S}$.*

Proof. Let $\mathbf{p}\in\overline{\mathbf{S}}$. If $\mathbf{p}\in\mathbf{S}$, actually, then for all i, $\mathbf{p}\in\mathbf{F}_i$. If $\mathbf{p}\in\overline{\mathbf{S}}\backslash\mathbf{S}$, then any ball about $\mathbf{p}$ must contain some points of $\mathbf{S}$, so $\mathbf{p}$ cannot be external to any $\mathbf{F}_i$. Hence, $\mathbf{p}\in\bigcap_i\mathbf{F}_i$ and, therefore, $\overline{\mathbf{S}}\subseteq\bigcap_i\mathbf{F}_i$.

Conversely, let $\mathbf{p}\in\bigcap_i\mathbf{F}_i$. Then $\mathbf{p}\in$ each $\mathbf{F}_i$, and so $\mathbf{p}\in\overline{\mathbf{S}}$ since $\overline{\mathbf{S}}$ is an $\mathbf{F}_i$ (Exercise 4.35(a)). Hence, $\bigcap_i\mathbf{F}_i\subseteq\overline{\mathbf{S}}$. The two set inclusions imply $\overline{\mathbf{S}}=\bigcap_i\mathbf{F}_i$. ■

4.7 COMPACTNESS; THE MAXIMUM–MINIMUM VALUE THEOREM

Theorem 2.9 and its two corollaries (Section 2.4) led up to a statement of the Bolzano-Weierstrass Theorem for sequences of points in $\mathbf{R}^m$ (Theorem 2.10). There, the space $\mathbf{R}^m$ was partitioned repeatedly into denumerably many congruent m-dimensional boxes. Here are two thoughts: (1) Make a slight change in viewpoint and replace the idea of *partitioning* space by one of *covering* space (with sets). (2) Instead of using infinitely many sets in this covering, strive for finitely many (if possible), as a simplification. We are led to make the following definitions.

Definition. *A collection of open sets* $\vartheta = \{\mathbf{O}_\alpha\}$ *is said to be an* **open cover** *of a set* $\mathbf{S}$ *in a metric space* $< \mathbf{M}, d >$ *if* $\mathbf{S} \subseteq \bigcup_\alpha \mathbf{O}_\alpha$. *If a finite subcollection of* $\mathbf{O}_\alpha$*'s can cover* $\mathbf{S}$, *then this subcollection is called a* **finite cover**.

Definition. *A set* $\mathbf{S}$ *in* $< \mathbf{M}, d >$ *is termed* **compact** *if each open cover of* $\mathbf{S}$ *can be reduced to a finite cover.*

In this section we will show that compact sets are one of the types of sets $\mathbf{S}$ whose nature is preserved by continuity in their image set $f(\mathbf{S})$. We will also prove one of the pillars of calculus, the Maximum–Minimum Value Theorem. It is possible to prove this by more elementary considerations (Fort, 1951; Jungck, 1963), but the use of compactness permits a more general proof.

■ Example 4.14

The set $\mathbf{S} = \{0\} \cup \{n^{-1}; n \in \mathbf{N}\}$ is compact. If $\vartheta = \{\mathbf{O}_i\}_{i=1}^{\infty}$ is an open cover of $\mathbf{S}$, there is some $\mathbf{O}_j$ that contains the zero. This $\mathbf{O}_j$ will automatically contain all but finitely many of the n^{-1}'s (because 0 is a cluster point for the sequence $\{n^{-1}\}_{n=1}^{\infty}$). Then for each $n^{-1} \notin \mathbf{O}_j$ choose an $\mathbf{O}_i$ that contains it. The union of $\mathbf{O}_j$ and all of these $\mathbf{O}_i$'s is then a finite cover of $\mathbf{S}$. ■

■ Example 4.15

The space $\mathbf{R}^1$ is not compact, for it is covered by $\vartheta = \{(n, n+1) \colon n \in \mathbf{Z}\}$, but by no finite subcollection from ϑ. ■

It is difficult to characterize compact sets in general metric or topological spaces. But in $\mathbf{R}^n$, they can be characterized quite simply (Theorem 4.8).

Lemma 4.7.1. *If, for any* $n \in \mathbf{N}$, $\mathbf{S} \subset \mathbf{R}^n$ *is compact, then* $\mathbf{S}$ *is closed.*

Proof. Assume, to the contrary, that $\mathbf{S}$ is not closed. Then by Theorem 4.7 there is at least one cluster point of $\mathbf{S}$ that does not lie in $\mathbf{S}$; call this cluster point $\mathbf{p}_0$. Now define the family of sets $\vartheta = \{\mathbf{O}_k\}_{k=1}^{\infty}$, where for each $k \in \mathbf{N}$,

$$\mathbf{O}_k = \left\{\mathbf{p} \in \mathbf{R}^n \colon d_n(\mathbf{p}, \mathbf{p}_0) > k^{-1}\right\}.$$

Each set $\mathbf{O}_k$ is open; since ϑ covers all of $\mathbf{R}^n$ except $\mathbf{p}_0$, it is automatically an open cover of $\mathbf{S}$. But $\mathbf{S}$ is compact, so a finite subcollection of ϑ will cover $\mathbf{S}$; that is, for some $K \in \mathbf{N}$ we have (Figure 4.10)

$$\mathbf{S} \subseteq \bigcup_{k=1}^{K} \mathbf{O}_k.$$

But this excludes all points $\mathbf{p}$ for which $d_n(\mathbf{p}, \mathbf{p}_0) \leq K^{-1}$, so no sequence in $\mathbf{S}$ can have $\mathbf{p}_0$ as a limit, a contradiction. It follows that $\mathbf{S}$ must be closed. ■

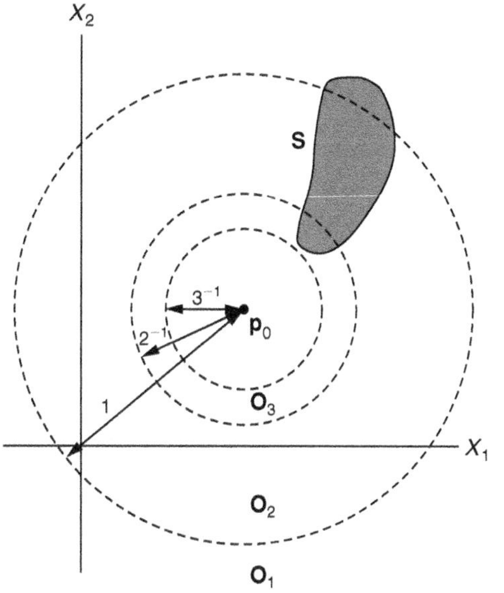

FIGURE 4.10
A set $\mathbf{S}$ *in* $\mathbf{R}^2$ *for which* $K = 3$*; that is,* $\mathbf{S} \subseteq \mathbf{O}_1 \cup \mathbf{O}_2 \cup \mathbf{O}_3$.

The **diameter** of a set $\mathbf{S}$ in a space $\mathbf{R}^n$, denoted diam($\mathbf{S}$), is defined by $\text{diam}(\mathbf{S}) = \sup_{\mathbf{p},\mathbf{q}\in\mathbf{S}} d_n(\mathbf{p},\mathbf{q})$. If diam($\mathbf{S}$) is finite, then $\mathbf{S}$ is said to be **bounded**. We note that in $\mathbf{R}^n$, where the metric is Euclidean, the diameter of a circle agrees with the usual notion of diameter, whereas the diameter of an ellipse is the length of the major axis.

Lemma 4.7.2. *If* $\mathbf{S} \subset \mathbf{R}^n$ *is compact, then* $\mathbf{S}$ *is bounded.*

Proof. Choose and fix a point $\mathbf{p} \in \mathbf{S}$ and let $\vartheta = \{\mathbf{B}_n(\mathbf{p}; k): k \in \mathbf{N}\}$ be an open cover of $\mathbf{S}$; use the definition of bounded. The proof is left to you. ■

Theorem 4.8 (Heine-Borel Theorem).[4] *For any* $n \in \mathbf{N}$, *a set in* $\mathbf{R}^n$ *is compact iff it is closed and bounded.*

Proof. ($\rightarrow$) This direction was covered in Lemmas 4.7.1 and 4.7.2.

($\leftarrow$) We prove this direction by contradiction. Suppose that $\mathbf{F}$ is closed and bounded but assume it is not contained in any finite number of sets of some open covering $\vartheta = \{\mathbf{O}_i\}_{i=1}^{\infty}$. There is a closed, bounded n-dimensional box $\mathfrak{B}_0$ that contains $\mathbf{F}$. Dissect $\mathfrak{B}_0$ into 2^n congruent subboxes, $\{\beta_k\}_{k=1}^{2^n}$. At least one of $\mathbf{F} \cap \beta_k$ must be nonempty and *not* contained in any finite subcollection from

---

[4]**Heinrich Eduard Heine** (1821–1881) first used the content of Theorem 4.8 for $\mathbf{R}^1$ in 1872, but the real importance of the theorem was pushed by **Emil Borel** (1871–1956) in 1895. Both writers selected finite sets from countably infinite covering sets. The extension of compactness to the case where a finite covering set is obtained from an uncountably infinite covering set was made by others soon after (Kline, 1972).

ϑ (denote it by $\mathbf{F} \cap \beta_j$), for if each of the $\mathbf{F} \cap \beta_k$'s were contained in a finite subcollection from ϑ, then $\mathbf{F} = \bigcup_{k=1}^{2^n} (\mathbf{F} \bigcap \beta_k)$ would also be contained in a finite subcollection from ϑ, contrary to assumption.

Relabel subbox β_j as $\mathfrak{B}_1$; dissect it into 2^n congruent subboxes. The reasoning continues with further dissections and leads to a nested sequence of closed, bounded, n-dimensional boxes, $\{\mathfrak{B}_k\}_{k=0}^{\infty}$. By Corollary 2.9.2 there is precisely one point $\mathbf{p}$ common to all the $\mathfrak{B}_k$'s. Point $\mathbf{p}$ is a cluster point of $\mathbf{F}$, and since $\mathbf{F}$ is closed, then $\mathbf{p} \in \mathbf{F}$. Further, $\mathbf{p}$ lies in some member of ϑ, say, $\mathbf{p} \in \mathbf{O}_j$. As $\mathbf{O}_j$ is open, then there is an $\varepsilon > 0$ such that $\mathbf{B}_n(\mathbf{p}; \varepsilon) \subseteq \mathbf{O}_j$. But as the diameter of the boxes approaches 0 as $k \to \infty$, then for large enough k' there is a box $\mathfrak{B}_{k'}$ such that $\mathfrak{B}_{k'} \subseteq \mathbf{B}_n(\mathbf{p}; \varepsilon)$. Hence, we see that $\mathbf{F} \cap \mathfrak{B}_{k'}$ is contained in $\mathbf{O}_j$, contrary to the method of construction of the subboxes $\mathfrak{B}_k$. So the initial assumption that $\mathbf{F}$ is not contained in any finite subcollection from ϑ is false, and $\mathbf{F}$ is then compact. ■

Note that Theorem 4.8 has been stated explicitly for real Euclidean vector spaces. There are some metric spaces in which closed, bounded sets may not be compact.

Theorem 4.9. *Let* $<\mathbf{M}, d_n'>$, $<\mathbf{R}^m, d_m>$ *be metric spaces, where* $\mathbf{M} \subset \mathbf{R}^n$ *is closed and bounded, and let* $\mathbf{f}: \mathbf{M} \to \mathbf{R}^m$ *be continuous on* $\mathbf{M}$. *Then* $\mathbf{f}(\mathbf{M})$ *is also closed and bounded.*

Proof. Let $\vartheta = \{\mathbf{O}_\alpha\}$ be a collection of open subsets of $\mathbf{R}^m$ that is an open cover of $\mathbf{f}(\mathbf{M})$. As $\mathbf{f}$ is continuous, then by Theorem 4.6 the sets $\mathbf{f}^{-1}(\mathbf{O}_\alpha)$ are open in $<\mathbf{M}, d_n'>$. Further, for any $\mathbf{x} \in \mathbf{M}$ it follows that $\mathbf{f}(\mathbf{x})$ belongs to some $\mathbf{O}_\beta$ and, thus, $\mathbf{x} \in \mathbf{f}^{-1}(\mathbf{O}_\beta)$. Hence, for the union of all $\mathbf{x} \in \mathbf{M}$, we have

$$\mathbf{M} = \bigcup_\alpha \mathbf{f}^{-1}(\mathbf{O}_\alpha).$$

Because $\mathbf{M}$ is closed and bounded, then from the Heine-Borel Theorem it is compact and, thus, has a finite open cover: $\mathbf{M} = \bigcup_{i=1}^{N} \mathbf{f}^{-1}(\mathbf{O}_i)$. We obtain

$$f(\mathbf{M}) = \mathbf{f}\left[\bigcup_{i=1}^{N} \mathbf{f}^{-1}(\mathbf{O}_i)\right] = \bigcup_{i=1}^{N} \mathbf{f}\left[\mathbf{f}^{-1}(\mathbf{O}_i)\right] \subseteq \bigcup_{i=1}^{N} \mathbf{O}_i,$$

and this says that $\mathbf{f}(\mathbf{M})$ is compact and, from the Heine-Borel Theorem again, is closed and bounded. ■

Theorem 4.9 is valid even in the trivial case where $\mathbf{f}(\mathbf{M})$ is just a point in $\mathbf{R}^m$. Note also that the proof would carry through with minor changes in wording for arbitrary metric spaces if "closed and bounded" were replaced throughout by "compact." Thus, we see (as promised) that continuity of a function on its domain preserves compactness in the image set. The Maximum–Minimum Value Theorem is an important consequence of Theorem 4.9.

Theorem 4.10 (Maximum–Minimum Value Theorem). *Let* $f : \mathbf{D}(f) \to \mathbf{R}^1$, $\mathbf{D}(f) \subset \mathbf{R}^n$ *be a continuous real-valued function and suppose that* $\mathbf{D}(f)$ *is compact. Then there are points* $\mathbf{p}_1, \mathbf{p}_2 \in \mathbf{D}(f)$ *at which* f *assumes absolute maximum, minimum values, respectively.*

Proof. By Theorem 4.9, the range $\mathbf{R}(f)$ is bounded, so the numbers $\sup_{\mathbf{p}\in\mathbf{D}(f)} f(\mathbf{p})$ and $\inf_{\mathbf{p}\in\mathbf{D}(f)} f(\mathbf{p})$ exist by the Completeness Axiom. If $\mathbf{R}(f)$ is a finite set, then $\sup_{\mathbf{p}\in\mathbf{D}(f)} f(\mathbf{p})$ and $\inf_{\mathbf{p}\in\mathbf{D}(f)} f(\mathbf{p})$ automatically belong to $\mathbf{R}(f)$ (Theorem 1.3 and Exercise 1.20), and we have $S = \sup_{\mathbf{p}\in\mathbf{D}(f)} f(\mathbf{p}) = f(\mathbf{p}_1)$ for some $\mathbf{p}_1 \in \mathbf{D}(f)$ and $s = \inf_{\mathbf{p}\in\mathbf{D}(f)} f(\mathbf{p}) = f(\mathbf{p}_2)$ for some $\mathbf{p}_2 \in \mathbf{D}(f)$.

If $\mathbf{R}(f)$ is an infinite set, then $\sup_{\mathbf{p}\in\mathbf{D}(f)} f(\mathbf{p})$ and $\inf_{\mathbf{p}\in\mathbf{D}(f)} f(\mathbf{p})$ are cluster points of $\mathbf{R}(f)$. Again, by Theorem 4.9, $\mathbf{R}(f)$ is closed, and from Theorem 4.7 it follows that $S = \sup_{\mathbf{p}\in\mathbf{D}(f)} f(\mathbf{p}) = f(\mathbf{p}_1)$ for some $\mathbf{p}_1 \in \mathbf{D}(f)$ and $s = \inf_{\mathbf{p}\in\mathbf{D}(f)} f(\mathbf{p}) = f(\mathbf{p}_2)$ for some $\mathbf{p}_2 \in \mathbf{D}(f)$. ■

■ Example 4.16

Let $f : \mathbf{D}(f) \to \mathbf{R}^1, \mathbf{D}(f) = [0, 4]$, be defined by $f(x) = \left(x^2 - 2x + 1\right) e^{-x} - 1$. By Theorem 4.10 this function must have an absolute maximum and absolute minimum at points $p_1, p_2 \in [0, 4]$, respectively. Relative extrema occur at $x_1 = 1, x_2 = 3$, where $f(x_1) = -1$ and $f(x_2) \approx -.801$. Additionally, $f(0) = 0$ and $f(4) \approx -.835$. Hence, an absolute maximum occurs at $p_1 = 0$ and an absolute minimum occurs at $p_2 = 1$. ■

■ Example 4.17

Let $f : \mathbf{D}(f) \to \mathbf{R}^1$ be the function defined by $f(\mathbf{x}) = f(x_1, x_2) = x_1^2 + 2x_2^2 - 4x_1 + 4x_2 - 3$ and $\mathbf{D}(f)$ is the set of points in and on the square $1 \le x_1 \le 3, -2 \le x_2 \le 0$. The function f is continuous on $\mathbf{D}(f)$ and $\mathbf{D}(f)$ is compact, so Theorem 4.10 applies. The necessary conditions for the existence of a relative extremum are[5]

$$\begin{cases} D_1 f = 2x_1 - 4 = 0 \\ D_2 f = 4x_1 + 4 = 0 \end{cases}$$

and the critical point here has a value of $f(2, -1) = -9$. It can be established that this is the value of a relative minimum. The boundary lines of $\mathbf{D}(f)$ must next be checked. Upon examination of the four sides of the square, we find for points x on the square that $-8 \le f(\mathbf{x}) \le -6$. Hence, an absolute maximum (of -6) occurs at various points $\mathbf{p}_1$ on the square and an absolute minimum (of -9) occurs at $\mathbf{p}_2 = (2, -1)$. ■

---

[5] The symbols $D_1 f, D_2 f$ denote the partial derivatives of f with respect to x_1, x_2, respectively.

4.8 CONNECTEDNESS; THE INTERMEDIATE-VALUE THEOREM

Informally, the notion of a set S being connected might mean that given any two points $p_1, p_2 \in S$, it is always possible to connect them by a curve that lies entirely in S. This imagery is appealing, but let us continue in the set-theoretic vein of the three previous sections (although, see later in Exercises 4.53, 4.54). We adopt the following standard definitions (Haaser and Sullivan, 1991; Sprecher, 1987).

Definition. *A set $S \subseteq M$ in a metric space $<M, d>$ is* ***disconnected*** *iff there are two nonempty, open sets $S_1, S_2 \subset M$ such that*

1. $S \subseteq S_1 \cup S_2$;
2. $S_1 \cap S_2 = \emptyset$;
3. $S \cap S_1 \neq \emptyset$, $S \cap S_2 \neq \emptyset$.

A set $S \subset M$ such that for no two nonempty, open sets $S_1, S_2 \subset M$ can all the above criteria be satisfied is called ***connected****.*

■ Example 4.18

The singleton set $S = \{(1, 2)\} \subset R^2$ is not disconnected because if nonempty, open $S_1, S_2 \subset R^2$ meet criteria (1) and (3), then explicitly $S \cap S_1 = S \cap S_2 = S$, and criterion (2) fails. Hence, S is connected. Clearly, the argument carries over to a singleton set in any metric space. ■

■ Example 4.19

However, the set $S = \{1, 2\} \subset R^1$ is disconnected, since if $S_1 = (0.9, 1.1)$ and $S_2 = (1.9, 2.1)$, then all three criteria are satisfied. ■

■ Example 4.20

The set $S = \{(x_1, x_2): x_1^2 - x_2^2 < 0\} \subset R^2$ is disconnected. Take $S_1 = \{(x_1, x_2): x_2 > |x_1| > 0\}$ and $S_2 = \{(x_1, x_2) : x_2 < -|x_1| < 0\}$. We have $S = S_1 \cup S_2$, $S \cap S_1 = S_1 \neq \emptyset$ and $S \cap S_2 = S_2 \neq \emptyset$, and $S_1 \cap S_2 = \emptyset$, so all three criteria are satisfied. ■

We will show in this section that connected sets are yet another type of set S whose nature is preserved by continuity in the image set $f(S)$. We will also prove another pillar of calculus, the Intermediate-Value Theorem.

A word about proofs of connectedness is useful. Because the definition of a connected set is framed as a negation, many proofs of connectedness run efficiently if indirect proof is used (Section 1.4).

Lemma 4.8.1. *In any metric space $<M, d>$, the empty set $\emptyset$ is connected.*

Lemma 4.8.2. *In any metric space $<M, d>$, M is connected iff there is no nonempty, proper subset of M that is both open and closed.*

Proof. Use the earlier hint for the proof in both directions. The proof is left to you. ■

Lemma 4.8.3. *Suppose that in the metric space $<M, d>$ the set $S \subset M$ is disconnected and is covered by nonempty, disjoint, open sets $S_1, S_2 \subset M$. Then if T is a connected subset of S, either $T \subset S_1$ or $T \subset S_2$.*

Proof. Proof by contrapositive; the proof is left to you. ■

The next result shows one way that new connected sets can be constructed from other connected sets.

Theorem 4.11. *If A, B are connected subsets of M in $<M, d>$ and $A \cap B \neq \emptyset$, then $A \cup B$ is connected.*

Proof. Suppose, to the contrary, that $A \cup B$ is disconnected; there are then disjoint, open sets $S_1, S_2 \subset M$ such that $(A \cup B) \subseteq (S_1 \cup S_2)$ and $(A \cup B) \cap S_1$, $(A \cup B) \cap S_2$ are nonempty. Let $p_0 \in A \cap B$; assume that $p_0 \in S_1$. Since A is connected in M, then by Lemma 4.8.3 A is contained in either S_1 or S_2. But A cannot be in S_2 because $p_0 \in (A \cap B) \subset A$ and $p_0 \in S_1$; hence, $A \subset S_1$. Likewise, since B is connected in M, we conclude that $B \subset S_1$. Therefore, it follows that $(A \cup B) \subset S_1$ and this contradicts $(A \cup B) \cap S_2$ being empty. It follows that $A \cup B$ must be connected. ■

Notice that Theorem 4.11 is not an iff-proposition. After you have read the next two results, you may be able to think of an example that illustrates the falsity of the converse of Theorem 4.11. Note that Theorem 4.12 is the converse of Lemma 4.8.4.

Lemma 4.8.4. *If S is a nonempty, connected, infinite subset of $\mathbf{R}^1$ in the metric space $< \mathbf{R}^1, d >$, then S is an interval.*

Proof. Suppose, contrapositively, that there are points x, y, z such that $x < y < z$, x and z lie in S, but $y \notin S$. Now look at $S \cap (-\infty, y)$ and $S \cap (y, \infty)$. The completion of the proof is left to you. ■

Theorem 4.12. *Any interval in $\mathbf{R}^1$ is connected.*

Proof. Let $S \subset \mathbf{R}^1$ be an interval in $< \mathbf{R}^1, d >$, and *assume* it is disconnected.

Then there are disjoint, open sets $S_1, S_2 \subset \mathbf{R}^1$ such that $S \subseteq S_1 \cup S_2$, $S \cap S_1 \neq \emptyset$, and $S \cap S_2 \neq \emptyset$. Choose $x \in S \cap S_1$ and $z \in S \cap S_2$. Clearly, $x \neq z$; by relabeling, if necessary, we may suppose that $x < z$.

Now define $y = \sup([x, z) \cap S_1)$; certainly, $x < y \leq z$, so $y \in S$ and, thus, $y \in S_1$ or $y \in S_2$, *but not both*. That is, for some $\varepsilon > 0$ either $(y - \varepsilon, y + \varepsilon) \subseteq S_1$ or

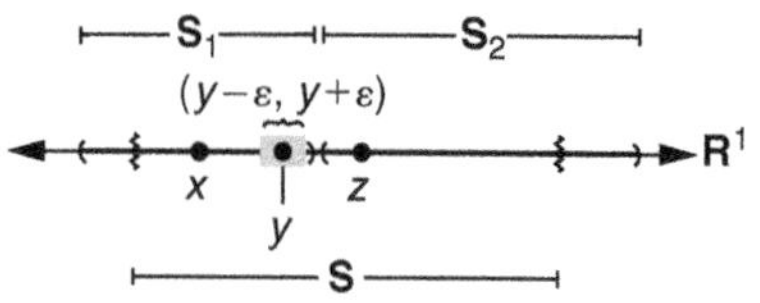

FIGURE 4.11

The contradiction for the case $y \in S_1$, assuming that S is disconnected.

$(y-\varepsilon, y+\varepsilon) \subseteq S_2$ (why?). If the former holds, then $y+\varepsilon/2 \in S_1$ and this contradicts y being an upper bound of $[x,z) \cap S_1$ (Figure 4.11).

Similarly, if $(y-\varepsilon, y+\varepsilon) \subseteq S_2$, then $S_1 \cap (y-\varepsilon, y] = \emptyset$, and this contradicts y being the supremum of $[x,z) \cap S_1$ because $y-\varepsilon$ is now a smaller upper bound of $[x,z) \cap S_1$ than is y. In either case, the contradiction obtained forces S to be connected. ■

■ Example 4.21

If $S \subset R^1$ is a compact set in $< R^1, d >$, then it may or may not be connected. If S is a closed, bounded interval, then S is compact from the Heine-Borel Theorem and S is connected from Theorem 4.12. But if S is the union of two disjoint, closed, bounded intervals, then it is still compact (verify!), but it is not connected. ■

The identification of connected sets in $< R^n, d_n >$ when $n > 1$ is a bit trickier than for $< R^1, d >$. We pursue one way to do this in the Exercises. Nevertheless, such sets in $< R^n, d_n >$ are abundant and we shall, for the present, accept this. Qualitatively, sets in $< R^n, d_n >$ that "look" connected, are connected.

Theorem 4.13. *Let $f: D(f) \to R(f), D(f) \subseteq M, R(f) \subseteq M'$, be a function defined from $D(f)$ in the metric space $< M, d >$ and onto $R(f)$ in the metric space $< M', d' >$. Suppose that f is continuous on $D(f)$ and that $D(f)$ is connected in $< M, d >$. Then $R(f)$ is connected in $< M', d' >$.*

Proof. Suppose that f is continuous on $D(f)$ and assume, contrapositively, that $R(f)$ is disconnected in $< M', d' >$. Then there are disjoint, open sets $S_1, S_2 \subset M$ such that $R(f) \subseteq S_1 \cup S_2$ and $R(f) \cap S_1, R(f) \cap S_2$ are nonempty. Let $T_1 = f^{-1}(S_1)$ and $T_2 = f^{-1}(S_2)$; by Theorem 4.6 these sets are open. They are also disjoint because $T_1 \cap T_2 = f^{-1}(S_1 \cap S_2) = f^{-1}(\emptyset) = \emptyset$, as each point in $D(f)$ has an image point in $R(f)$. Further, we see that $[f^{-1}(R(f))] \cap T_1 = f^{-1}[R(f) \cap S_1] \neq \emptyset$, and similarly for T_2, since every point in $R(f)$ has an inverse image in $D(f)$.

Finally, we have

$$f^{-1}(\mathbf{R}(f)) \subseteq f^{-1}(\mathbf{S}_1 \cup \mathbf{S}_2)$$

or equivalently, $\mathbf{D}(f) \subseteq f^{-1}(\mathbf{S}_1) \cup f^{-1}(\mathbf{S}_2)$

$$= \mathbf{T}_1 \cup \mathbf{T}_2,$$

so sets $\mathbf{T}_1, \mathbf{T}_2$ cover $\mathbf{D}(f)$. All the above equations show, therefore, that by definition $\mathbf{D}(f)$ is disconnected in $< \mathbf{M}, d >$. ■

■ Example 4.22

Consider the function $f: [0, 1) \to \mathbf{R}^1$, where $f(x) = 1/\sqrt{1 - \sqrt{x}}$. The function is continuous everywhere on $[0, 1)$, which is a connected set (Theorem 4.12). By Theorem 4.13 and Lemma 4.8.4, $\mathbf{R}(f)$ is an interval. ■

Theorem 4.13, as with Theorem 4.9, is automatically valid when $\mathbf{R}(f)$ is just a point in $\mathbf{M}'$. Thus, we see, as also promised, that continuity of a function on its domain preserves connectedness in the image set if the domain is connected. The Intermediate-Value Theorem now follows as an almost trivial consequence of this statement. The theorem is highly plausible, but it defied rigorous proof by the mathematicians of the eighteenth century. The first substantially correct proof was supplied in 1817 by Bolzano, while he was on the faculty of the University of Prague (Russ, 1980). This and other mathematical work done by Bolzano was largely ignored for fifty years. Early proofs of the Intermediate-Value Theorem used a bisection technique similar to that in our proof of Corollary 2.9.2 of the Nested Intervals Theorem (Theorem 2.9).

Theorem 4.14 (Intermediate-Value Theorem). *Let $f: \mathbf{D}(f) \to \mathbf{R}^1$ be a continuous function from $\mathbf{D}(f)$ in a metric space $< \mathbf{M}, d >$ into $\mathbf{R}^1$. If $\mathbf{p}_1, \mathbf{p}_2$ lie in a connected subset $\mathbf{S} \subseteq \mathbf{D}(f)$ and $f(\mathbf{p}_1) < f(\mathbf{p}_2)$, then for any y in the interval $(f(\mathbf{p}_1), f(\mathbf{p}_2))$ there is a point $\mathbf{p} \in \mathbf{S}$ such that $f(\mathbf{p}) = y$.*

Proof. Since f is continuous on $\mathbf{S}$ and $\mathbf{S}$ is connected, then by Theorem 4.13 $f(\mathbf{S})$ is connected in $< \mathbf{R}^1, d >$, and by Theorem 4.12 $f(\mathbf{S})$ is an interval. Thus, as $f(\mathbf{p}_1)$ and $f(\mathbf{p}_2)$ belong to $f(\mathbf{S})$, then $(f(\mathbf{p}_1), f(\mathbf{p}_2)) \subseteq f(\mathbf{S})$. ■

You have undoubtedly made implicit use of the Intermediate-Value Theorem in earlier mathematics in connection with the estimation of roots of equations (Figure 4.12). The idea is to bracket a real root x_0 of an equation $f(x) = 0$ in successively smaller and smaller intervals $[a_n, b_n]$, where $f(a_n), f(b_n)$ are of opposite signs. The following example is illustrative; the technique is called the **bisection method** because successive intervals are halved.

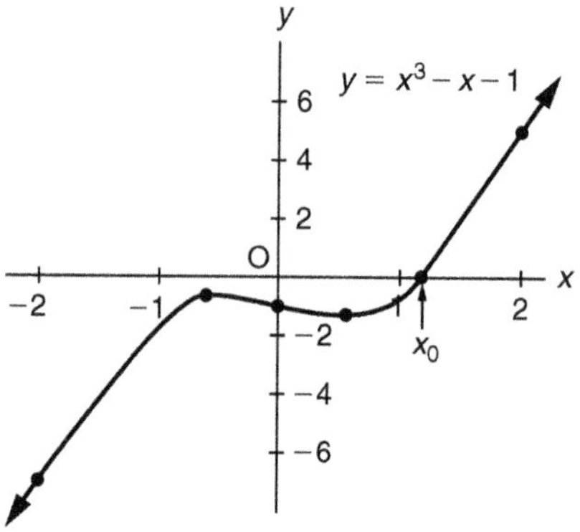

FIGURE 4.12
Illustration of the Intermediate-Value Theorem.

Table 4.2 Selected Data for Example 4.23

n	$f(a_{n-1})$	$f(c_{n-1})$	$f(b_{n-1})$	a_n	b_n	c_n
1	−1	0.875	5	1	1.5	1.25
2	−1	−.296875	0.875	1.25	1.5	1.375
3	−.296875	0.224609	0.875	1.25	1.375	1.3125
4	−.296875	−.051514	0.224609	1.3125	1.375	1.34375
5	−.051514	0.082611	0.224609	1.3125	1.34375	1.328125
6	−.051514	0.014576	0.082611	1.3125	1.328125	1.320313
7	−.051514	−.018708	0.014576	1.320313	1.328125	1.324219
8	−.018708	−.002127	0.014576	1.324219	1.328125	1.326172

■ Example 4.23

Figure 4.12 shows that $x_0 \in [a_0, b_0] = [1, 2]$; letting $f(x) = x^3 - x - 1$, we have $f(a_0) = -1$ and $f(b_0) = 5$. Denote the midpoint of any $[a_n, b_n]$ by c_n. Then $c_0 = 3/2$ and $f(c_0) = 7/8$. Hence, we choose $a_1 = a_0$ and $b_1 = c_0 = 3/2$. Next, we obtain $f(a_1) = -1, f(b_1) = 7/8, f(c_1) = f(5/4) = -19/64$. So now we choose $a_2 = c_1 = 5/4$ and $b_2 = b_1 = 3/2$. The continuation of the work is arranged in a table (Table 4.2). ■

The estimated real root of $x^3 - x - 1 = 0$ after eight cycles (done on a graphing calculator) is 1.326. The work is best carried out, of course, by writing and running a short computer program.

One consequence of the Intermediate-Value Theorem (IVT) is, like the IVT itself, so highly plausible that earlier mathematicians (who did not use the IVT) stumbled on its rigorous proof. Nevertheless, Theorem 4.15 is a useful characterization of some functions. Please review in Section 1.6 the concept of an injection.

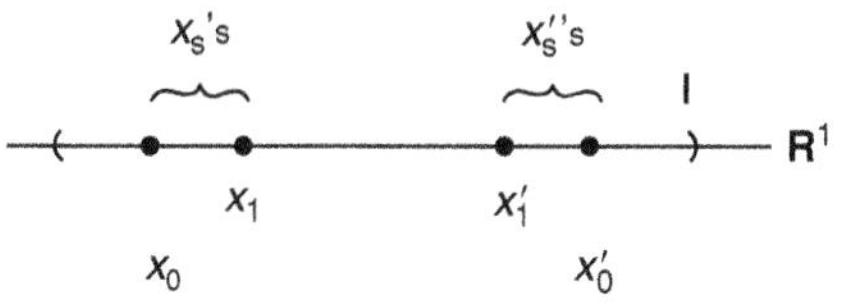

FIGURE 4.13
A disposition of some points in the proof of Theorem 4.15.

Theorem 4.15. *Suppose that the function $f : \mathbf{I} \to \mathbf{R}^1$, is an injection and is continuous on the interval* **I**. *Then f is strictly monotonic on* **I**.

Proof. Let $x_0 < x_0'$ be in **I** and assume $f(x_0) < f(x_0')$. Let $x_1 < x_1'$ be arbitrary and also in **I**. We now define for all $s \in [0, 1]$ the following two functions of s that are continuous on $[0, 1]$:

$$\begin{cases} x_s = (1-s)x_0 + sx_1 \\ x_s' = (1-s)x_0' + sx_1'. \end{cases}$$

All values of x_s, x_s' reside in **I** (Figure 4.13).

We have that

$$\begin{aligned} x_s' - x_s &= (1-s)x_0' + sx_1' - \{(1-s)x_0 + sx_1\} \\ &= (1-s)(x_0' - x_0) + s(x_1' - x_1) > 0. \end{aligned}$$

The function F, defined by

$$F(s) = f(x_s') - f(x_s),$$

is continuous on $[0, 1]$ by Theorem 4.3 because each term is continuous on $[0, 1]$. Also, F is never 0 because $x_s' > x_s$ and f is an injection. It is impossible that F could be of two signs on $[0, 1]$ because F is continuous there, and by Theorem 4.14 there would be an $s \in (0, 1)$ such that $F(s) = 0$. Hence, F is of one sign throughout $[0, 1]$. As $F(0) > 0$ was assumed, then $F(s) > 0$ for all $s \in [0, 1]$. Since x_1, x_1' were arbitrary in **I**, then $x < y$ implies $f(x) < f(y)$ for all pairs $\{x, y\}$ in **I**, and thus f is strictly increasing on **I**. The argument is analogous if $f(x_0) > f(x_0')$ is assumed initially. ■

■ Example 4.24

Consider f: $[3, 100] \to \mathbf{R}^1$, where $f(x) = x^3 - 3x^2 - 6x + 2$. If $x_1 < x_2$ exist in $[3, 100]$ and are such that $f(x_1) = f(x_2)$, then we obtain

$$1 = \frac{3}{S} + \frac{6 + x_1x_2}{S^2}, S = x_1 + x_2$$

(verify!). But $S > 6$, so $3/S < \frac{1}{2}$, and $(6 + x_1x_2)/S^2 \geq \frac{1}{2}$ iff $x_1^2 + x_2^2 \leq 12$, which is impossible. Hence, f is an injection, and since $-16 = f(3) < f(4) = -6$, then by Theorem 4.15 f is increasing on $[3, 100]$. ■

Other proofs of Theorem 4.15 also usually rely upon the IVT (Bayne, Joseph, Kwack and Lawson, 2002; Buck, 1978). You are invited to compare the mechanics in these proofs with that in ours.

4.9 UNIFORM CONTINUITY

We recall now that if f is a function from the metric space $<\mathbf{M}, d>$ into the metric space $<\mathbf{M}', d'>$ and if f is continuous on some subset $\mathbf{S} \subset \mathbf{M}$, then at a given point $\mathbf{p}_0 \in \mathbf{S}$ and for any $\varepsilon > 0$, there is a $\delta > 0$ (dependent upon ε) such that

$$\mathbf{p} \in \mathbf{B}_d(\mathbf{p}_0; \delta) \cap \mathbf{S} \to f(\mathbf{p}) \in \mathbf{B}_{d'}(f(\mathbf{p}_0); \varepsilon).$$

There is no reason to suppose from this definition that for a given $\varepsilon > 0$ the same $\delta > 0$ could suffice for all $\mathbf{p}_0 \in \mathbf{S}$.

■ Example 4.25

Consider f: $\mathbf{D}(f) \to \mathbf{R}(f)$, where $f(x) = \cos(x^{-1})$, and take $\mathbf{S} \subset \mathbf{D}(f)$ to be $(0, 3/2]$. Let $\varepsilon = \frac{1}{2}$ be given and fix $\delta = 0.1$, arbitrarily. Figure 4.14 shows that this δ will clearly work at $x_0 = 1$.

However, this δ clearly will not work at $x_0 = \frac{1}{4}$. In fact, given $\varepsilon = \frac{1}{2}$, if $1 > \delta > 0$ is fixed arbitrarily and an odd natural number N is chosen so that $[8/(3\pi N)] < \delta$, then the points $x_0 = 4/(\pi N), x_1 = 4/(3\pi N)$ are such that $|x_0 - x_1| = 8/(3\pi N)$; that is, $x_1 \in \mathbf{B}(x_0; \delta)$, but $|f(x_0) - f(x_1)| = |\cos(\pi N/4) - \cos(3\pi N/4)| = \sqrt{2} > \varepsilon$, so $f(x_1) \notin \mathbf{B}(f(x_0); \varepsilon)$. Thus, for $\varepsilon = \frac{1}{2}$, no matter how small is δ, there will always be a point $x_0 \in (0, 3/2]$ at which δ will not work. Of course, from Theorem 4.4, it is apparent that f is continuous on $\mathbf{S}$. ■

It could happen in some cases, however, that for some function f the same value of $\delta > 0$, for fixed $\varepsilon > 0$, would work for every $\mathbf{p}_0$ in the subset $\mathbf{S} \subset \mathbf{M}$ of interest. This would indicate the existence of a special type of continuity.

Definition. *Let $f : \mathbf{M} \to \mathbf{M}'$ be a function from the metric space $< \mathbf{M}, d >$ into the metric space $< \mathbf{M}', d' >$, and let $\mathbf{S} \subset \mathbf{M}$. Then f is **uniformly continuous** on $\mathbf{S}$ iff, for every $\varepsilon > 0$, there is a $\delta > 0$, depending only upon ε, such that for each $\mathbf{p}_0 \in \mathbf{S}$, we have*

$$\mathbf{p} \in \mathbf{B}_d(\mathbf{p}_0; \delta) \cap \mathbf{S} \to f(\mathbf{p}) \in \mathbf{B}_{d'}(f(\mathbf{p}_0); \varepsilon).$$

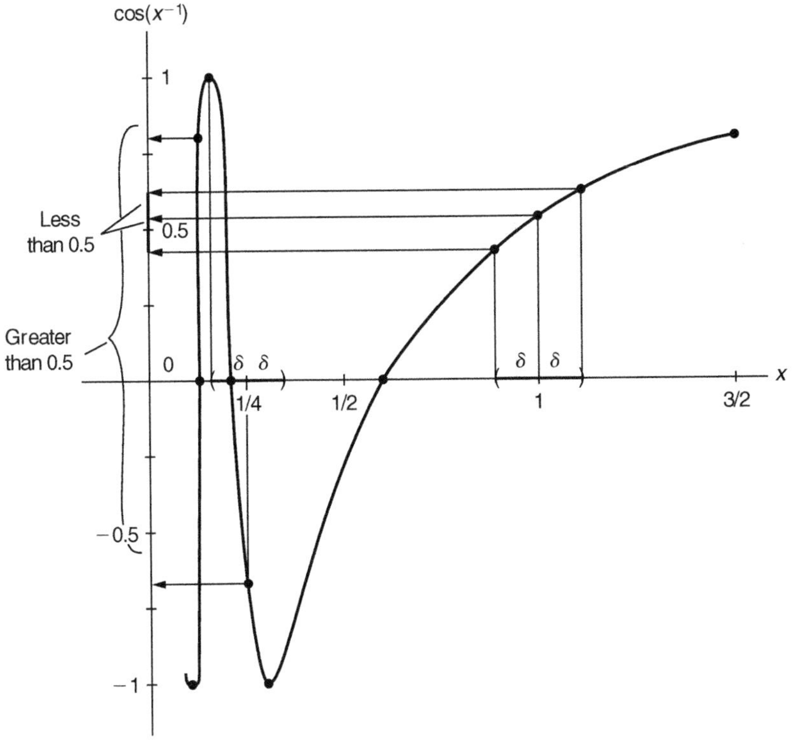

FIGURE 4.14
Illustration of the dependence of δ upon x_0 at fixed ε.

■ Example 4.26

Consider $f : \mathbf{D}(f) \to \mathbf{R}^1$, where $f(x) = \sqrt{x}$, and this time take $\mathbf{S} \subset \mathbf{D}(f)$ to be $(1/2, 3/2]$. Let $\varepsilon > 0$ be given and choose $x_0 \in \mathbf{S}$, arbitrarily. If x_1 is any point in $\mathbf{S}$ within δ units of x_0, then

$$\left|\sqrt{x_0} - \sqrt{x_1}\right| = \frac{|x_0 - x_1|}{\sqrt{x_0} + \sqrt{x_1}} < \frac{\delta}{\sqrt{1/2} + \sqrt{1/2}} = \frac{\delta}{\sqrt{2}} \leq \varepsilon,$$

if $\delta = 7\varepsilon/5$. Thus, for any $x_0 \in \mathbf{S}$ we have

$$x \in \mathbf{B}(\mathbf{x}_0; 7\varepsilon/5) \cap \mathbf{S} \to f(x) \in \mathbf{B}(f(x_0); \varepsilon),$$

so f is uniformly continuous on $\mathbf{S}$. ■

■ Example 4.27

Consider $\mathbf{f}: \mathbf{S} \to \mathbf{R}^2$, where $\mathbf{S} = \{(x_1, x_2) : 0 < x_1 < 1, 0 < x_2 < 2\}$ and $\mathbf{f}(\mathbf{x}) = f[(x_1, x_2)] = (x_1^2, 2 - x_2)$. Let $\varepsilon > 0$ be given and fix $\mathbf{x} \in \mathbf{S}$ arbitrarily.

If $\mathbf{y}$ is any point in $\mathbf{S}$ such that $0 < ||\mathbf{x} - \mathbf{y}|| = \left\{|x_1 - y_1|^2 + |x_2 - y_2|^2\right\}^{1/2} < \delta$, then

$$\begin{aligned}||\mathbf{f}(\mathbf{x}) - \mathbf{f}(\mathbf{y})|| &= \left\{|x_1^2 - y_1^2|^2 + [(2 - x_2) - (2 - y_2)]^2\right\}^{1/2} \\ &< 3\left\{|x_1 - y_1|^2 + |x_2 - y_2|^2\right\}^{1/2} \text{ (verify!)} \\ &< 3\delta \\ &< \varepsilon.\end{aligned}$$

Hence, if we take $\delta = \varepsilon/3$, then for *any* $\mathbf{x}_0 \in \mathbf{S}$ we have

$$\mathbf{x} \in \mathbf{B}_2(\mathbf{x}_0; \varepsilon/3) \cap \mathbf{S} \to \mathbf{f}(\mathbf{x}) \in \mathbf{B}_2(\mathbf{f}(\mathbf{x}_0); \varepsilon),$$

so $\mathbf{f}$ is uniformly continuous on $\mathbf{S}$. ■

The basic theorem (see Theorem 4.16) on uniform continuity was worked out in 1872 in simplified form by Heine. A modern version uses the concept of compactness explicitly, and extends Heine's version to general metric spaces. It is interesting that the weaker condition of continuity implies here the stronger condition of uniform continuity.

Theorem 4.16. *Let f be a function from $< \mathbf{M}, d >$ into $< \mathbf{M}', d' >$ and suppose that f is continuous on the compact set $\mathbf{S} \subset \mathbf{M}$. Then f is uniformly continuous on $\mathbf{S}$.*

Proof. The continuity of f at an $\mathbf{x}_0 \in \mathbf{S}$ implies that, given any $\varepsilon > 0$, there is a $\delta > 0$, dependent upon ε (and, possibly, $\mathbf{x}_0$), such that

$$\mathbf{x} \in \mathbf{B}_d(\mathbf{x}_0; \delta) \cap \mathbf{S} \to f(\mathbf{x}) \in \mathbf{B}_{d'}(f(\mathbf{x}_0); \varepsilon).$$

We now imagine that all points $\mathbf{p} \in \mathbf{S}$ are contained in balls of half the radii of those above. This collection of balls is an open cover of $\mathbf{S}$, and since $\mathbf{S}$ is compact, then there is a finite subcover:

$$\mathbf{S} \subseteq \bigcup_{k=1}^{r} \mathbf{B}_d(\mathbf{p}_k; \delta_k/2).$$

Let $\delta = \min\{\delta_1/2, \delta_2/2, \ldots, \delta_r/2\}$.

Now let $\mathbf{p}_0 \in \mathbf{S}$ be arbitrary and let $\mathbf{p}$ be any point in $\mathbf{S}$ such that $0 < d(\mathbf{p}_0, \mathbf{p}) < \delta$. There is a ball $\mathbf{B}_d(\mathbf{p}_k; \delta_k/2)$ that contains $\mathbf{p}_0$, so the continuity of f there implies $d'(f(\mathbf{p}_0), f(\mathbf{p}_k)) < \varepsilon/2$. Then from the Triangle Inequality,

$$\begin{aligned}d(\mathbf{p}_k, \mathbf{p}) &\leq d(\mathbf{p}_0, \mathbf{p}) + d(\mathbf{p}_k, \mathbf{p}_0) < \delta + \frac{\delta_k}{2} \\ &\leq \frac{\delta_k}{2} + \frac{\delta_k}{2} = \delta_k,\end{aligned}$$

so $\mathbf{p} \in \mathbf{B}_d(\mathbf{p}_k; \delta_k) \cap \mathbf{S}$. It follows from the opening lines that $f(\mathbf{p}) \in \mathbf{B}_{d'}(f(\mathbf{p}_k); \varepsilon/2)$. Finally, use of the Triangle Inequality a second time gives

$$\begin{aligned} d'(f(\mathbf{p}_0), f(\mathbf{p})) &\leq d'(f(\mathbf{p}_0), f(\mathbf{p}_k)) + d'(f(\mathbf{p}_k), f(\mathbf{p})) \\ &< \varepsilon/2 + \varepsilon/2 \\ &= \varepsilon, \end{aligned}$$

that is, given $\varepsilon > 0$, there is a $\delta > 0$, dependent *only* upon ε, such that for each $\mathbf{p}_0 \in \mathbf{S}$, we have

$$\mathbf{p} \in \mathbf{B}_d(\mathbf{p}_0; \delta) \rightarrow f(\mathbf{p}) \in \mathbf{B}_{d'}(f(\mathbf{p}_0; \varepsilon)).$$

■

■ Example 4.28

In Example 4.26 let f be as indicated, but take $\mathbf{S}$ to be $[1/2, 3/2]$. Then f is uniformly continuous on $\mathbf{S}$. As that example shows, Theorem 4.16 is a sufficient but not necessary condition for f to be uniformly continuous on a given set. ■

As expected, uniformly continuous functions possess several nice features. Some of these include (a) function addition preserves uniform continuity (Exercise 4.56), (b) function composition preserves uniform continuity, and (c) Cauchy sequences are mapped into Cauchy sequences by uniformly continuous functions (Exercise 4.60). We shall make use of uniform continuity in the important Theorem 6.13 when we come to integration theory.

4.10* THE CONTRACTION MAPPING THEOREM[6]

A special class of uniformly continuous functions are the Lipschitz functions, which are useful in various numerical settings.

Definition. *Let $f : \mathbf{M} \rightarrow \mathbf{M}'$ be a function from the metric space $< \mathbf{M}, d >$ into the space $< \mathbf{M}', d' >$. Then f is called a* **Lipschitz function**[7] *iff there is a positive number λ, the* **Lipschitz constant**, *such that $d'(f(x), f(y)) \leq \lambda d(x, y)$ for all $x, y \in \mathbf{M}$. If $\lambda < 1$, then f is termed a* **contraction** *of $\mathbf{M}$ into $\mathbf{M}'$.*

Contractions from a metric space into itself were the subject of work (1911) by the Dutch topologist **Luitzen E. J. Brouwer** (1882–1966) and work (1922) by the Polish topologist **Stefan Banach** (1892–1945). Such contractions have the very interesting property that is described next. We note that if **M** were an infinite

---

[6]The symbol * denotes that this section can be skipped without interruption of the continuity of the text.

[7]After the German mathematician **Rudolf Otto Lipschitz** (1832–1903), who worked in differential equations and differential geometry.

subset of $\mathbf{R}^m$ that is closed and bounded, then the metric space $< \mathbf{M}, d >$ would be complete (why?), which is highly desirable. In order to include more general metric spaces for consideration, it is necessary to stipulate the completeness beforehand.

Theorem 4.17 (Contraction Mapping Theorem). *If f is a contraction of $\mathbf{M}$ into $\mathbf{M}$ in the complete metric space $< \mathbf{M}, d >$, then there exists a point $\mathbf{x} \in \mathbf{M}$ such that $f(\mathbf{x}) = \mathbf{x}$.*

Proof. To get started, define the following sequence inductively on $\mathbf{M}$:

$$x_n = \begin{cases} x_0 & n = 0 \\ f(x_{n-1}) & n \geq 1, \end{cases}$$

where $\mathbf{x}_0 \in \mathbf{M}$ is arbitrary. Recognize next that f is a Lipschitz function, and for any integer $K > k$, use the Triangle Inequality to obtain

$$d(\mathbf{x}_k, \mathbf{x}_K) \leq \lambda^k \left(1 - \lambda^{K-k}\right) (1 - \lambda)^{-1} d(\mathbf{x}_0, \mathbf{x}_1).$$

Finally, make use of the sequence definition of continuity.

The completion of the proof is left to you. ■

The point $\mathbf{x}$ in the theorem, incidentally, is unique (Exercise 4.66). There are many applications of Theorem 4.17; here is an elementary one. Consider the equation $e^x - 2x - 2 = 0$. It can be deduced easily from the IVT (Theorem 4.14) that a real root lies in $[1, 2]$. We proceed to rearrange the equation so that it assumes the form $x = f(x)$, which is needed for the application of Theorem 4.17. If we rearrange the equation as

$$x = \frac{e^x - 2}{2} = f(x),$$

then $f(x)$ is clearly increasing, but $f(1) \approx 0.36 < 1$ and $f(2) \approx 2.69 > 2$, so this rearrangement will not be useful. Figure 4.15(a) shows what can happen after a few iterations.

In an alternative rearrangement,

$$x = \ln(2 + 2x) = g(x),$$

the function $g(x)$ is also increasing; further, $g(1) \approx 1.38 \in [1, 2]$ and $g(2) \approx 1.79 \in [1, 2]$. Thus, if we define $x_{n+1} = \ln(2 + 2x_n)$, it follows that $x_n \in [1, 2]$ for all $n \in \mathbf{N}$ (Figure 4.15(b)).

To show that we actually have a contraction, we borrow from calculus (Section 5.7) the Taylor series representation for $\ln(1 + u), -1 < u \leq 1$. This

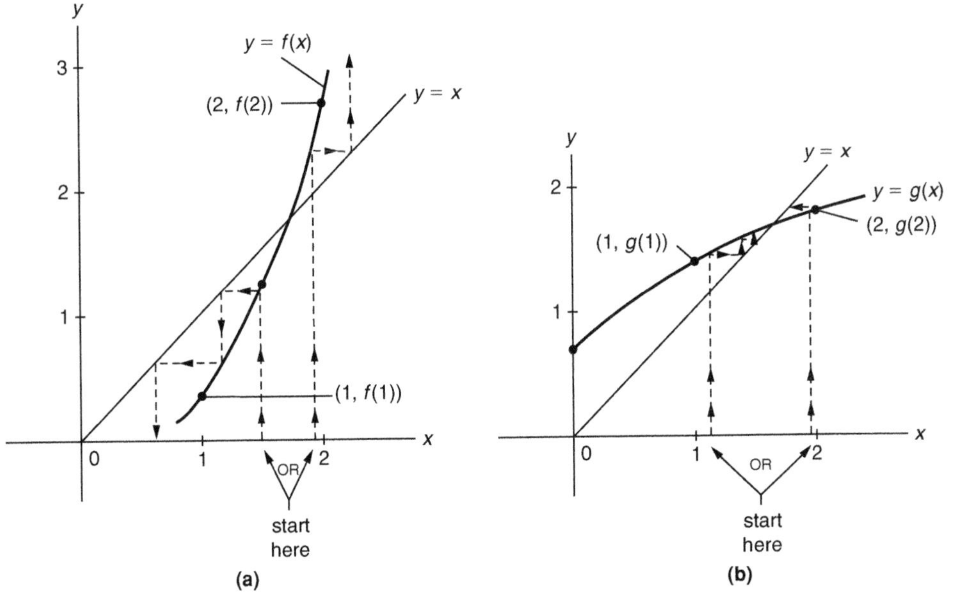

FIGURE 4.15
An application of the Contraction Mapping Theorem that will (a) fail (b) succeed.

series is

$$\ln(1+u) = u - (u^2/2) + (u^3/3) - (u^4/4) + \cdots = \sum_{k=1}^{\infty}(-1)^{k+1}(u^k/k).$$

From the proof of Theorem 3.10, we have (since u will be positive and the terms in the series decrease in magnitude)

$$u - (u^2/2) < \ln(1+u) < u. \qquad (*)$$

In Figure 4.15(b) we obtain for arbitrary $n \in \mathbf{N}$,

$$\begin{aligned}
|x_{n+2} - x_{n+1}| &= |g(x_{n+1}) - g(x_n)| \\
&= |\ln(2 + 2x_{n+1}) - \ln(2 + 2x_n)| \\
&= \ln\left[1 + \frac{|x_{n+1} - x_n|}{1 + x_n}\right] \\
&< \frac{|x_{n+1} - x_n|}{1 + x_n} \quad \text{from } (*) \\
&\le \frac{1}{2}|x_{n+1} - x_n| \quad \text{since } x_n \ge 1.
\end{aligned}$$

Table 4.3 Selected Data for the Solution of the Equation $x = \ln(2 + 2x)$

n	x_n	n	x_n
0	1.5	12	1.678345607
4	1.674689025	16	1.678346963
8	1.678275830	20	1.678346989

Thus, we have a contraction with Lipschitz constant $\lambda = \frac{1}{2}$.

The midpoint of $[1, 2]$ is a good place to begin. The work for the problem can be done on calculator; the results are shown in Table 4.3.

The estimated root of the original equation $e^x - 2x - 2 = 0$ is, after 20 iterations, 1.6783470.

EXERCISES

Section 4.1

4.1. Prove, using the balls-definition in Section 4.1, that (a) $\sqrt{x}$ is continuous at $x = 3$ and (b) $\frac{2x+3}{x-4}$ is continuous at $x = 1$.

4.2. Suppose that $\mathbf{f} : \mathbf{D}(\mathbf{f}) \to \mathbf{R}^m, \mathbf{D}(\mathbf{f}) \subseteq \mathbf{R}^n$, is a function. Prove that if $\mathbf{f}(\mathbf{x})$ is continuous at $\mathbf{x} = \mathbf{x}_0$, then so is $||\mathbf{f}(\mathbf{x})||$.

4.3. A function $f : \mathbf{R}^2 \to \mathbf{R}^1$ is defined for $\mathbf{x} = (x_1, x_2) \in \mathbf{R}^2$ by

$$f(\mathbf{x}) = \begin{cases} \frac{2(x_1-1)(x_2-1)}{(x_1-1)^2+(x_2-1)^2} & \mathbf{x} \neq (1, 1) \\ 1 & \mathbf{x} = (1, 1). \end{cases}$$

Suppose that in $\mathbf{R}^2$ the point $(1, 1)$ is approached along the line $x_1 = x_2$. Then along that line $f(\mathbf{x}) = 2(x_2 - 1)^2/[(x_1 - 1)^2 + (x_2 - 1)^2] = 1$, so $\lim_{\mathbf{x}\to(1,1)} f(\mathbf{x}) = \lim_{x_2\to 1} f(\mathbf{x}) = \lim_{x_2\to 1} 1 = 1 = f[(1, 1)]$. A student then argues that f is continuous at $(1, 1)$. Comment on this.

4.4. Write out the proof of the equivalence of these two statements:

(a) $\mathbf{f}$ is continuous at the interior point $\mathbf{a}$ of $\mathbf{D}(\mathbf{f})$.

(b) Given any $\varepsilon > 0$, there is a $\delta > 0$ such that $\mathbf{f}[\mathbf{B}_n(\mathbf{a}; \delta)] \subseteq \mathbf{B}_m(\mathbf{f}(\mathbf{a}); \varepsilon)$, where $\mathbf{f} : \mathbf{D}(\mathbf{f}) \to \mathbf{R}^m, \mathbf{D}(\mathbf{f}) \subseteq \mathbf{R}^n$.

4.5. Let $f : \mathbf{D}(f) \to \mathbf{R}^1, \mathbf{D}(f) \subseteq \mathbf{R}^2$, be defined by $y = f(\mathbf{x}) = f[(x_1, x_2)] = 1 + x_1 - x_2$, and $\mathbf{D}(f)$ is the set of points inside the circle $(x_1 - 2)^2 + (x_2 - 2)^2 = 1$. Use the balls definition to prove that f is continuous at $\mathbf{a} = (2, 2)$.

Section 4.2

4.6. Identify which of the following sentences are predicates, and in those cases indicate the predicate variable(s).

(a) A fourth-degree polynomial $P(x)$: $\mathbf{R}^1 \to \mathbf{R}^1$ has n relative maxima and m relative minima, $0 \le n, m \le 3$.

(b) The three perpendicular bisectors of an acute triangle meet inside the triangle.

(c) The function f, which is differentiable at $x = 1$, is continuous there.

(d) For real x, b, $\int_0^x \sqrt{1+bt}\, dt = 2\sqrt{(1+bt)^3}/(3b)\big|_0^x$.

4.7. Translate the logical symbolism into prose; the universes of discourse for the variables are given at the right.

(a) $(\exists x)$: $[(x > 0)$ AND $(\forall y)(xy \neq 1)]$; $(x, y \in \mathbf{R})$

(b) $(\forall x)(\exists y)$: $(x + 2y = 0)$; $(x, y \in \mathbf{C}\backslash\{0\})$

(c) $(\exists x)$: $(\forall y)(x + 2y = 0)$; $(x, y \in \mathbf{C}\backslash\{0\})$

(d) $(\forall \varepsilon)[\varepsilon > 0 \to (\exists \delta)$: $\{(\delta > 0$ AND $(\forall \mathbf{p})[(\mathbf{p} \in \mathbf{B}_n(\mathbf{p}_0; \delta) \cap \mathbf{D}(\mathbf{f})) \to \mathbf{f}(\mathbf{p}) \in \mathbf{B}_m(\mathbf{f}(\mathbf{p}_0); \varepsilon)]\}]$ $(\varepsilon, \delta > 0; \mathbf{p} \in \mathbf{R}^n; \mathbf{f}(\mathbf{p}) \in \mathbf{R}^m$, $\mathbf{f}$ is continuous at $\mathbf{p}_0)$.

Which, if any, of these statements are true?

4.8. Translate the prose in each case into logical symbolism.

(a) Some dogs are black and friendly.

(b) For every natural number there is a real number that is the logarithm of that natural number.

(c) There is a prime p such that $2^p - 1$ is not prime.

(d) For each $\varepsilon > 0$ there is a $\delta > 0$ such that if $|x - a|$ is positive but less than δ, then $|f(x) - L|$ is less than ε, iff the limit of f as x approaches a is L.

Which, if any, of these statements are false?

4.9. Work out the negations of each indicated sentence.

(a) $(\exists u)$: $[u > 0$ AND $(\forall v)(uv \neq 1)]$; $(u, v \in \mathbf{R})$

(b) $(\forall x)[x > 0 \to (\exists \varepsilon)$: $(\varepsilon > 0 \to x \geq \varepsilon)]$; $(x, \varepsilon > 0)$

(c) Exercise 4.7(c);

(d) $(\exists x)$: $\{(\exists \varepsilon)$: $[\varepsilon > 0$ AND $(\forall y)(y \in \mathbf{R}\backslash\mathbf{Q} \to y \notin (x - \varepsilon, x + \varepsilon))]\}$; $(x, y \in \mathbf{R}; \varepsilon > 0)$

(e) Exercise 4.7(d). Now return to Theorem 4.1.

Section 4.3

4.10. Let f: $[0, 1] \to \mathbf{R}^1$ be defined by $f(x) = \begin{cases} x & x \in \mathbf{Q} \\ 0 & x \notin \mathbf{Q}. \end{cases}$

Is this function continuous anywhere? Justify your answer by using the sequence approach.

4.11. The **signum function**, $\operatorname{sgn}(x)$, is defined as

$$\operatorname{sgn}(x) = \begin{cases} x/|x| & x \neq 0 \\ 0 & x = 0. \end{cases}$$

Use the sequence definition of continuity to show that $\operatorname{sgn}(x)$ is discontinuous at $x = 0$.

4.12. Identify the points of discontinuity, if there are any, for each function and tell whether these points are removable or essential.

(a) $f(x) = \begin{cases} (x-1)^3/(x-1) & x \neq 1 \\ 0 & x = 1; \end{cases}$

(b) $f(x) = \begin{cases} x+1 & x < 2 \\ 4-x & x > 2; \end{cases}$

(c) $f(\mathbf{x}) = f(x_1, x_2) = \begin{cases} x_1x_2/\sqrt{x_1^4 + x_2^4} & \mathbf{x} \neq \mathbf{0} \\ 1/\sqrt{2} & \mathbf{x} = \mathbf{0}; \end{cases}$

(d) $f(x) = \begin{cases} x\cos(1/x) & x \in (0, 1] \\ 1 & x = 0; \end{cases}$

(e) $f(x) = \begin{cases} \sin(\pi x/6) & x < 1 \\ x^2 - x + (1/2) & x \geq 1; \end{cases}$

(f) f: $[0, 1] \to [0, 1]$, where

$$f(x) = \begin{cases} 0 & x \notin \mathbf{Q} \\ 1/q & x = p/q, \quad p, q \text{ relatively prime} \\ 1 & x = 0 \text{ or } 1. \end{cases}$$

(the **Ruler Function**)

4.13. Suppose that functions $f(x), g(x)$ are both discontinuous at $x = x_0$. Must $(f \oplus g)(x)$ also be discontinuous at $x = x_0$? Discuss.

Section 4.4

4.14. Prove the analog of Theorem 4.2 for the case $f(\mathbf{a}) < 0$.

4.15. Prove parts (a) and (b) of Theorem 4.3.

4.16. In the proof of Theorem 4.3(c), prove that there is a $\delta_2 > 0$ such that $\mathbf{x} \in \mathbf{B}_n(\mathbf{a}; \delta_2) \cap \mathbf{D}$ implies $0 < \frac{1}{|g(\mathbf{x})|} < \frac{2}{|g(\mathbf{a})|}$.

4.17. Prove Theorem 4.4 via the sequence approach (as suggested).

4.18. Let $f(x) = \sqrt{\ln(x)}$ and $g(x) = \ln(\sqrt{x})$; it is known that $g(56) > f(56)$. Show that there exists a $\delta > 0$ such that $0 < |x - 56| < \delta$ implies that $g(x) > f(x)$.

4.19. A **group** $< \mathbf{G}, * >$ is a set $\mathbf{G}$ of elements, together with a binary operation (*) on $\mathbf{G}$, such that (a) there is an **identity element** $\mathbf{I} \in \mathbf{G}$ such that for any $\mathbf{g} \in \mathbf{G}$ we have $\mathbf{I}^*\mathbf{g} = \mathbf{g}^*\mathbf{I} = \mathbf{g}$, (b) to each $\mathbf{g} \in \mathbf{G}$ there corresponds an **inverse element** $\mathbf{h} \in \mathbf{G}$ such that $\mathbf{h}^*\mathbf{g} = \mathbf{g}^*\mathbf{h} = \mathbf{I}$, (c) for any $\mathbf{g}, \mathbf{h}, \mathbf{k} \in \mathbf{G}$ we have $\mathbf{g}^* (\mathbf{h}^*\mathbf{k}) = (\mathbf{g}^*\mathbf{h})^*\mathbf{k}$; that is, the **Associative Law** holds.

Now fix a natural number $n > 1$; let $\mathbf{G}$ be the set of all real-valued functions $f(\mathbf{x})$ defined on and continuous on the n-dimensional unit cube centered about the origin in $\mathbf{R}^n$, and let **(*)** be function addition. Prove that $< \mathbf{G}, ^* >$ is a group.

Section 4.5

4.20. Discuss where the following functions are continuous on their natural (or indicated) domains.

(a) $f(x) = (\lceil x \rceil - x)(x - \lfloor x \rfloor)$;

(b) $f(x) = \begin{cases} x\sin(x^{-1}) & x \neq 0 \\ 0 & x = 0; \end{cases}$

(c) $f(x) = \sqrt[4]{(x-1)(2-x)[\ln(x/2)]}$;
(d) $f(x) = \tan(2 \sin x)$;
(e) $f(\mathbf{x}) = f[(x_1, x_2)] = \frac{x_1^2+1}{x_2^3+2x_2-5} \quad (\mathbf{x} \in \mathbf{R}^2)$.

4.21. Let f be a real-valued function defined on some interval $\mathbf{I}$, and let $k > 0$. Suppose that for each $x_1, x_2 \in \mathbf{I}$ we have

$$|f(x_1) - f(x_2)| \leq k|x_1 - x_2|.$$

Such a function f is called a **Lipschitz function**. Show that f is continuous on $\mathbf{I}$.

4.22. Consider f: $[0, 2\pi] \to [-1, 1]$, where $f(x) = \sin(x)$. Prove that f is continuous on $[0, 2\pi]$.

4.23. Prove all statements in Lemma 4.5.1.

4.24. Prove Lemma 4.5.2. Then sketch what the ball $\mathbf{B}_2(\mathbf{x}_0; 1), \mathbf{x}_0 = (2, 3)$, would look like in $\mathbf{R}^2$ if the metric were $d(\mathbf{x}, \mathbf{x}') = \max\{|x_1 - x_1'|, |x_2 - x_2'|\}$ instead of the usual Euclidean metric. Here, in $\mathbf{R}^2$, $\mathbf{x} = (x_1, x_2)$ and $\mathbf{x}' = (x_1', x_2')$.

4.25. Another common metric in $\mathbf{R}^2$ is the so-called **taxicab metric** (MacG. Dawson, 2007):

$$d(\mathbf{x}, \mathbf{x}') = |x_1 - x_1'| + |x_2 - x_2'|,$$

with $\mathbf{x}, \mathbf{x}'$ given as in Exercise 4.24.

(a) Sketch the "unit circle" in $\mathbf{R}^2$ with center at $\mathbf{0}$.
(b) Show how to draw three "equilateral" triangles of side 2 that share a common side. For each triangle label the coordinates of the vertices.

4.26. (a) Prove that the intersection of a *finite* number of open subsets of $\mathbf{M}$ in the metric space $< \mathbf{M}, d >$ is itself an open set.
(b) Show by example that the result in (a) may not hold if "finite" is replaced by "infinite."

4.27. Prove that in any metric space $< \mathbf{M}, d >$, an open set $\mathbf{T} \subseteq \mathbf{M}$ is the union of a collection of (open) balls.[8]

4.28. Suppose that $\mathbf{S}$ is a nonempty, open subset of $\mathbf{R}^2$. If this $\mathbf{S}$ is now viewed as a subset of $\mathbf{R}^3$, is $\mathbf{S}$ still open? Discuss.

4.29. Let $\mathbf{S}$ be a nonempty subset of $\mathbf{M}$ in the metric space $< \mathbf{M}, d >$. Prove that $\text{Int}(\mathbf{S})$ is the union of all open subsets of $\mathbf{S}$. To get started, denote by $\vartheta = \{\mathbf{O}_\alpha\}$ the set all open subsets of $\mathbf{S}$. You then need to show that if $\mathbf{p} \in \bigcup_\alpha \mathbf{O}_\alpha$, then $\mathbf{p} \in \text{Int}(\mathbf{S})$, and conversely.

4.30. In the proof of Theorem 4.6,
(a) show that $f(\mathbf{B}(\mathbf{p}; \delta)) \subseteq \mathbf{Y}$ implies that $\mathbf{B}(\mathbf{p}; \delta) \subseteq f^{-1}[f(\mathbf{B}(\mathbf{p}; \delta))] \subseteq f^{-1}(\mathbf{Y})$, where $f: \mathbf{M} \to \mathbf{M}'$ is continuous;
(b) show that $\mathbf{B}(\mathbf{p}; \delta) \subseteq f^{-1}(\mathbf{B}'(f(\mathbf{p}); \varepsilon))$ implies that $f(\mathbf{B}(\mathbf{p}; \delta)) \subseteq f[f^{-1}(\mathbf{B}'(f(\mathbf{p}); \varepsilon))] \subseteq \mathbf{B}'(f(\mathbf{p}); \varepsilon)$.

---

[8] **Lindelöf's Covering Theorem** (not discussed in this book) guarantees that for *any* subset of $\mathbf{R}^n$ the collection can be taken to be (at worst) countably infinite.

4.31. In the metric space $\mathbf{R}^1$ let $\mathbf{S}_1, \mathbf{S}_2$ be two nonempty, disjoint sets that contain all of their boundary points. Further, let $\mathbf{T} = \{x : x \in \mathbf{R}^1, d(x, \mathbf{S}_1) < d(x, \mathbf{S}_2)\}$, where $d(x, \mathbf{S}_i)$ means the distance from x to the set $\mathbf{S}_i$, and is given by

$$d(x, \mathbf{S}_i) = \inf_{z \in \mathbf{S}_i} |x - z|.$$

In the case when $x \in \mathbf{S}_i$, this formula gives $d(x, \mathbf{S}_i) = 0$, which is uninteresting; henceforth, we ignore this possibility.

(a) Prove that the functions g_i: $\mathbf{R}^1 \to \mathbf{R}^1$, given by $g_i(x) = d(x, \mathbf{S}_i)$, $i = 1, 2$, are continuous on $\mathbf{R}^1 \backslash \mathbf{S}_i$.

(b) In view of part (a), use Theorem 4.6 to show that $\mathbf{T}$ is an open set (Haaser and Sullivan, 1991).

Section 4.6

4.32. Prove Lemma 4.6.1.

4.33. Prove Lemma 4.6.2. Statement (b) in the sketch of the proof is a version of one of **de Morgan's Laws**; the standard approach to it is to show that each side is a subset of the other.

4.34. Show that the union of a *finite* collection $\mathfrak{C}$ of closed subsets of $\mathbf{M}$ in the metric space $< \mathbf{M}, d >$ is itself a closed set. Can "finite" be replaced by "infinite" here?

4.35. **(a)** Write out the proof that $\mathbf{S} \subset \mathbf{M}$ in the metric space $< \mathbf{M}, d >$ is closed iff $\mathbf{S} = \overline{\mathbf{S}}$.

(b) The analogous theorem for open sets is that $\mathbf{S} \subset \mathbf{M}$ in the metric space $< \mathbf{M}, d >$ is open iff $\mathbf{S} = \text{Int}(\mathbf{S})$. Prove it.

4.36. Refer to Example 4.11.

(a) If $\mathbf{H}$ is viewed as a subset of $\mathbf{R}^1$, what is its complete set of boundary points?

(b) If $\mathbf{H}$ is viewed as a subset of $\mathbf{S}$, then explain why $\mathbf{H}$ is open.

4.37. The following is a sequence defined in $\mathbf{S} = [0, 3] \cap \mathbf{Q}$, a subspace of $< \mathbf{R}^1, d >$:

$$x_k = \begin{cases} 2 & k = 1 \\ x_{k-1} + \frac{1}{k!} & k \geq 2. \end{cases}$$

(a) Confirm that each $x_k \in \mathbf{S}$. How do you know that $\lim_{k \to \infty} x_k$ exists in $\mathbf{R}^1$?

(b) Assume that this limit is rational, say, p/q. Now multiply $\frac{p}{q} - \sum_{k=0}^{q} \frac{1}{k!}$ by $q!$ and deduce a contradiction.

(c) What do you conclude about $\lim_{k \to \infty} x_k$, and what does this imply about $\mathbf{S}$, in view of Theorem 4.7?

4.38. In the follow-up to Theorem 4.7:

(a) Why was the phrase "$\mathbf{S}$ is not closed" in Example 4.12 not written instead as "$\mathbf{S}$ is open"?

(b) Prove the Corollary 4.7.1.

(c) Explain why, in the statement of Corollary 4.7.1, it is necessary for $\mathbf{F}$ to be an infinite subset of $\mathbf{M}$.

Section 4.7

4.39. Prove Lemma 4.7.2 along the line suggested. Could Lemmas 4.7.1 and 4.7.2 have been stated for arbitrary metric spaces?

4.40. Let $\mathbf{S}$ be the compact set in $\mathbf{R}^2$ that consists of all points contained in the region bounded by the line segment $y = 4, -2 \le x \le 2$, and by the graph of $y = x^2, -2 \le x \le 2$. Conjecture the value of $\operatorname{diam}(\mathbf{S})$.

4.41. In the proof of Theorem 4.9, show that $\bigcup_{i=1}^{N} f\left[f^{-1}(\mathbf{O}_i)\right] \subseteq \bigcup_{i=1}^{N} \mathbf{O}_i$.

4.42. In view of Theorem 4.9, is it possible for a continuous function defined on a non-compact set in $\mathbf{R}^1$ to have a range in $\mathbf{R}^1$ that is a compact set? Prove that it is not, or provide an example that it is possible.

4.43. Let $f: \mathbf{D}(f) \to \mathbf{R}^1$ be the function $f(\mathbf{x}) = f(x_1, x_2) = (-x_1 - 2x_2 + 2)e^{-x_1 - 2x_2^2}$, where $\mathbf{D}(f)$ is the set of points in and on the square $0 \le x_1 \le 1, 0 \le x_2 \le 1$. Present an analysis in the manner of Example 4.17, showing all details and concluding with the values of the absolute maximum and minimum.

4.44. In the papers (Bernau, 1967; Pennington, 1960; Fort, Jr., 1951) three different proofs are given for the existence of an extremum of a real-valued function defined on a closed, bounded interval. Compare and contrast the three approaches and write up a summary of the reading.

Section 4.8

4.45. Prove Lemmas 4.8.1 and 4.8.2.

4.46. Prove Lemma 4.8.3.

4.47. **(a)** If $\mathbf{A}, \mathbf{B}$ are as in Theorem 4.11, then find an example in $\mathbf{R}^2$ where $\mathbf{A} \cap \mathbf{B}$ is disconnected.
(b) Complete the proof of Lemma 4.8.4.

4.48. Regarding Theorem 4.12:
(a) Why is it, as indicated, that for some $\varepsilon > 0$ either $(y - \varepsilon, y + \varepsilon) \subseteq \mathbf{S}_1$, or $(y - \varepsilon, y + \varepsilon) \subseteq \mathbf{S}_2$ must hold?
(b) In what way does the Theorem depend upon the all-important Axiom of Completeness?

4.49. The sets $\mathbf{S} = (0, 1), \mathbf{S}' = [0, 1]\backslash\mathbf{Q}$ are both uncountably infinite subsets in $\mathbf{R}^1$. Nevertheless, prove there is no continuous function f such that $f(\mathbf{S}) = \mathbf{S}'$.

4.50. Write a short program to use the bisection method to find all real roots of the following equations:
(a) $5\sqrt{x} - 7\ln(x) = 0$;
(b) $x^5 - x^4 - x^2 - 5 = 0$;
(c) $3x = 2\cosh(x)$;
(d) $x = 1 + \sum_{k=1}^{3} \sin(kx)$.

4.51. Let g, h be functions continuous on $[0, 1]$, and suppose that $g(0) < h(0)$ and $g(1) > h(1)$. Prove that $g(x) = h(x)$ for some $x \in (0, 1)$.

4.52. **Theorem 4.18.** Let $f: \mathbf{I} \to \mathbf{R}^1$ be a continuous injection on the bounded interval $\mathbf{I}$, and let $\mathbf{R}(f)$ denote the range of f. Then the inverse function $f^{-1}: \mathbf{R}(f) \to \mathbf{I}$ exists, is an injection, is strictly monotonic on $\mathbf{R}(f)$, and is continuous on $\mathbf{R}(f)$.
(a) Use definitions and prove first that f^{-1}, as a *function*, exists.
(b) Again, using definitions, prove that f^{-1} is an injection.
(c) Next, establish that f^{-1} is strictly monotonic on $\mathbf{R}(f)$.
(d) Finally, prove that f^{-1} is continuous on $\mathbf{R}(f)$.
Theorem 4.18 was alluded to in Approach 1 in Section 4.4.

4.53. Definition. A continuous mapping of the form $\gamma: \mathbf{I} \to \mathbf{R}^k$, where $\mathbf{I} \subseteq \mathbf{R}^1$ is an interval, is called a **path**. The image set $\gamma(\mathbf{I})$ is a **curve** in $\mathbf{R}^k$. If γ is an injection, then $\gamma(\mathbf{I})$ is called an **arc**, and if $\mathbf{I} = [a, b]$ and $\gamma(a) = \gamma(b)$, then $\gamma(\mathbf{I})$ is said to be a **closed curve**.
Sketch the curve corresponding to the path $\gamma: [0, \pi/4] \to \mathbf{R}^2, \gamma(t) = (\cos t\sqrt{\cos 2t}, \sin t\sqrt{\cos 2t})$. Use calculus to determine at what value of t the curve has a maximum.

4.54. The curve in $\mathbf{R}^2$ in Exercise 4.53 is clearly connected. More generally, a subset $\mathbf{S} \subset \mathbf{M}$ in the metric space $< \mathbf{M}, d >$ is **path-connected** iff for each $\mathbf{p}_1, \mathbf{p}_2 \in \mathbf{S}$ there is a path $\gamma: [a, b] \to \mathbf{S}$ such that $\gamma(a) = \mathbf{p}_1, \gamma(b) = \mathbf{p}_2$. The following useful theorem emerges.

Theorem 4.19. If $\mathbf{S} \subset \mathbf{M}$ in the metric space $< \mathbf{M}, d >$ is path-connected, then $\mathbf{S}$ is a connected set.
Assume, to the contrary, that $\mathbf{S}$ is disconnected; choose arbitrary $\mathbf{p}_1 \in \mathbf{S} \cap \mathbf{S}_1, \mathbf{p}_2 \in \mathbf{S} \cap \mathbf{S}_2$, where $\mathbf{S}_1, \mathbf{S}_2$ are disjoint and $\mathbf{S} \subseteq \mathbf{S}_1 \cup \mathbf{S}_2$. Let $\gamma: [a, b] \to \mathbf{S}$ be a path such that $\gamma(a) = \mathbf{p}_1$ and $\gamma(b) = \mathbf{p}_2$. Now examine the set $\mathbf{Y} = \gamma([a, b])$; use Theorems 4.12 and 4.13 and reach a contradiction. Write out the proof in full detail.

Sections 4.9 and 4.10*

4.55. Prove that $f: [1/2, \infty) \to \mathbf{R}^1$, where $f(x) = 3/x$, is uniformly continuous on its domain.

4.56. Suppose that functions f, g from $< \mathbf{M}, d >$ into $< \mathbf{M}', d' >$ are uniformly continuous on a common domain $\mathbf{D} \subseteq \mathbf{M}$. Prove that $f \oplus g$ is uniformly continuous on $\mathbf{D}$.

4.57. Show by an example that if f, g are as in Exercise 4.56 and $\mathbf{M} = \mathbf{M}' = \mathbf{R}^1$, then fg may not be uniformly continuous on $\mathbf{D}$.

4.58. If the function $\mathbf{f}: \mathbf{S} \to \mathbf{R}^m, \mathbf{S} \subset \mathbf{R}^n$, is uniformly continuous on the bounded set $\mathbf{S}$, then show that $\mathbf{f}$ is bounded on $\mathbf{S}$.

4.59. Refer to Example 4.27.
(a) Account for the inequality following $||\mathbf{f}(\mathbf{x}) - \mathbf{f}(\mathbf{y})||$.
(b) Determine whether or not the composite function $\mathbf{f}[\mathbf{f}]$ is uniformly continuous on $\mathbf{S}$.

4.60. Suppose that $f : \mathbf{S} \to \mathbf{M}', \mathbf{S} \subset \mathbf{M}$, is uniformly continuous on $\mathbf{S}$ in the metric space $< \mathbf{M}, d >$, and that $\{\mathbf{p}_n\}_{n=1}^{\infty}$ is a Cauchy sequence in $\mathbf{S}$. Prove that $\{f(\mathbf{p}_n)\}_{n=1}^{\infty}$ is a Cauchy sequence in the metric space $< \mathbf{M}', d' >$.

4.61. Extend Exercise 4.21 by showing that if f is a Lipschitz function from the metric space $< \mathbf{M}, d >$ into the space $< \mathbf{M}', d' >$, then f is uniformly continuous on $\mathbf{M}$.

4.62. Complete the proof of Theorem 4.17 along the lines suggested in the text.

4.63. Do this experiment: Enter any number $x_0 \in [0, 2]$ into your calculator; compute $x_1 = \sqrt[3]{1 + x_0}$. Then compute in similar fashion $x_2, x_3, \ldots, x_{12}$ from $x_{n+1} = \sqrt[3]{1 + x_n}$. What do you get? Discuss, in the context of Theorem 4.17 and with justifications of all details, what is going on here. Indicate the limiting result that should be obtained in the experiment.

4.64. Read Drager and Foote (1986) and Kolodner (1967), and compare their presentations of the Contraction Mapping Theorem to that in Theorem 4.17 (and Exercise 4.62).

4.65. Obtain a value for the Lipschitz constant λ for each function on the indicated interval; then solve $x - f(x) = 0$.
(a) $f(x) = (x^3 + 3)/5, \quad [0, 1]$;
(b) $f(x) = (x^4/1296) - (11x^2/60), \ [-12, -9.6]$.

4.66. Prove that the point $\mathbf{x}$ in the statement of Theorem 4.17 is uniquely determined.

4.67. Part of the work in Exercise 4.63 could have been facilitated if we had employed the **Bounded Derivative Condition**. Read what this is in Wagner (1982); also read the author's application of the Bounded Derivative Condition to the well-known Newton-Raphson Method. Write up a summary of your reading, and include some sample applications of your own.

REFERENCES

Cited Literature

Baker, C.W., *Introduction to Topology*, Krieger Publishing Co., Malabar, FL, 1997. A well-written, short (155 pp.), introduction to the subject; good for one's first read in the area.

Bayne, R.E., Joseph, J.E., Kwack, M.H. and Lawson, T.H., "Alternative Proofs of Some Results from Elementary Analysis," *Missouri J. Math. Sci.*, **14**, 19–27 (2002). This interesting article has alternative proofs of the Bolzano-Weierstrass Theorem, the Heine-Borel Theorem, and our Theorem 4.15.

Bernau, S.J., "The Bounds of a Continuous Function," *Amer. Math. Monthly*, **74**, 1082 (1967). An alternative proof of a version of our Theorem 4.10. Cited in our Exercise 4.44.

Buck, R.C., *Advanced Calculus*, 3rd ed., McGraw-Hill, New York, 1978, p. 96. Still another proof of our Theorem 4.15; a model of brevity.

Copi, I.M., *Introduction to Logic*, 7th ed., Macmillan, NY, 1986. A rather full text that will appeal to readers who enjoy logical argumentation.

Drager, L.D. and Foote, R.L., "The Contraction Mapping Lemma and the Inverse Function Theorem in Advanced Calculus," *Amer. Math. Monthly*, **93**, 52–54 (1986). The authors prove the Contraction Mapping Theorem for a function $\mathbf{f} : \mathbf{R}^k \to \mathbf{R}^k$ by explicitly using the result that a continuous, real-valued function defined on a nonempty, closed, bounded subset of $\mathbf{R}^k$ attains a maximum and a minimum on that set (our Theorem 4.10). Cited in our Exercise 4.64.

Fort, Jr., M.K., "The Maximum Value of a Continuous Function," *Amer. Math. Monthly*, **58**, 32–33 (1951). This author's proof is based on the application of the Completeness Axiom to an increasing sequence of real numbers that is bounded above. Cited in our Exercise 4.44.

Gamelin, T.W. and Greene, R.E., *Introduction to Topology*, 2nd ed., Dover Publications, NY, 1999. A bit more advanced than Baker; see Chapters I and II on metric spaces and topological spaces, respectively.

Haaser, N.B. and Sullivan, J.A., *Real Analysis*, Dover Publications, NY, 1991, pp. 64–65, 66, 88–91. The first citation is to point continuity of a function from one metric space into another metric space, the second citation is our source for Exercise 4.32, and the third citation is to the Intermediate-Value Theorem for mappings from a connected metric space into $\mathbf{R}^1$. This book is worth owning. Cited in our Exercise 4.31.

Jungck, G., "The Extreme Value Theorem," *Amer. Math. Monthly*, **70**, 864–865 (1963). A proof of the existence of extrema of a real-valued, continuous function on a closed, bounded interval that uses the Completeness Axiom.

Kline, M., *Mathematical Thought from Ancient to Modern Times*, Oxford University Press, New York, 1972, pp. 949–954. A thumbnail sketch of some nineteenth century developments in function theory, including contributions by Heine and Borel. For a peek at still later developments (in what became known as functional analysis), see Kline's Chapter 46 (pp. 1076–1095).

Kolodner, I.I., "On the Proof of the Contraction Mapping Theorem," *Amer. Math. Monthly*, **74**, 1212–1213 (1967). A more general proof than that of Drager and Foote; interesting. Cited in our Exercise 4.64.

Lipschutz, S., *General Topology*, Schaum's Outline Series, McGraw-Hill, NY, 1965. Very useful study guide; recommended.

MacG. Dawson, R.J., "Crackpot Angle Bisectors!," *Math. Mag.*, **80**, 59–64 (2007). Very interesting article on some of the geometry in a two-dimensional space that uses the taxicab metric. Cited in our Exercise 4.25.

Messer, R. and Straffin, P., *Topology Now!*, Mathematical Association of America, Washington, D.C., 2006. Very accessible, without requiring extensive mathematical prerequisites.

Pennington, W.B., "Existence of a Maximum of a Continuous Function," *Amer. Math. Monthly*, **67**, 892–893 (1960). The author's proof uses the bisection technique; the author alludes to the famous **Heine-Borel Theorem** as something we would use for functions defined on sets more general than intervals. Cited in our Exercise 4.44.

Russ, S.B., "A Translation of Bolzano's Paper on the Intermediate Value Theorem," *Hist. Math.*, **7**, 156–185 (1980). This is the first such translation into English by a Bolzano specialist of this famous paper.

Simmons, G.F., *Introduction to Topology and Modern Analysis*, Krieger Publishing Co., Malabar, FL, 1983, pp. 49–90. These pages give a sound summary of material on metric spaces. Chapter 4 (pp. 110–128) and Chapter 6 (pp. 142–152) cover compactness and connectedness, respectively.

Sprecher, D.A., *Elements of Real Analysis*, Dover Publications, NY, 1987, pp. 146–148. A brief section on connected sets; like the Haaser-Sullivan book, this one by Sprecher is worth owning.

Wagner, C.H., "A Generic Approach to Iterative Methods: Variations on a Theme by Stefan Banach," *Math. Mag.*, **55**, 259–273 (1982). Very nice article, packed with lots of good mathematics. Cited in our Exercise 4.67. Recommended.

Wolf, R.S., *A Tour Through Mathematical Logic*, Carus Mathematical Monograph No. 30, Mathematical Association of America, Washington, D.C., 2005. Very accessible; written for mathematicians rather than logicians (as is true of Copi's book).

Additional Literature

Beardon, A.F., "On the Continuity of Monotonic Functions," *Amer. Math. Monthly*, **74**, 314–315 (1967).

Beardon, A.F., "Contractions of the Real Line," *Amer. Math. Monthly*, **113**, 557–558 (2006).

Berberian, S.K., *A First Course in Real Analysis*, Springer-Verlag, NY, 1994, pp. 84–95 (continuity).

Gemignani, M.C., "On Properties Preserved by Continuous Functions," *Math. Mag.*, **41**, 181–183 (1968).

Heuer, G.A., "Functions Continuous at the Irrationals and Discontinuous at the Rationals," *Amer. Math. Monthly*, **72**, 370–373 (1965).

Hewitt, E., "The Role of Compactness in Analysis," *Amer. Math. Monthly*, **67**, 499–516 (1960).

Olsen, L., "A New Proof of Darboux's Theorem," *Amer. Math. Monthly*, **111**, 713–715 (2004) (the IVT applied to derivatives).

Osler, T.J. and Stugard, N., "A Collection of Numbers Whose Proof of Irrationality is Like That of the Number e," *Math. Comp. Educ.*, **40** (2), 103–107 (2006).

Rosenlicht, M., *Introduction to Analysis*, Dover Publications, NY, 1986, pp. 67–95 (Chapter IV. "Continuous Functions").

Rudin, W., *Principles of Mathematical Analysis*, 3rd ed., McGraw-Hill, NY, 1976, pp. 220–223 (the Contraction Mapping and Inverse Function Theorems).

Rusnock, P. and Kerr-Lawson, A., "Bolzano and Uniform Continuity," *Hist. Math.*, **32**, 303–311 (2005).

Staib, J.H., "A Sequence-Approach to Uniform Continuity," *Math. Mag.*, **40**, 270–273 (1967).

Staib, J.H., "Sequences vs Neighborhoods," *Math. Mag.*, **44**, 145–146 (1971) (the Ruler Function).

Walk, S.M., "Mind Your ∀'s and ∃'s," *Coll. Math. J.*, **35**, 362–369 (2004).

CHAPTER 5

Differentiation

"As long as the function $f(x)$ and its derivative $f'(x)$ remain continuous between the limits $x = x_0$ and $x = x_0 + h$, we have in general $[f(x+h) - f(x)]/h = f'(x + \theta h)$."

Augustin-Louis Cauchy

CONTENTS

Reviewed in this chapter	Cauchy's definition of the derivative; first- and second-derivative tests; Rolle's Theorem; l'Hôpital's Rule (statement).

New in this chapter	Carathéodory's definition; Mean-Value Theorem; Taylor's Theorem; radius of convergence; exponential and logarithmic functions; l'Hôpital's Rule (proof); differentiation in $\mathbf{R}^n$.

5.1 CAUCHY'S DEFINITION

The derivative is the most important concept in the differential calculus, and its clear definition in Cauchy's 1823 book *Calcul infinitésimal* (Grabiner, 1981, 1983) was a powerful stimulant for mathematics. Cauchy's definition of derivative is now, after more than 185 years, so familiar to us that we scarcely give it a second thought. But we should give it the reverence that it deserves.

Definition. *Let a be an interior point of a set* $\mathbf{S} \subseteq \mathbf{R}^1$ *and let* $f\colon \mathbf{S} \to \mathbf{R}^1$ *be a function. Then the value of the* **derivative** *of* $y = f(x)$ *at* $x = a$ *is* $f'(a) = \lim_{x\to a} \frac{f(x)-f(a)}{x-a}$, *provided that this limit exists (in* $\mathbf{R}^1$*).*

The notation $f'(a)$ is the most prevalent notation in use. Other notational forms for the value of the derivative of f at a include $Df(a)$ and the Leibniz notations $\frac{df}{dx}(a)$ or $\left.\frac{dy}{dx}\right|_{x=a}$. In any case, the definition shows clearly that the derivative of a constant is 0.

Since "sided" limits can be defined at endpoints of intervals, derivatives can be defined there also. In the language of balls, $f'(a)$ is the value of the derivative of f at a iff, given any $\varepsilon > 0$, there is a $\delta > 0$ such that $x \in [\mathbf{B}(a;\delta)\backslash\{a\}] \cap \mathbf{S}$ implies that

$$\frac{f(x)-f(a)}{x-a} \in \mathbf{B}(f'(a);\varepsilon).$$

You are familiar with the derivatives of several standard functions (Table 5.1). Their rigorous development by the pioneers involved quite a lot of good mathematics.

Table 5.1 A Brief Selection of Derivatives

$f(x)$	$f'(x)$	$f(x)$	$f'(x)$
1. x^c	cx^{c-1}	6. $\tan x$	$\sec^2 x$
2. e^x	e^x	7. $\mathrm{Sin}^{-1}\, x^a$	$1/\sqrt{1-x^2}$
3. $\ln x$	$1/x$	8. $\mathrm{Tan}^{-1}\, x^a$	$1/(1+x^2)$
4. $\sin x$	$\cos x$	9. $\sinh x$	$\cosh x$
5. $\cos x$	$-\sin x$	10. $\cosh x$	$\sinh x$

[a] *Principal values:* $-\pi/2 \le \mathrm{Sin}^{-1}x \le \pi/2, -\pi/2 < \mathrm{Tan}^{-1}x < \pi/2$

We shall put some of these functions on a secure footing later. Formula 1, for example, requires some attention. A special case of it, when $c = n \in \mathbf{N}$, is easy to deal with.

■ Example 5.1

In formula 1, with $c = n \in \mathbf{N}$, let $h = x - a$, where x is variable and a is any constant. Then

$$f'(a) = \lim_{x \to a} \frac{x^n - a^n}{x - a} = \lim_{h \to 0} \frac{(a+h)^n - a^n}{h}.$$

Newton's Binomial Theorem (Exercise 2.6) gives

$$\begin{aligned}(a+h)^n &= \sum_{k=0}^{n} \binom{n}{k} a^{n-k} h^k \\ &= a^n + \sum_{k=1}^{n} \binom{n}{k} a^{n-k} h^k,\end{aligned}$$

and so

$$\begin{aligned}f'(a) &= \lim_{h \to 0} \left[\frac{1}{h} \sum_{k=1}^{n} \binom{n}{k} a^{n-k} h^k \right] \\ &= \lim_{h \to 0} \left[\binom{n}{1} a^{n-1} + \sum_{k=2}^{n} \binom{n}{k} a^{n-k} h^{k-1} \right] \\ &= na^{n-1} + \sum_{k=2}^{n} \left[\lim_{h \to 0} \binom{n}{k} a^{n-k} h^{k-1} \right] \\ &= na^{n-1}.\end{aligned}$$

■

Formulas 4 through 6, especially, generally are not handled satisfactorily in introductory texts because of the reliance upon certain intuitive geometric notions (Hardy, 1967; Spiegel, 1956). We will deal with the trigonometric functions in Exercises 6.38, 6.50, and 6.51. Nevertheless, a "half-rigorous" proof of formula 6 would be useful.

■ Example 5.2

Let $f(a) = \tan a, a \neq (2k+1)\pi/2, k \in \mathbf{Z}$. Then

$$\tan(a+h) - \tan a = \frac{\sin(a+h)}{\cos(a+h)} - \frac{\sin a}{\cos a}$$

$$= \frac{\sin(a+h)\cos a - \cos(a+h)\sin a}{\cos(a+h)\cos a}$$
$$= \frac{\sin h}{\cos(a+h)\cos a},$$

after making use of two standard identities. Hence,

$$\frac{\tan(a+h) - \tan a}{h} = \frac{1}{\cos a}\left[\frac{1}{\cos(a+h)}\frac{\sin h}{h}\right]$$

and

$$f'(a) = \frac{1}{\cos a}\left[\lim_{h\to 0}\frac{1}{\cos(a+h)}\right]\left[\lim_{h\to 0}\frac{\sin h}{h}\right]$$
$$= \frac{1}{\cos^2 a}\lim_{h\to 0}\frac{\sin h}{h},$$

if the limit exists. To obtain expeditiously the indicated limit, we note that the perimeter $P(n)$ of a regular n-gon inscribed in a circle of radius r is $P(n) = 2nr\sin(\pi/n)$ (verify!). If we accept that $\lim_{n\to\infty} P(n) = 2\pi r$, then upon setting $\pi/n = h$ and $P(n) = H(h)$, we obtain

$$\lim_{n\to\infty} P(n) = \lim_{h\to 0} H(h) = \lim_{h\to 0} 2\pi r\left(\frac{\sin h}{h}\right) = 2\pi r,$$

and so, $\lim_{h\to 0}(\sin h)/h = 1$. It follows that $f'(a) = \sec^2 a$. ■

Several elementary theorems follow directly from Cauchy's definition. The first (see next) makes contact with Chapter 4; this fundamental theorem has a false converse (Exercise 5.2).

Theorem 5.1. *If f has a derivative at a, then f is continuous at a.*

Proof. Suppose that the domain of f contains $[a-c, a+c]$, $c > 0$, and let h satisfy $0 < |h| \le c$. Then

$$f(a+h) - f(a) = [f(a+h) - f(a)]\left(\frac{h}{h}\right).$$

The completion of the proof is left to you. ■

■ Example 5.3

Refer to the signum function in Exercise 4.11. As sgn(x) is discontinuous at $x = 0$, then it cannot have a derivative there. ■

Theorem 5.1 has had a colorful history. Many mathematicians of the nineteenth century, including luminaries of the day, such as S.-F. Lacroix, A.M. Ampère, and J.L.F. Bertrand, claimed to have "proved" that a function has a derivative at *all* points where it is continuous (Kline, 1980). All of these "proofs" were wrong![1] On the other hand, a wild example of a function that has a derivative *nowhere*, but is continuous at each irrational x and discontinuous at each rational x (see Exercise 4.13(f)), is provided by the Ruler Function on [0,1] (Dunham, 2003). Still zanier examples are functions that are continuous *everywhere* but also have a derivative nowhere (Hildebrandt, 1933; McCarthy, 1953; Wen, 2000). We will look at an example in Section 7.6.

Theorem 5.2. *If f, g have derivatives at a and if $c \in \mathbf{R}$, then*

(a) $[cf + g]'(a) = cf'(a) + g'(a)$;

(b) $(fg)'(a) = f(a)g'(a) + f'(a)g(a)$;

(c) if $g(x) \neq 0$ in some ball $\mathbf{B}(a;\delta)$, then

$$(f/g)'(a) = \left[g(a)f'(a) - f(a)g'(a)\right] / [g(a)]^2 .$$

Proof. We sketch the proof of (b) and leave the other parts as exercises.

Elementary function algebra gives

$$\begin{aligned}(fg)(a+h) - (fg)(a) &= f(a+h)g(a+h) \\ &\quad + [f(a+h)g(a) - f(a+h)g(a)] - f(a)g(a) \\ &= f(a+h)[g(a+h) - g(a)] + g(a)[f(a+h) - f(a)].\end{aligned}$$

Now divide by $h \neq 0$; the completion of the proof is left to you. ■

■ Example 5.4

We proceed to work out formula 5 in Table 5.1, for which we call upon the work in Example 5.2. Write $1 + \tan^2\theta = 1/[f(\theta)]^2$, whenever $f(\theta) = \cos\theta \neq 0$. The derivatives of both sides of this identity lead to a new identity. From Theorem 5.2 (b, c) and Example 5.2 we obtain

$$2\tan\theta\sec^2\theta = \frac{-2f'(\theta)}{\cos^3\theta}, \quad (\cos\theta \neq 0)$$

or

$$f'(\theta) = -\sin\theta. \tag{*}$$

[1] As in other matters mathematical, Bolzano was ahead of his time here also. He gave in 1834 an example of a continuous function with a derivative nowhere, thus anticipating the more widely recognized examples due to Riemann and Weierstrass 20 to 40 years later. Bolzano's example went unnoticed. For details, see Kowalewski (1923).

On the other hand, Cauchy's definition gives

$$
\begin{aligned}
f'(\theta) &= \lim_{h\to 0} \frac{\cos(\theta+h)-\cos\theta}{h} \\
&= \lim_{h\to 0} \frac{\cos\theta\cos h-\sin\theta\sin h-\cos\theta}{h} \\
&= \cos\theta\left[\lim_{h\to 0}\frac{\cos h-1}{h}\right]-\sin\theta\left[\lim_{h\to 0}\frac{\sin h}{h}\right] \\
&= \cos\theta\left[\lim_{h\to 0}\frac{\cos h-1}{h}\right]-\sin\theta \qquad (**)
\end{aligned}
$$

from Example 5.2. The limit in the brackets is independent of the choice of θ. Whenever $\cos\theta \neq 0$, then we obtain from $(*)$

$$
-\sin\theta = \cos\theta\left[\lim_{h\to 0}\frac{\cos h-1}{h}\right]-\sin\theta,
$$

and, thus, the limit is 0. It follows from $(**)$ that $f'(\theta) = -\sin\theta$ for all θ. ■

The ability to obtain derivatives is increased enormously by use of the famous Chain Rule. This deep theorem can be proved from Cauchy's definition. We quote the Theorem now but delay its proof until the next section, where we shall present an alternative proof and an enlightening perspective on derivatives.

Theorem 5.3 (Chain Rule). *If f has a derivative at a and g has a derivative at $b = f(a)$, then the composite function $g[f]$ has a derivative at a, given by $g'(b)f'(a)$.*

5.2 CARATHÉODORY'S DEFINITION

The German-born, Greek mathematician **Constantin Carathéodory** (1873–1950) gave a definition of the derivative that does not focus on limits of quotients of vanishing differences and is not, therefore, linked to geometric notions. Instead, the focus is on continuity of a particular function (Kuhn, 1991).

Definition. *The function f: $\mathbf{S} \to \mathbf{R}^1$, has a* ***derivative*** *at $a \in \mathbf{S}$ iff there exists a function ϕ_a that is continuous at a and satisfies the relation $f(x) - f(a) = \phi_a(x)$ $(x - a)$ for all $x \in \mathbf{S}$.*

The notation $\phi_a(x)$ emphasizes that ϕ depends upon the point a. We have immediately the following.

Theorem 5.4. *In Carathéodory's formulation, if $f'(a)$ exists, then it is $\phi_a(a)$.*

Proof. The proof is left to you. ■

■ Example 5.5

Consider $f\colon \mathbf{R}^1 \to \mathbf{R}^1$, where $f(x) = x^3$. Then Carathéodory's definition gives at $a = 3$

$$\phi_3(x) = \begin{cases} \dfrac{x^3 - 3^3}{x - 3} = x^2 + 3x + 9 & x \neq 3 \\ 27 & x = 3. \end{cases}$$

This function is continuous at a. By Theorem 5.4 we have $f'(a) = \phi_3(3) = 27$. ■

Carathéodory's definition handles the proofs of the three parts of Theorem 5.2 with ease. A better test of the definition is its use in the proof of the Chain Rule.

Theorem 5.3 (Chain Rule).

Proof. By Carathéodory's definition there are functions β_a, γ_b defined on open intervals $\mathbf{I}_a, \mathbf{I}_b$ about $x = a, x = b$, respectively, and continuous at these points, such that

$$\begin{cases} f(x) - f(a) = \beta_a(x)(x - a) & (*) \\ g(x) - g(b) = \gamma_b(x)(x - b). & (**) \end{cases}$$

Accordingly, we have

$$\begin{aligned} h(x) - h(a) &= g[f(x)] - g[f(a)] \\ &= \{\gamma_b[f(x)]\}\,(f(x) - f(a)) && \text{from } (**) \\ &= (\gamma_b[f])(x)\,[\beta_a(x)(x - a)] && \text{from } (*) \end{aligned}$$

for all $x \in \mathbf{I}_a$ such that $f(x) \in \mathbf{I}_b$. Since $a \in \mathbf{I}_a$ and $b = f(a) \in \mathbf{I}_b$, then by Theorem 4.4 $\gamma_b[f]$ is continuous at a. An application of Theorem 4.3(b) leads to the conclusion that $(\gamma_b[f])(x)\beta_a(x)$ is also continuous at a. From Carathéodory's definition we conclude that $h = g[f]$ has a derivative at a, which is

$$\begin{aligned} h'(a) = (g[f])'(a) &= (\gamma_b[f])(a)\,\beta_a(a) \\ &= (\gamma_b[f])(a)\,f'(a) \\ &= [\gamma_b(b)]\,f'(a) \\ &= g'(b)\,f'(a). \end{aligned}$$ ■

By induction, the Chain Rule is extendable to compositions of more than two functions. Carathéodory's definition of the derivative offers an elegant approach to the rule.

■ Example 5.6

Here's a simple example. Let $f(x) = \sqrt[3]{x}$; Example 5.1 cannot be used to obtain $f'(x)$ because $1/3 \notin \mathbf{N}$. Nor can we use formula 1 in its generality because we have not yet established this. But if we let $g(x) = x^3$, then $x = g[f(x)]$ and if we set the derivatives of the two sides equal, then Theorem 5.3 yields

$$1 = g'[f(x)]f'(x) = 3[f(x)]^2 f'(x).$$

Hence, for any $x \neq 0$ we obtain

$$f'(x) = \frac{1}{3[f(x)]^2} = \frac{1}{3\sqrt[3]{x^2}}.$$

This example is a particular case of the Inverse Function Theorem, which is itself provable from Carathéodory's definition (Exercise 5.12). ■

■ Example 5.7

On the interval $(0, \pi)$ $\cos x$ is continuous (how do you know this?). It is also strictly decreasing (Corollary 5.7.1) and is, therefore, an injection. Hence, by Exercise 4.52 it has an inverse function (denoted $\cos^{-1} x$) that is also continuous and strictly decreasing. Additionally, for any $x \in (-1, 1)$ we have from Exercise 1.43(b) the identity

$$\cos[\text{Cos}^{-1} x] = x.$$

The derivatives of the two sides must be equal, so application of the Chain Rule gives (verify!)

$$\left(\text{Cos}^{-1} x\right)' = \frac{-1}{\sqrt{1 - x^2}} \qquad (-1 < x < 1).$$

■

5.3 ROLLE'S THEOREM

You recall that in introductory calculus maxima and minima of functions were located by setting their derivatives equal to zero. Let us be more precise about this.

Definition. *A function $f : \mathbf{I} \to \mathbf{R}^1$ has a **relative maximum (minimum)** at a point c in the interval $\mathbf{I}$ iff there is a ball $\mathbf{B}(c; \delta)$ such that $f(x) \leq f(c) (f(x) \geq f(c))$ for all $x \in \mathbf{B}(c; \delta) \cap \mathbf{I}$.*

A relative maximum or minimum is called a **relative extremum**. In some cases a relative extremum may correspond to an absolute maximum or minimum on $\mathbf{I}$ (Figure 5.1(a)), whereas in other cases a relative extremum will be neither an absolute maximum nor an absolute minimum (Figure 5.1(b)). A function may not even have a relative or an absolute extremum on some interval $\mathbf{I}$ (Figure 5.1(c)). Also note from the definition that a relative extremum can occur at an endpoint of an interval.

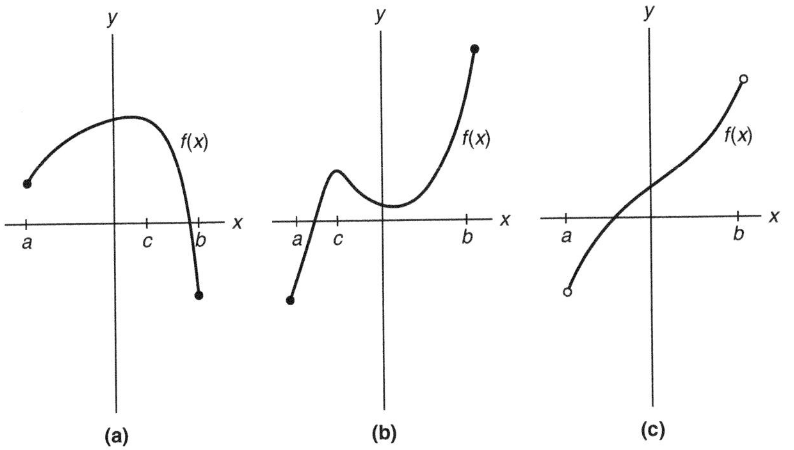

FIGURE 5.1
Illustration of relative and absolute extrema on a bounded interval.

Theorem 5.5. *If c is an interior point of an interval* **I** *(finite or infinite), if a function $f: \mathbf{I} \to \mathbf{R}^1$ is known to have a relative minimum at c, and if f has a derivative at c, then $f'(c) = 0$.*

Proof. By Carathéodory's definition and by hypothesis, there is a ball $\mathbf{B}(c; \delta) \subset \mathbf{I}$ and a function $\phi_c(x)$, continuous at c, such that $x \in \mathbf{B}(c; \delta)$ implies

$$f(x) - f(c) = \phi_c(x)(x - c).$$

If $x < c$, then $f(x) - f(c) \geq 0$ by hypothesis; $x - c < 0$ then implies that $\phi_c(x) \leq 0$. But if $x > c$, then $f(x) - f(c) \geq 0$ again holds and $x - c > 0$ now implies that $\phi_c(x) \geq 0$. The continuity of $\phi_c(x)$ at c forces $\phi_c(c) = 0$. By Theorem 5.4, we then have $f'(c) = 0$. ■

An analogous theorem holds for functions f with a relative maximum at c. Note that so long as f has a derivative at c, the condition $f'(c) = 0$ is only a necessary condition and not a sufficient condition for f to have a relative extremum at c (Exercise 5.17). However, see (later) Theorem 5.9.

■ Example 5.8

Let $f: [-1, 2] \to \mathbf{R}^1$ be defined by $f(x) = x^3 e^{-2x}$. This function has a derivative everywhere on its domain.[2] We find $f'(x) = x^2(3 - 2x)e^{-2x}$ and, hence, $f'(x) = 0$ when $x = c = 0, 3/2$. The *potential* locations of relative extrema are, thus, $x = -1, 0, 3/2, 2$, but further analysis is needed (Section 5.5). However, what does your intuition tell you? ■

---

[2]We borrow from Section 5.7 (or Table 5.1) the fact that $(e^{ax})' = ae^{ax}$, as well as any other basic properties of the exponential function.

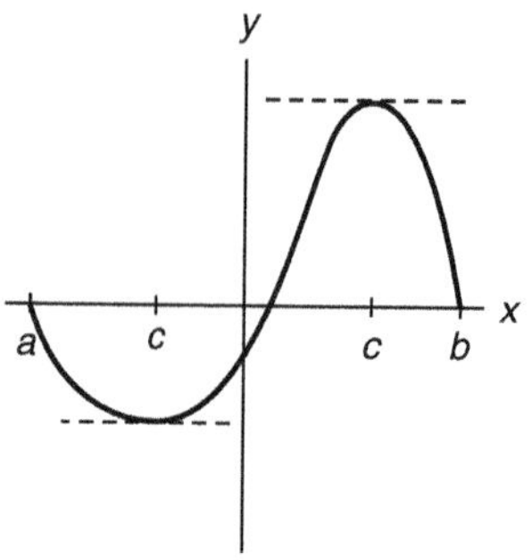

FIGURE 5.2
Illustration of Rolle's Theorem.

We now come to the main result in this section, due to the French mathematician **Michel Rolle** (1652–1719), who presented his Theorem in 1691 but did not prove it.

Theorem 5.6 (Rolle's Theorem). *If f is continuous on the closed, bounded interval $[a, b]$ and has a derivative on (a, b), and if $f(a) = f(b) = 0$, then there is at least one interior point c at which $f'(c) = 0$.*

Proof. Let us assume that f is not the zero function on $[a, b]$. Then some values of f are positive, or negative, or possibly both exist (Figure 5.2). By Theorems 4.8 and 4.10, f attains absolute extrema on $[a, b]$. As f is not constant, then at least one of the absolute extrema must occur at an interior point c. By Theorem 5.5 (or its analog), we then have $f'(c) = 0$. ■

■ Example 5.9

Let $f: \mathbf{R}^1 \to \mathbf{R}^1$ be defined by $f(x) = x^5 + 3x + 4$; we show that $f(x) = 0$ has only one real root.

We have $f(-2) = -34$ and $f(0) = 4$, so by the Intermediate-Value Theorem (Theorem 4.14) there is a number b_1 such that $-2 < b_1 < 0$ and $f(b_1) = 0$. The function f has a derivative everywhere. If there were a second point, $x = b_2$, at which $f(b_2) = 0$, then by Rolle's Theorem there should be a point $c \in (b_1, b_2)$ at which $f'(c) = 5c^4 + 3 = 0$. But this latter is impossible, so no such b_2 exists. ■

5.4 THE MEAN-VALUE THEOREM

The Mean-Value Theorem is one of the most famous theorems in calculus. Despite attempts to downplay its importance (Boas, Jr., 1981), its place in the calculus curriculum is secure (Swann, 1997).

Although the Mean-Value Theorem was known to J.-L. Lagrange and A.-M. Ampère, the Theorem was proved by Cauchy from his definition of the derivative (see opening quotation) and was then used to make several deductions (Grabiner, 1981). We shall present two proofs of it, one a rather traditionally structured proof, and the other a proof with some linear algebra content that is a warm-up for a later exposition of Taylor's Theorem (Theorem 5.11).

Theorem 5.7 (Mean-Value Theorem). *Suppose that $f'(x)$ exists at each point in (a, b) and that $L_1 = \lim_{x \to a^+} f(x), L_2 = \lim_{x \to b^-} f(x)$ are real. Then there is a point $c \in (a, b)$ such that $L_2 - L_1 = f'(c)\,(b - a)$.*

The content of the Theorem is interpretable by the geometry displayed in Figure 5.3. We shall prove a slightly more general theorem, and then deduce the Mean-Value Theorem as a consequence of it.

In the proof of Theorem 5.8, we shall require a function $D(t)$ to be constructed from two other functions, $F(t)$ and $G(t)$, and to satisfy the hypotheses of Rolle's Theorem on the interval $[a, b]$. The geometric picture is similar to that in Figure 5.3, except that in the xy-plane the coordinate variables are now functions of a parameter t (Figure 5.4):

$$\begin{cases} x = G(t) \\ y = F(t). \end{cases}$$

A simple construction for $D(t)$ is a linear combination of $F(t)$, $G(t)$.

$$D(t) = k_1 F(t) + k_2 G(t) + k_3$$

We desire $D(a) = D(b) = 0$ in order to use Rolle's Theorem. Solving the resulting pair of linear equations,

$$\begin{cases} k_1 F(a) + k_2 G(a) + k_3 = 0 \\ k_1 F(b) + k_2 G(b) + k_3 = 0, \end{cases}$$

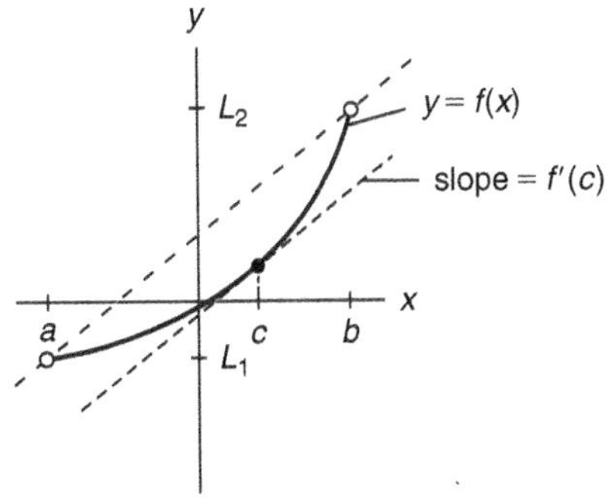

FIGURE 5.3
Illustration of the Mean-Value Theorem.

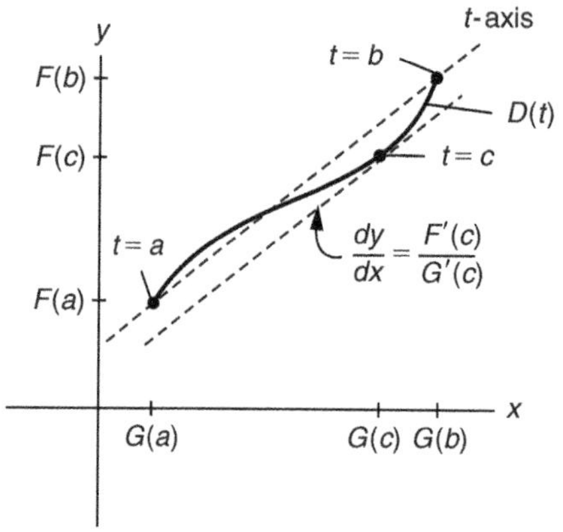

FIGURE 5.4
A special function is used in the Generalized Mean-Value Theorem.

we can obtain $k_1 = G(b) - G(a)$ and $k_2 = F(a) - F(b)$. These solutions are not unique, but that will not matter here. Finally, k_3 can be found by back substitution of k_1, k_2. The final result is

$$D(t) = F(t)[G(b) - G(a)] - G(t)\left[F(b) - F(a)\right] - \left[F(a)G(b) - G(a)F(b)\right].$$

We note also that the derivative (slope of a tangent line) at any point $t = c, a < c < b$, on the curve $D(t)$ is given by

$$\frac{dy}{dx} = \lim_{h \to 0} \frac{F(c+h) - F(c)}{G(c+h) - G(c)} = \lim_{h \to 0} \frac{\left[F(c+h) - F(c)\right]/h}{\left[G(c+h) - G(c)\right]/h} = \frac{F'(c)}{G'(c)},$$

provided that $G'(c) \neq 0$.

Theorem 5.8 (Generalized Mean-Value Theorem). *Suppose that f, g have real derivatives at each $t \in (a, b)$; suppose also that $L_1 = \lim_{t \to a^+} f(t)$, $L_2 = \lim_{t \to b^-} f(t)$, $l_1 = \lim_{t \to a^+} g(t)$, $l_2 = \lim_{t \to b^-} g(t)$ are real. Then there exists a point $c \in (a, b)$ such that $g'(c)\left[L_2 - L_1\right] = f'(c)\left(l_2 - l_1\right)$.*

Proof. Let functions F, G be defined on $[a, b]$ as follows:

$$F(t) = \begin{cases} L_1 & t = a \\ f(t) & a < t < b \\ L_2 & t = b \end{cases} \qquad G(t) = \begin{cases} l_1 & t = a \\ g(t) & a < t < b \\ l_2 & t = b. \end{cases}$$

We now define the special function

$$D(t) = F(t)\left[G(b) - G(a)\right] - G(t)\left[F(b) - F(a)\right] - \left[F(a)G(b) - G(a)F(b)\right].$$

This has a derivative at each $t \in (a, b)$ and is continuous on $[a, b]$, in particular, at the endpoints, by virtue of the definitions made earlier.

We also observe that $D(a) = L_1(l_2 - l_1) - l_1(L_2 - L_1) - (L_1 l_2 - l_1 L_2) = 0$, and similarly, $D(b) = 0$. Thus, by Rolle's Theorem, there is a $c \in (a, b)$ such that $D'(c) = 0$. Since c is not an endpoint, then the equation $D'(c) = 0$ reduces to

$$g'(c)[L_2 - L_1] = f'(c)(l_2 - l_1). \qquad \blacksquare$$

Theorem 5.7 now follows immediately if $g(t) = t$ for all $t \in [a, b]$. Under these conditions, $g'(c) = 1$ and $l_2 - l_1 = b - a$. The Mean-Value Theorem is mainly an *existence* theorem. In most applications of it, as in the corollaries given later in this section, we are not interested in the particular value of c. Also, much subsequent work has been inspired by the Mean-Value Theorem and a number of generalizations of it are known (Goodner, 1962; Reich, 1969; Sanderson, 1972).

We now give an alternative route to the Mean-Value Theorem (Barrett and Jacobson, 1960). Figure 5.5(a) shows a parallelogram ABCD. An arbitrary subset of three of its vertices has coordinates A: $(x(a), y(a))$, B: $(x(s), y(s))$, C: $(x(b), y(b))$, where x, y are functions of the parameter t. When $t = a, s, b$, respectively, the corresponding values of x, y are the coordinates of the vertices A, B, C.

In Figure 5.5(b) a triangular region of ABCD has been translated so as to give a new parallelogram AOO′D of identical area but with an altitude of length h that is parallel to the y-axis. This is a geometric simplification, and we leave it to you to show that the areas K of the parallelograms are given by the absolute

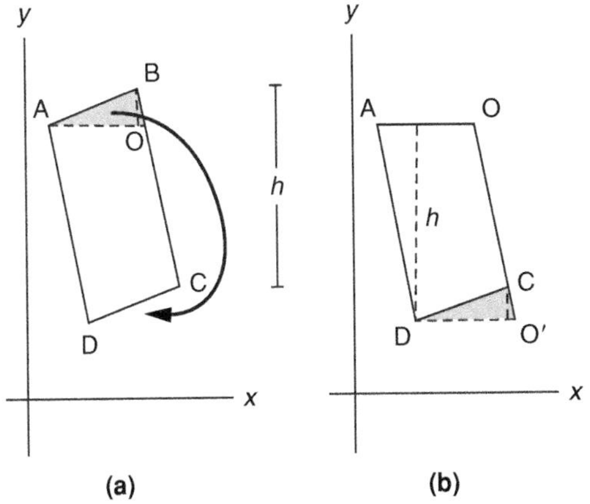

FIGURE 5.5
Computation of the area of a parallelogram.

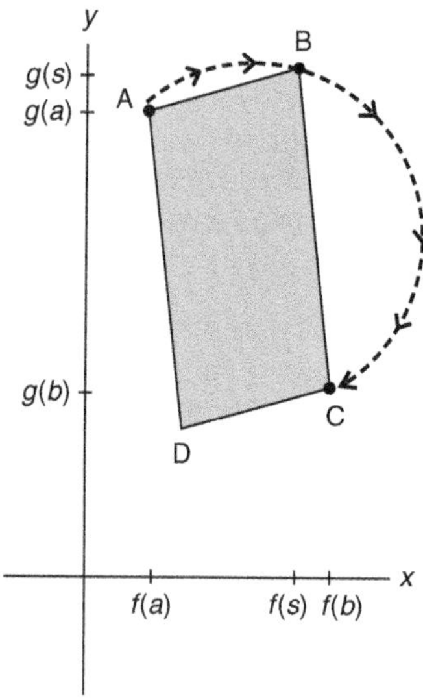

FIGURE 5.6
Parameterization of the area of a parallelogram.

determinant (Exercise 5.26)

$$\pm K = \begin{vmatrix} x(a) & y(a) & 1 \\ x(s) & y(s) & 1 \\ x(b) & y(b) & 1 \end{vmatrix}.$$

Now suppose that in Figure 5.6 vertices A, C are fixed, but B, D are variable (the figure always being a parallelogram, however). A parameterized curve $\mathfrak{C}$

$$\mathfrak{C} : x = f(t), y = g(t) \quad (a \leq t \leq b)$$

passes through A, B, C. Since A, C have fixed values of t, the area of ABCD depends only upon the value of t for vertex B:

$$\pm K(t) = \begin{vmatrix} f(a) & g(a) & 1 \\ f(t) & g(t) & 1 \\ f(b) & g(b) & 1 \end{vmatrix}.$$

We stipulate that f, g are continuous on $[a, b]$ and have derivatives everywhere on (a, b). We can then show from determinant theory and from Rolle's Theorem that there is a $c \in (a, b)$ such that $K'(c) = 0$; that is (Exercise 5.26),

$$\begin{vmatrix} f(a) & g(a) & 1 \\ f'(c) & g'(c) & 0 \\ f(b) & g(b) & 1 \end{vmatrix} = 0.$$

Expansion yields $g'(c)[f(b) - f(a)] = f'(c)[g(b) - g(a)]$ and, thus, we have recovered the Generalized Mean-Value Theorem; Theorem 5.7 again follows immediately.

To show the usefulness of the Mean-Value Theorem, we give now three corollaries of the theorem. Recall from Section 2.1 that a real-valued function f defined on an interval $\mathbf{I}$ is an **increasing (decreasing)** function on $\mathbf{I}$ iff for every pair $x_1, x_2 \in \mathbf{I}$, $x_1 < x_2$, we have $f(x_1) \le f(x_2)$ $(f(x_1) \ge f(x_2))$.

Corollary 5.7.1. *If $f'(x) \ge 0$ $(f'(x) \le 0)$ for all $x \in (a, b)$, then f is increasing (decreasing) on (a, b).*

Proof. Use Theorem 5.7; the proof is left to you. ■

An especially interesting application of the Mean-Value Theorem leads to the well-known **Arithmetic Mean–Geometric Mean Inequality** (Schaumberger, 1985). The **arithmetic mean** (AM) and the **geometric mean** (GM) of a set of positive numbers $\{a_1, a_2, \ldots, a_n\}$ are defined so as to generalize definitions given in Exercise 1.22:

$$\mathrm{AM} = \frac{a_1 + a_2 + \cdots + a_n}{n} \qquad \mathrm{GM} = \sqrt[n]{a_1 a_2 \ldots a_n}.$$

Sample calculations in Table 5.2 suggest that GM $\le$ AM. The definition of GM indicates that it may be useful to work with logarithms; we assume familiarity with them.

From the Mean-Value Theorem we can deduce (Exercise 5.28)

$$\ln x \le x - 1$$

for any $x > 0$, with equality only when $x = 1$. Given a set $\{a_1, a_2, \ldots, a_n\}$ of positive numbers, we define $x_i = a_i/\mathrm{GM}$ for each $i = 1, 2, \ldots, n$. Inserting these

Table 5.2 Illustrative Data for the AM-GM Inequality

	{0.5, 0.75, e}	{10, 10, 17}	{3, 3.2, 3.4, 3.6, 3.8}	{0.1, 9, 10, 11, 12, 13}	{5, 5, 5}
AM	1.323	12.333	3.400	9.183	5
GM	1.006	11.935	3.388	4.990	5

into the logarithmic inequality and summing over i, we obtain

$$\sum_{i=1}^{n} \ln\left(\frac{a_i}{\mathrm{GM}}\right) \le \sum_{i=1}^{n}\left[\frac{a_i}{\mathrm{GM}} - 1\right],$$

or equivalently,

$$0 \le \frac{1}{\mathrm{GM}}\sum_{i=1}^{n} a_i - n.$$

A slight rearrangement then gives the following.

Corollary 5.7.2 (Arithmetic Mean–Geometric Mean Inequality). *If $\{a_1, a_2, \ldots, a_n\}$ is a set of positive numbers, then $\mathrm{GM} \le \mathrm{AM}$, with equality holding iff $a_1 = a_2 = \cdots = a_n$.*

The importance of Corollary 5.7.2 cannot be overestimated. It is the starting point for several other inequalities that are used in analysis (Borden, 1998).

Theorem 1.7 in Section 1.4 gave a version of the useful Bernoulli's Inequality restricted to positive, integral exponents. We now extend this to rational exponents.

Corollary 5.7.3 (Bernoulli's Inequality). *If $x > -1, x \neq 0$, and α is a nonintegral rational number that exceeds 1, then $(1+x)^\alpha > 1 + \alpha x$.*

Proof. From Exercise 5.14, it is immediately clear that if $f(x) = (1+x)^\alpha$ and $\alpha > 1$ is a nonintegral rational, then $f'(x) = \alpha(1+x)^{\alpha-1}$. There are then two cases.

CASE 1. $x > 0$. The Mean-Value Theorem on the interval $[0, x]$ yields

$$(1+x)^\alpha - 1^\alpha = \alpha(1+c)^{\alpha-1}(x-0)$$

for some $c \in (0, x)$. Since $1 + c > 1$ and α, x are positive, we have, after a slight rearrangement,

$$(1+x)^\alpha > 1 + \alpha x.$$

CASE 2. $-1 < x < 0$. This time we have on the interval $[x, 0]$

$$1^\alpha - (1+x)^\alpha = \alpha(1+c)^{\alpha-1}(0-x)$$

for some $c \in (x, 0)$. The right-hand side is positive and $1 + c < 1$, so $1 - (1+x)^\alpha < -\alpha x$. The desired result follows by a slight rearrangement. ■

■ Example 5.10

If θ satisfies $-\pi/2 < \theta < 0$, then from Corollary 5.7.3 we have $(1+\sin\theta)^{3/2} > 1 + (3/2)\sin\theta$. For example, at $\theta = -\pi/18$, we have $0.7512 > 0.7395$. ■

5.5 MORE ON RELATIVE EXTREMA

In Theorem 5.5 we gave a *necessary* condition for a function $f : \mathbf{I} \to \mathbf{R}^1$ to have a relative minimum at an interior point $c \in \mathbf{I}$. We now give a *sufficient* condition for such a minimum to occur. We make use of the Mean-Value Theorem.

Theorem 5.9 (First Derivative Test). *Let f be continuous on $[a, b] \subseteq \mathbf{I}$ and have a derivative everywhere on $(a, c) \cup (c, b)$, where $c \in (a, b)$. If $f'(x) \leq 0$ for all $x \in (a, c)$ and $f'(x) \geq 0$ for all $x \in (c, b)$, then f has a relative minimum at c.*

Proof. If $x \in (a, c)$, then by the Mean-Value Theorem

$$f(c) - f(x) = (c - x)f'(\alpha),$$

for some $\alpha \in (x, c)$. As $c - x > 0$ and $f'(\alpha) \leq 0$, then $f(x) \geq f(c)$.

On the other hand, if $x \in (c, b)$, then we have

$$f(c) - f(x) = (c - x)f'(\beta),$$

for some $\beta \in (c, x)$. As $c - x < 0$ and $f'(\beta) \geq 0$, then $f(x) \geq f(c)$ again holds. Thus, by definition, c is the location of a relative minimum of f. ■

Note from the hypotheses that f is not required to have a derivative at c. However, where in the proof did we use the fact that f is at least continuous at c? You can frame and prove an analogous theorem for relative maxima.

The First Derivative Test requires us to examine the sign of $f'(x)$ at *all* points in a sufficiently small deleted ball about c. In contrast, the standard Second Derivative Test necessitates examination of the sign of $f''(x)$ at *only one* point.[3]

We pause briefly to recall from introductory calculus that a derivative of a function may itself possess a derivative at some points. Thus, if $f : \mathbf{S} \to \mathbf{R}^1$ is given and there is a $\delta > 0$ such that $f'(x)$ exists at all $x \in \mathbf{B}(a; \delta) \cap \mathbf{S}$, then f has a second derivative at a whose value is given by

$$\lim_{x \to a} \frac{f'(x) - f'(a)}{x - a},$$

provided that this limit exists (in $\mathbf{R}^1$). We denote the value of this derivative by $f''(a)$. Clearly, $f'(a)$ must exist (in $\mathbf{R}^1$) if $f''(a)$ has any hope of also existing. Similarly, third-, fourth-, and nth-order derivatives at a, denoted commonly by $f'''(a)$, $f^{(4)}(a)$, $f^{(n)}(a)$, may also exist for some function f. Some practice with higher-order derivatives appears later (Exercises 5.35, 5.36).

[3]An alternative Second Derivative Test can be formulated that requires examination of the sign of $f''(x)$ at all points in sufficiently small intervals to the left and right of c, but does not demand that $f''(c)$ exist (Creighton, 1975).

Theorem 5.10 (Second Derivative Test). *Suppose that $f'(x)$ is continuous on $[a, b]$ and that $f'(c) = 0$, where $c \in (a, b)$. If $f''(c) > 0$, then f has a relative minimum at c.*

Proof. We have $f''(c) = \lim_{h\to 0} \frac{f'(c+h)-f'(c)}{h} = \lim_{h\to 0} \frac{f'(c+h)}{h} > 0$. Since f' is continuous on $[a, b]$, then the function F

$$F(h) = \begin{cases} \frac{f'(c+h)}{h} & h \neq 0 \\ f''(c) & h = 0 \end{cases} \qquad (*)$$

is continuous at all $h \in [a - c, b - c]$. Now call upon Theorem 4.2. The completion of the proof is left to you. ■

You can frame and prove a theorem analogous to Theorem 5.10 for relative maxima. It may be noted that Theorem 5.10 provides no information on the nature of c if $f''(c) = 0$. The point c could be the location of a relative maximum, of a relative minimum, or of neither.

■ Example 5.11

Line $\mathcal{L}$, normal to the parabola $y = x^2$ at $P = (a, a^2)$, is extended so as to intersect the parabola again at Q (Figure 5.7). Find $a > 0$ that minimizes the y-coordinate of the midpoint M of line segment PQ (Hall, 2003).

The tangent at P has slope $2a$, so slope $\mathcal{L} = -1/2a$. Denote $Q = (x, y)$; then

$$\frac{y - a^2}{x - a} = \frac{x^2 - a^2}{x - a} = x + a = -1/2a.$$

Solving for x, we obtain $x = \left(-2a^2 - 1\right)/2a, y = x^2 = a^2 + 1 + (1/4a^2)$, and the y-coordinate of M is then

$$y_M = \frac{1}{2}\left(y + a^2\right) = a^2 + (1/2) + \left(1/8a^2\right) = f(a).$$

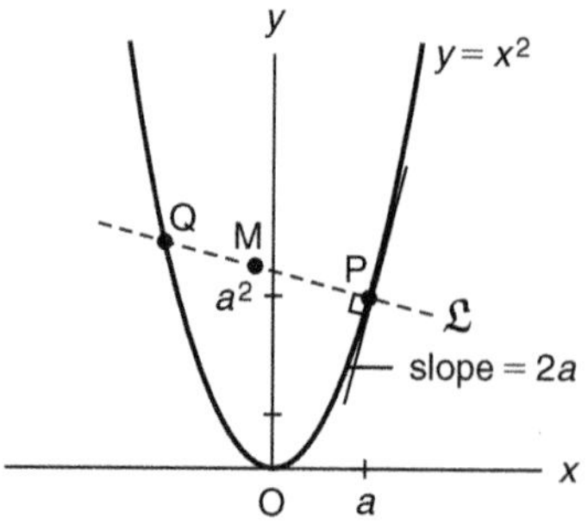

FIGURE 5.7
Diagram for Example 5.11.

Applying Theorem 5.9 first, we find (remembering that $a > 0$)

$$f'(a) = 2a - \left(1/4a^3\right),$$

and $f'(a) < 0$ for all $a \in (0, 1/\sqrt[4]{8})$ and $f'(a) > 0$ for all $a \in (1/\sqrt[4]{8}, \infty)$. Hence, y_M has a relative minimum at $1/\sqrt[4]{8}$.

But Theorem 5.5 says that at any point a_M that is not an endpoint and where f has both a derivative and a relative minimum, $f'(a_M) = 0$ must hold. Hence, the only relative minimum of y_M occurs at $a_M = 1/\sqrt[4]{8}$.

■ Example 5.12

In the previous Example we find

$$f''(a_M) = \left[2 + \left(3/4a^4\right)\right]_{a=a_M} = 8 > 0,$$

so by Theorem 5.10 we again conclude that y_M has a relative minimum at $a_M = 1/\sqrt[4]{8}$. ■

The second derivative has other uses besides the classification of relative extrema. For example, it is useful in the discussion of the concavity of curves (Taylor, 1942).

5.6 TAYLOR'S THEOREM

In Section 3.6 we discussed briefly the notion of a series of functions, commonly a power series, that converges pointwise on some domain $\mathbf{D} \subseteq \mathbf{R}^1$ to a function f. The following related definition is important.

Definition. *A pointwise convergent series of functions whose value at any $x \in \mathbf{D}$ is given by $\sum_{n=0}^{\infty} f_n(x)$* **represents** *a given function f iff at each $x \in \mathbf{D}$ the series converges to the value $f(x)$.*

Taylor's Theorem is concerned with the use of power series to represent functions that are more complex than simple polynomials. There are several approaches to Taylor's Theorem, depending upon which ideas we assume at the outset (Kalman, 1985; Kountourogiannis and Loya, 2003). Our discussion is abstracted from Blumenthal (1926).[4]

We draw on the core idea in our alternative route to the Mean-Value Theorem (Section 5.4), but we throw away all of the geometry. Let $\mathbf{I}$ be an open interval about the point a and let $x \neq a$ be in $\mathbf{I}$. Also, let F, G, H be functions, each defined on and having a derivative on $[a, x] \subset \mathbf{I}$. For any $t \in [a, x]$ we write, as

---

[4]The authors are indebted to Professor Troy Hicks (University of Missouri-Rolla) for drawing their attention to this paper.

before (the $\pm$ is now irrelevant),

$$K(t) = \begin{vmatrix} F(a) & G(a) & H(a) \\ F(t) & G(t) & H(t) \\ F(x) & G(x) & H(x) \end{vmatrix}.$$

Since F, G are automatically continuous on $[a, x]$, then there is a $c \in (a, x)$ such that[5]

$$K'(c) = \begin{vmatrix} F(a) & G(a) & H(a) \\ F'(c) & G'(c) & H'(c) \\ F(x) & G(x) & H(x) \end{vmatrix} = 0. \qquad (*)$$

The usefulness of equation $(*)$ hinges upon the substitutions that we might make for the entries in the determinant.

Now let $f: \mathbf{I} \to \mathbf{R}^1$ have a derivative on $[a, x]$. Theorem 5.7 tells us that $f(x) - f(a) - (x - a)f'(c) = 0$ for some $c \in (a, x)$. In general, we have $f(x) - f(a) - (x - a)f'(a) \neq 0$. We denote it as $R_1(x)$, the **remainder at x**, because the highest-order derivative that has been included is the first. We next define $F(t) = f(x) - f(t) - (x - t)f'(t)$ and assume that f'' exists on $[a, x]$. Then $F(x) = 0$, $F(a) = R_1(x)$, and $F'(t) = -(x - t)f''(t)$ for all $t \in [a, x]$. Making these substitutions in $(*)$, we obtain

$$K'(c) = \begin{vmatrix} R_1(x) & G(a) & H(a) \\ -(x - c)f''(c) & G'(c) & H'(c) \\ 0 & G(x) & H(x) \end{vmatrix} = 0$$

for some $c \in (a, x)$. The determinant is expanded and solved for $R_1(x)$:

$$R_1(x) = (x - c)f''(c)\frac{G(x)H(a) - G(a)H(x)}{G'(c)H(x) - G(x)H'(c)},$$

where G, H must be chosen so that the denominator is nonzero.

To this end, let $G(t) = (x - t)^2$, $H(t) = k \neq 0$, $t \in [a, x]$. Substitutions into the expression for $R_1(x)$ lead to

$$R_1(x) = \frac{1}{2}f''(c)(x - a)^2,$$

[5] In order to avoid any misunderstanding, we adopt the convention in this section and the next two that intervals $[a, b]$ or (a, b) can mean either $b > a$ or $b < a$, and that $c \in (a, b)$ (for example) can mean either $a < c < b$ or $a > c > b$.

and so we have

$$f(x) = f(a) + (x-a)f'(a) + \frac{1}{2}f''(c)(x-a)^2.$$

We have thereby extended the Mean-Value Theorem out to second-order.

The way to continue is now clear. We proceed to define $R_2(x) = f(x) - f(a) - (x-a)f'(a) - \frac{1}{2}(x-a)^2 f''(a)$, $F(t) = f(x) - f(t) - (x-t)f'(t) - \frac{1}{2}(x-t)^2 f''(t)$, $G(t) = (x-t)^3$, $H(t) = k \neq 0$, and require f''' to exist on $[a, x]$. These stipulations lead to an expression for $R_2(x)$. The final result, upon n-fold extension of the sequence of steps, is (Exercise 5.41) as follows.

Theorem 5.11 (Taylor's Theorem). *Let* **I** *be an open interval about the point* a *and let* $x \neq a$ *be in* **I**. *For* $f : \mathbf{I} \to \mathbf{R}^1$, *suppose that* $f^{(n+1)}$ *exists on* $[a, x]$. *Then there is a* $c \in (a, x)$, *dependent upon* a, x, *and* n, *such that*

$$f(x) = f(a) + f'(a)(x-a) + \frac{f''(a)}{2!}(x-a)^2 + \cdots + \frac{f^{(n)}(a)}{n!}(x-a)^n + R_n(x)$$

$$= T_n(x) + R_n(x),$$

where $T_n(x) = f(a) + \sum_{k=1}^{n} f^{(k)}(a)(x-a)^k/k!$ *is the* ***nth-order Taylor polynomial*** *for* f *at* a, *and the remainder is*

$$R_n(x) = \frac{f^{(n+1)}(c)}{(n+1)!}(x-a)^{n+1}.$$

■

The form of the remainder term in Theorem 5.11 is due to the Italian-born French mathematician **Joseph-Louis Lagrange** (1736–1813). Other forms for $R_n(x)$ are obtainable (for example, see Theorem 5.14). Theorem 5.11, but without the remainder term, was published in 1715 by the English mathematician **Brook Taylor** (1685–1731), although the theorem was known to others before him. If $a = 0$ in Taylor's Theorem, then the name MacLaurin often is attached to the result, even though this special case of Taylor's Theorem was presented in MacLaurin's own book some 27 years *after* Taylor.

■ Example 5.13

From Examples 5.2 and 5.4 we can establish that the derivative of $\sin\theta$ is $\cos\theta$ (Exercise 5.25(a)). Then from this and Taylor's Theorem, we obtain

$$\sin\theta = \sin a + \frac{\cos a}{1!}(\theta - a) - \frac{\sin a}{2!}(\theta - a)^2 - \frac{\cos a}{3!}(\theta - a)^3$$

$$+ \frac{\sin a}{4!}(\theta - a)^4 + \frac{\cos a}{5!}(\theta - a)^5 + \left[\frac{-\sin c}{6!}(\theta - a)^6\right],$$

for some $c \in (a, \theta)$. Choose, arbitrarily, $\theta = 2\pi/5$ and $a = \pi/2$.

Then

$$|R_5(2\pi/5)| = \left| \frac{-\sin c}{6!} \left(\frac{2\pi}{5} - \frac{\pi}{2} \right)^6 \right| = \frac{\pi^6 |\sin c|}{7.20 \times 10^8} < 1.335263 \times 10^{-6}.$$

Indeed, the absolute error in estimating $\sin(2\pi/5)$ by $T_5(2\pi/5)$ is calculated to be $\left| 0.951056516 \text{ (calculator)} - \{1 - \frac{1}{2}(-\pi/10)^2 + \frac{1}{24}(-\pi/10)^4\} \right| = 1.332995 \times 10^{-6}$, which is just less than 1.335263×10^{-6}. ■

We now make the important connection between Taylor's Theorem and the representation of a function f by a power series. Whereas $T_n(x)$ is always a polynomial, $f(x) - T_n(x) = R_n(x)$ is not generally a polynomial, so any $T_n(x)$ is an inadequate representation of f, and an infinite series is needed. Thus, as stressed by Cauchy, we have the following.

Corollary 5.11.1. *Suppose that f is defined on the open interval* **I** *and that f has derivatives of all orders at each $x \in [a, b] \subseteq$* **I**. *Then the* **Taylor series** $f(a) + \sum_{n=1}^{\infty} \frac{f^{(n)}(a)}{n!}(x-a)^n$ *represents f on $[a, b]$ iff $\lim_{n\to\infty} R_n(x) = 0$ at each $x \in [a, b]$.*

■ Example 5.14

In Example 5.13, let a be arbitrary; then $R_n(\theta)$ satisfies

$$|R_n(\theta)| = \left| \frac{f^{(n)}(c)}{(n+1)!}(\theta - a)^{n+1} \right| = \frac{\left| \begin{Bmatrix} \sin c \\ \cos c \end{Bmatrix} \right|}{(n+1)!} |\theta - a|^{n+1} < \frac{|\theta - a|^{n+1}}{(n+1)!},$$

for some $c \in (a, \theta)$. After some experimentation and a short inductive proof (Exercise 5.43), we can show that for any $n \in \mathbf{N}$

$$\left(\frac{n+1}{3} \right)^{n+1} < (n+1)!.$$

Hence, if $\theta \neq a$ is also arbitrary and if we set $K = |\theta - a|$, then

$$0 < \frac{|\theta - a|^{n+1}}{(n+1)!} = \frac{K^{n+1}}{(n+1)!} < \frac{K^{n+1}}{\left(\frac{n+1}{3}\right)^{n+1}} = \left(\frac{3K}{n+1} \right)^{n+1}.$$

Then for all $n > 6K - 1$ the quantity $3K/(n+1)$ is less than $1/2$, and $0 < |\theta - a|^{n+1}/(n+1)! < (1/2)^{n+1}$. It follows (Squeeze Theorem) that $\lim_{n\to\infty} |\theta - a|^{n+1}/(n+1)! = 0$ and $\lim_{n\to\infty} R_n(\theta) = 0$, so from Corollary 5.11.1 the Taylor series for $f(\theta) = \sin\theta$ at any a represents f for all θ. ■

It may be hard sometimes to work with the limit in Corollary 5.11.1 because of the uncertain location of the point c in the formula for $R_n(x)$. However, the representation of f by its Taylor series also follows from a slightly different set of hypotheses.

Corollary 5.11.2. *If f is as in Corollary 5.11.1 and if there is an $M > 0$ (dependent upon a or b) such that $|f^n(x)| \le M^n$ for all n and all $x \in [a, b]$, then f is represented on $[a, b]$ by the series*

$$f(a) + \sum_{n=1}^{\infty} \frac{f^{(n)}(a)}{n!}(x-a)^n.$$

Proof. The proof is left to you. ■

5.7 THE EXPONENTIAL AND LOGARITHMIC FUNCTIONS

At various places in the book we have assumed minimal familiarity with the exponential and logarithmic functions. We are now in a position to put them on a secure footing. In Example 2.9 it was stated that the sequence $\{(1+k^{-1})^k\}_{k=1}^{\infty}$ converges to the irrational number universally designated as e. What happens if k^{-1} is replaced by x/k, where $x \neq 0$ is fixed but arbitrary ($x = 0$ is uninteresting)?

For any $x \neq 0$, we can establish (Exercise 5.50) that

$$\left[1 + \frac{x}{k+1}\right]^{k+1} > \left[1 + \frac{x}{k}\right]^k$$

for all $k > |x|$. Let k_0 be the smallest natural number larger than $|x|$. Then for any $x \neq 0$ the sequence S_1 given by

$$\left\{\left(1 + \frac{x}{k}\right)^k\right\}_{k=k_0}^{\infty}$$

is an increasing sequence of positive numbers. Upon replacing x by $-x$ and taking reciprocals, we have that the corresponding, positive sequence S_2 given by

$$\left\{\left(1 - \frac{x}{k}\right)^{-k}\right\}_{k=k_0}^{\infty}$$

is a decreasing sequence. As S_2 is bounded below by 0, then by the Completeness Axiom (Theorem 2.2, specifically) it follows that S_2 converges.

Additionally, we have

$$\left(1-\frac{x}{k}\right)^{-k}-\left(1+\frac{x}{k}\right)^{k}=\left(1-\frac{x}{k}\right)^{-k}\left[1-\left(1-\frac{x^2}{k^2}\right)^{k}\right]$$
$$<\left(1-\frac{x}{k}\right)^{-k}\left(\frac{x^2}{k}\right) \qquad \text{if } |x|<k$$

from Bernoulli's Inequality (Theorem 1.7). Since the first factor on the right-hand side is positive and bounded and $\lim_{k\to\infty}(x^2/k)=0$, then we see that each term of $\mathbf{S}_2$ lies above the corresponding term of $\mathbf{S}_1$, and the difference between the two sequences approaches 0. Hence, $\mathbf{S}_1$ is bounded above by the terms of $\mathbf{S}_2$, so $\mathbf{S}_1$ also converges.

The following properties of $\exp(x)$, the common limit of $\mathbf{S}_1$ and $\mathbf{S}_2$, now emerge and will serve to give us a clearer picture of $\exp(x)$:

1. $\exp(x)$ is a *function* of x (because limits are unique).
2. Since no restrictions on x were necessary in the development, then the domain of $\exp(x)$ is $\mathbf{R}^1$.
3. In particular, $\exp(0)=1$ and $\exp(1)=e$.
4. $\exp(x)$ is an *increasing*, positive function; this follows from the definition of $\exp(x)$ and Theorem 2.8.
5. For $0<|h|<1$, the following inequalities hold for all $k\in\mathbf{N}$:

$$1+h<\left(1+\frac{h}{k}\right)^{k}<\exp(h)<\left(1-\frac{h}{k}\right)^{-k}<(1-h)^{-1}.$$

By the Squeeze Theorem (Exercise 2.22), we conclude that $\exp(h)$ is continuous at $h=0$.

6. For any $x,y\in\mathbf{R}^1$, we have for all $k>|x+y|$

$$\frac{\left(1+\frac{h}{k}\right)^{k}\left(1+\frac{y}{k}\right)^{k}}{(1+(x+y)/k)^{k}}=\left[1+\frac{xy/(k+x+y)}{k}\right]^{k}. \qquad (*)$$

Choose k so large that $|h|=|xy/(k+x+y)|<1$. Then from property (5) the bracketed expression in equation (*) is sandwiched between $1+h$ and $\exp(h)$. In view of the continuity of $\exp(h)$ at $h=0$, we conclude that the bracketed expression approaches 1 as $h\to 0$; that is, as $k\to\infty$.

Hence, we obtain, finally, the functional equation for $\exp(x)$:

$$\exp(x) \times \exp(y) = \exp(x + y). \qquad (**)$$

7. In particular, if $y = -x$, then from property (3) and equation (**) we obtain $\exp(-x) = 1/\exp(x)$.

It follows that in property (4) "an *increasing*" can be replaced by "a strictly *increasing*."

8. Properties (3), (6), and (7) are those to be expected of any exponential function a^x; from (3) we obtain $\exp(1) = a^1 = a = e$ and $\exp(x) = e^x$.

9. Use of Cauchy's definition of the derivative now yields

$$(e^x)' = \lim_{h\to 0} \frac{e^{x+h} - e^x}{h} = e^x \lim_{h\to 0} \frac{e^h - 1}{h},$$

from property (6), provided the limit exists. But from property (5) we deduce if $0 < h < 1$, then $\frac{(1+h)-1}{h} < \frac{e^h-1}{h} < \frac{(1-h)^{-1}-1}{h}$, or equivalently, $1 < \frac{e^h-1}{h} < \frac{1}{1-h}$. All inequalities are reversed if $0 > h > -1$. Since $\lim_{h\to 0} [1/(1-h)] = 1$, then by the Squeeze Theorem it follows that $\lim_{h\to 0} (e^h - 1)/h = 1$ and $(e^x)' = e^x$. Clearly, e^x has derivatives of all orders.

We can now investigate the representation of e^x by a Taylor series. Let $a = 0$ in Theorem 5.11; then properties (3) and (8) give

$$e^x = 1 + x + \frac{x^2}{2!} + \cdots + \frac{x^n}{n!} + \frac{e^c}{(n+1)!} x^{n+1},$$

for some $c \in (0, x)$. Let this interval be contained in $[b, d]$, $d > b$. From property (4), the function e^x assumes its maximum value on $[b, d]$ at d. As in the proof of Corollary 5.11.2, $\frac{|x|^{n+1}}{(n+1)!}$ is the $(n+1)$st term of a convergent infinite series (how do you know this?). Hence, we have

$$\lim_{n\to\infty} e^c \frac{|x|^{n+1}}{(n+1)!} \le \lim_{n\to\infty} e^d \frac{|x|^{n+1}}{(n+1)!} = e^d \lim_{n\to\infty} \frac{|x|^{n+1}}{(n+1)!} = e^d \times 0 = 0,$$

and by Corollary 5.11.1, the Taylor series $1 + \sum_{n=1}^{\infty} \frac{x^n}{n!}$ represents e^x on $[b, d]$. Since b, d were arbitrary, then e^x can be so represented everywhere on its domain $\mathbf{R}^1$.

■ Example 5.15

For the routine estimation of the value of e, the use of the Taylor series is better than use of the limit definition of exp(1). We have

$$e^{1/3} - \left[1 + \sum_{k=1}^{6} \frac{1}{3^k k!}\right] \begin{matrix} > \\ < \end{matrix} \begin{cases} 1/(3^7 7!) \\ e^{1/3}/(3^7 7!) \end{cases}$$

or

$$e^{1/3} \begin{matrix} > \\ < \end{matrix} \begin{cases} 1.395612330 + 9.1 \times 10^{-8} \\ \dfrac{1.395612330}{1 - (9.1 \times 10^{-8})}. \end{cases}$$

Finally, we obtain $2.718281805 < e < 2.718282066$, in which the first uncertain decimal place is the sixth. In contrast, $\left(1 + k^{-1}\right)^k$ has a value of about 2.71881 if $k = 20$, which is correct only to three decimal places. ■

Since e^x is an increasing function on $\mathbf{R}^1$, then the function is an injection and by Exercise 1.43(a) the inverse function, the **logarithmic function**, $\ln x$, exists. Its domain is the range of e^x (Figure 5.8).

Some of the properties of $\ln x$ are as follows:

1. $\ln x$ is a strictly increasing function on $(0, \infty)$.
2. $\ln 1 = 0$; $\ln e = 1$.
3. If $x, y > 0$, then the functional equation for the logarithmic function is $\ln(xy) = \ln x + \ln y$.

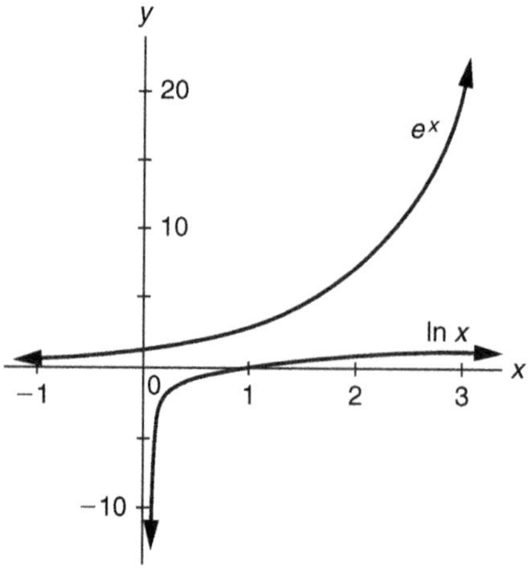

FIGURE 5.8

Partial graphs of the exponential function and the logarithmic function.

4. $\lim_{x\to\infty} \ln x = \infty$; $\lim_{x\to 0^+} \ln x = -\infty$.

5. Use of Cauchy's definition of the derivative gives

$$(\ln x') = \lim_{h\to 0} \frac{\ln(x+h) - \ln(x)}{h}$$

$$= \frac{1}{x}\lim_{y\to 0} \frac{\ln(1+y)}{y} \quad (h/x = y)$$

from property (3), provided the limit exists. If $y \in (-0.20, 0) \cup (0, 0.20)$, then from the Taylor series representation of e^y it follows that $1 + y < e^y$, and by property (1) and Exercise 1.43(d), we obtain $\ln(1+y) < y$.

Additionally, we can establish (Exercise 5.52(d)) that if $y \in (-0.20, 0) \cup (0, 0.20)$, then $e^{y-2y^2} < 1 + y$ holds. Consequently, $y - 2y^2 < \ln(1+y)$, and so we have (after a division by $y \neq 0$)

$$1 - 2y < \frac{\ln(1+y)}{y} < 1$$

when $y \in (0, 0.20)$, and the inequalities are reversed when $y \in (-0.20, 0)$. The Squeeze Theorem then gives $\lim_{y\to 0^+} \frac{\ln(1+y)}{y} = 1$, and we conclude that $(\ln x)' = 1/x$ for any $x \in (0, \infty)$. All the higher derivatives of $\ln x$ follow from Exercise 5.13.

We cannot represent $\ln x$ as a Taylor series about $a = 0$ because of property (4). Instead, we let $f(x) = \ln(1+x)$ and for this function we choose $a = 0$. By induction, we have for $k \in \mathbf{N}$

$$f^{(k)}(x) = \frac{(-1)^{k-1}(k-1)!}{(1+x)^k},$$

and then from Taylor's Theorem

$$\ln(1+x) = f(0) + \sum_{k=1}^{n} \frac{f^{(k)}(0)}{k!}x^k + \frac{f^{(n+1)}(c)}{(n+1)!}x^{n+1}$$

$$= 0 + \sum_{k=1}^{n} \frac{(-1)^{k-1}(k-1)!}{k!}x^k + \frac{(-1)^n n!/(1+c)^{n+1}}{(n+1)!}x^{n+1}$$

$$= \sum_{k=1}^{n} \frac{(-1)^{k-1}}{k}x^k + \frac{(-1)^n}{(n+1)(1+c)^{n+1}}x^{n+1},$$

for some $c \in (0, x)$.

The expansion is trivial for $x = 0$ because $\ln 1 = 0$. If $0 < x < 1$ holds, then $\lim_{n\to\infty} x^{n+1} = \lim_{n\to\infty} [1/(n+1)] = 0$ and $1/(1+c)^{n+1}$ is bounded from above by 1. Thus, $\lim_{n\to\infty} R_n(x) = 0$; the same conclusion also holds at $x = 1$ (why?). By Corollary 5.11.1, we conclude that the Taylor series for $\ln(1+x)$ represents this function for all $x \in [0, 1]$. We are not quite finished, however, with our analysis of $\ln(1+x)$; we shall complete it in the next section.

■ Example 5.16

Direct use of the Taylor series as a way to estimate ln 2 is not very good because that series for $\ln(1+x)$ converges very slowly for $x = 1$. It is known that $\ln 2 \approx 0.693147$; Taylor polynomials, $T_n(1)$, of various degrees give the results shown in Table 5.3.

Table 5.3 Estimations of ln 2

n	$T_n(1)$	| error |
3	0.833333	0.140186
6	0.616667	0.076480
9	0.745635	0.052488
12	0.653211	0.039936

On the other hand, arithmetic gives $\ln 2 = 2\ln(10/9) + 2\ln(6/5) + \ln(9/8)$. Approximating each individual logarithm on the right-hand side by corresponding Taylor polynomials $T_6(x)$, we obtain

$$\begin{aligned}\ln 2 &\approx 2T_6(1/9) + 2T_6(1/5) + T_6(1/8)\\ &= 2(0.1053605) + 2(0.1823199) + 0.1177830\\ &= 0.693144.\end{aligned}$$

The magnitude of our error is now 3×10^{-6}, which is much better. ■

5.8 RADIUS OF CONVERGENCE

In the consideration of the Taylor series representation of a function $f(x)$, there are always two general issues:

1. Determine for which set S of values of x the Taylor series about some point a converges.
2. Determine for which set $S' \subseteq S$ of values of x the remainder $R_n(x)$ approaches 0 as $n \to \infty$.

Both issues must be resolved because it is possible for a Taylor series for $f(x)$ to converge at some $x = x_0$, but not to the value $f(x_0)$ (Exercise 5.68). We need to examine convergence of power series a little more closely.

Theorem 5.12 (Cauchy-Hadamard[6] Theorem). *For the power series* $\sum_{n=0}^{\infty} c_n(x-a)^n$

(a) *convergence of the series at $x = x_0 \neq a$ implies absolute convergence of the series at all x such that $|x-a| < |x_0 - a|$;*

(b) *divergence of the series at $x = x'$ implies divergence of the series at all x such that $|x-a| > |x'-a|$.*

Proof.

(a) By Theorem 3.3 the sequence $\{c_n(x_n - a)^n\}_{n=0}^{\infty}$ converges to 0, and by Theorem 2.4 this sequence is bounded: $|c_n(x_0 - a)^n| < L$, for some $L > 0$ and for all $n \in \mathbf{N} \cup \{0\}$. Selecting any x such that $|x-a| < |x_0 - a|$ and defining $r = |x-a|/|x_0 - a| < 1$, we obtain

$$\left|c_n(x-a)^n\right| = \left|c_n(x_0-a)^n\right| \left|\frac{x-a}{x_0-a}\right|^n < Lr^n.$$

By the Comparison Test (Theorem 3.4), we then see that $\sum_{n=0}^{\infty} |c_n(x-a)^n|$ converges because $\sum_{n=0}^{\infty} Lr^n = L/(1-r)$.

(b) Suppose, to the contrary, that there is an x such that $|x-a| > |x'-a|$, and yet $\sum_{n=0}^{\infty} c_n(x-a)^n$ converges. But then, by part (a), $\sum_{n=0}^{\infty} |c_n(x'-a)^n|$ would have to be convergent, and by Theorem 3.8, so would $\sum_{n=0}^{\infty} c_n(x'-a)^n$, a contradiction of hypothesis. Hence, $\sum_{n=0}^{\infty} c_n(x-a)^n$ must diverge. ■

Part (a) of the theorem indicates that the power series converges for those x in the shaded region in Figure 5.9, which is an interval. The theorem says, in short, that convergence is restricted *only* to intervals. Any such interval may be just a point, or an interval of nonzero, finite length, or all of $\mathbf{R}^1$. We are led, therefore, to the following key definition:

---

[6] After the versatile French mathematician **Jacques S. Hadamard** (1865–1963), one of two individuals who first proved the Prime Number Theorem (in 1896). Theorem 5.12, and especially the formula in Theorem 5.13, were given by Hadamard in his doctoral dissertation of 1892 (Maz'ya and Shaposhnikova, 1998).

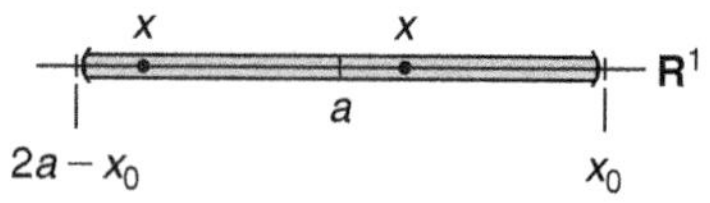

FIGURE 5.9
An interval of convergence of a power series.

Definition. *If R is the largest real number such that $\sum_{n=0}^{\infty} c_n(x-a)^n$ converges for all x that satisfy $-R < x - a < R$, then R is called the* **radius of convergence** *of the series and $(-R, R)$ is the* **interval of convergence**. *If the series converges only for $x = a$, we define $R = 0$, and if the series converges for all real x, we define $R = \infty$.*

The definition makes no statement about the possible convergence of $\sum_{n=0}^{\infty} c_n(x-a)^n$ at the endpoints when $R > 0$. Anything can happen there, so the endpoints must always be investigated separately (Exercise 5.57).

Theorem 5.13. *For the power series $\sum_{n=0}^{\infty} c_n(x-a)^n$, let $\rho = \limsup_{n\to\infty} |c_n|^{1/n}$. Then the radius of convergence is given by*

$$R = \begin{cases} 1/\rho & 0 < \rho < \infty \\ 0 & \rho = \infty \\ \infty & \rho = 0. \end{cases}$$

Proof. Since the general term of the power series is $c_n(x-a)^n$, we have

$$\limsup_{n\to\infty} |c_n(x-a)^n|^{1/n} = |x-a| \limsup_{n\to\infty} |c_n|^{1/n} = \rho|x-a|.$$

The completion of the proof is left to you. ■

■ Example 5.17

For the power series $1 + (x-1) + \frac{(x-1)^2}{4} + \frac{(x-1)^3}{9} + \cdots$, where $c_n = 1/n^2$, $n > 0$, we have

$$\rho = \limsup_{n\to\infty} |c_n|^{1/n} = \lim_{n\to\infty} |c_n|^{1/n} = \lim_{n\to\infty} n^{-2/n}.$$

Define the sequence $\{p_n\}_{n=1}^{\infty}$ by $p_n = [-2\ln n]/n$. We can establish (Exercise 5.58) that $\lim_{n\to\infty} p_n = 0$. Since e^x is continuous at $x = 0$, then it follows from Theorem 4.1 that the sequence $\{n^{-2/n}\}_{n=1}^{\infty} = \{e^{p_n}\}_{n=1}^{\infty}$ converges to $e^0 = 1$. Finally, from Theorem 5.13, we obtain $R = 1/\rho = 1$. ■

■ Example 5.18

In the Taylor series about $a = 0$ for $\ln(1+x)$, we have $c_n = (-1)^{n-1}/n, n > 0$. Then

$$\rho = \limsup_{n\to\infty} |c_n|^{1/n} = \lim_{n\to\infty} |c_n|^{1/n} = \lim_{n\to\infty} n^{-1/n} = 1,$$

by reasoning practically identical to that in the previous example. Thus, $R = 1$ and the interval of convergence of the Taylor series for $\ln(1+x)$ is $(-1, 1)$. It is, therefore, pointless to investigate the behavior of the remainder $R_n(x)$ for any $x > 1$. We already know that $\lim_{n\to\infty} R_n(1) = 0$. ■

We still need to deal with the remainder in the Taylor series expansion of $\ln(1+x)$ when $-1 < x < 0$ because the Lagrange form of the remainder is inadequate for this purpose. Another form of the remainder will do what we need.

Theorem 5.14 (Cauchy's Form of the Remainder).[7] *Under the same hypotheses as in Theorem 5.11, the expansion of f is nearly identical except that the remainder is given by*

$$R_n(x) = \frac{(x-c)^n}{n!}(x-a)f^{(n+1)}(c)$$

for some $c \in (a, x)$.

Proof. In (*) in Section 5.6, let $F(t) = f(x) - f(t) - \sum_{k=1}^{n} \frac{(x-t)^k}{k!} f^{(k)}(t)$, $G(t) = x - t$, and $H(t) = h \neq 0$; the completion of the proof is left to you. ■

For $f(x) = \ln(1+x)$, we have $f^{(n+1)}(x) = (-1)^n n!/(1+x)^{n+1}$, so when $-1 < x < c < 0$ holds, Cauchy's form of the remainder becomes

$$|R_n(x)| = \frac{|x||x-c|}{(1+c)^{n+1}} = \frac{|x||x|^n(1-cx^{-1})^n}{(1+c)(1+c)^n} < \frac{|x|^n}{1+x}\left(\frac{1-cx^{-1}}{1+c}\right)^n.$$

We can show, additionally, that

$$0 < \frac{1-cx^{-1}}{1+c} < 1,$$

so $|R_n(x)| < |x|^n/(1+x)$, and as $|x| < 1$, then $\lim_{n\to\infty} |R_n(x)| = \lim_{n\to\infty} |x|^n = 0$. We conclude, finally, that the Taylor series for $\ln(1+x)$ represents the function precisely for all $x \in (-1, 1]$.

---

[7] So-called because Cauchy presented this form in his 1826 book *Exercices de mathématiques*.

5.9 l'HÔPITAL'S RULE

Elucidation of the limit in property (9) for e^x (Section 5.7), of the limit in property (5) for $\ln x$, and of the limit in Example 5.17 (Section 5.8) would have been easier if we had employed the best-known theorem in all of calculus. The theorem is associated with the name of the French marquis **Guillaume F.A. de l'Hôpital** (1661–1704), but is now known to have originated with Johann Bernoulli (Truesdell, 1958). We split the theorem, known universally as **l'Hôpital's Rule**, into two parts and prove each part differently in order to provide additional applications of some topics already covered in the text.

Theorem 5.15 (l'Hôpital's Rule). *Suppose that f, g are continuous on $[a, b]$ and have derivatives everywhere on (a, b), that $g'(x) \neq 0$ on (a, b), that $f(b) = g(b) = 0$, and that $\lim_{x \to b^-} [f'(x)/g'(x)] = L \in \mathbf{R}^1$. Then we have $\lim_{x \to b^-} [f(x)/g(x)] = L$.*

Proof. By hypothesis, if $\varepsilon > 0$ is given, then there is a $\delta > 0$ such that $b - \delta < x < b$ implies $\left|\frac{f'(x)}{g'(x)} - L\right| < \varepsilon$. At the same time, for each such x the Generalized Mean-Value Theorem (Theorem 5.8) applies:

$$[f(b) - f(x)]\, g'(c_x) = [g(b) - g(x)]\, f'(c_x),$$

for some c_x that satisfies $x < c_x < b$. Setting $f(b) = g(b) = 0$, we obtain $\frac{f(x)}{g(x)} = \frac{f'(c_x)}{g'(c_x)}$. But $b - \delta < x < b$ implies $b - \delta < c_x < b$, so we also have $\left|\frac{f(x)}{g(x)} - L\right| < \varepsilon$. As ε is arbitrary, this says that $\lim_{x \to b^-} \frac{f(x)}{g(x)} = L$. ■

It is essential that f and g *both* approach 0 as $x \to b^-$; therefore, these must be checked ahead of time. Also, note that $f(x)/g(x)$ is defined throughout (a, b) (how do you know this?), so we are entitled to seek its limit as $x \to b^-$ (Exercise 5.63).

■ Example 5.19

Find $\lim_{x \to 0^-} \frac{\tan x - (x/2)}{\sin x}$.

The numerator and denominator each approach 0 as $x \to 0^-$. Designating $f(x) = \tan x - (x/2)$ and $g(x) = \sin x$, we have

$$\frac{f'(x)}{g'(x)} = \frac{\sec^2 x - (1/2)}{\cos x}.$$

The derivatives exist everywhere on $(-1, 0)$ and $\cos x \neq 0$ on this interval. Then $\lim_{x \to 0^-} f'(x) = 1/2$ and $\lim_{x \to 0^-} \cos x = 1$, and Theorem 5.15 gives

$$\lim_{x \to 0^-} \frac{f(x)}{g(x)} = \frac{1}{2}.$$

■

■ Example 5.20

From property (5) of $\ln x$ in Section 5.7, we obtain

$$\lim_{y\to 0^+} \frac{\ln(1+y)}{y} = \lim_{y\to 0^+} \frac{1/(1+y)}{1} = 1.$$

■

Corollary 5.15.1. *Suppose for some $a > 0$ that f and g are continuous on $[a,\infty)$ and have derivatives on (a,∞). Suppose also that $g(x) \neq 0$ and $g'(x) \neq 0$ on (a,∞), that $\lim_{x\to\infty} f(x) = \lim_{x\to\infty} g(x) = 0$, and that $\lim_{x\to\infty} [f'(x)/g'(x)] = L \in \mathbf{R}^1$. Then we have*

$$\lim_{x\to\infty} \frac{f(x)}{g(x)} = L.$$

Proof. Let $z = x^{-1}$; then $z \to 0^+$ as $x \to \infty$, and Theorem 5.15 (together with Theorem 5.3) can be applied. The completion of the proof is left to you. ■

In the second part of l'Hôpital's Rule we do not make any use of a mean-value theorem. However, we do call upon continuity explicitly, and some elementary manipulations with inequalities are needed (Boas, Jr., 1969; Hartig, 1991).

Theorem 5.16 (l'Hôpital's Rule). *Suppose that f, g have continuous derivatives on (a,b), that $g'(x) \neq 0$ on (a,b), that $\lim_{x\to b^-} f(x) = \lim_{x\to b^-} g(x) = \infty$, and that $\lim_{x\to b^-} [f'(x)/g'(x)] = L$. Then we have*

$$\lim_{x\to b^-} \frac{f(x)}{g(x)} = L.$$

Proof. We do the case where b and L are both finite; only minor changes in the proof are needed for other cases.

Given $\varepsilon > 0$, there exists a δ, $0 < \delta < (b-a)$, such that $x \in (b-\delta, b)$ implies

$$-\varepsilon/2 < \left[f'(x)/g'(x) - L\right] < \varepsilon/2.$$

Since g' is continuous and is never 0 on (a,b), it follows from the Intermediate-Value Theorem that g' is of fixed sign. Without loss of generality, we suppose that $g'(x) > 0$. These two inequalities are equivalent to

$$-\frac{1}{2}\varepsilon g'(x) < f'(x) - Lg'(x) < \frac{1}{2}\varepsilon g'(x). \qquad (*)$$

The right-hand inequality in equation (*) is equivalent to

$$f'(x) - [(\varepsilon/2) + L]\, g'(x) < 0$$

for all $x \in (b-\delta, b)$. But if a continuous function of negative value everywhere on a finite interval $[t_1, t_2]$, $t_1 < t_2$, is integrated, the result is a negative number. Hence, we obtain for $b - \delta \leq t_1 < t_2 < b$

$$\int_{t_1}^{t_2} \left\{f'(x) - [(\varepsilon/2) + L]\, g'(x)\right\} dx$$
$$= f(t_2) - f(t_1) - [(\varepsilon/2) + L]\,[g(t_2) - g(t_1)] < 0.$$

For t_2 sufficiently close to b we have $g(t_2) > 0$, since we assumed $g'(x) > 0$. The previous inequality, upon rearrangement and division by the positive $g(t_2)$, becomes

$$\frac{f(t_2)}{g(t_2)} - \left(\frac{\varepsilon}{2} + L\right) < \frac{f(t_1) - \left(\frac{\varepsilon}{2} + L\right) g(t_1)}{g(t_2)}.$$

Fix t_1; then for t_2 sufficiently close to b, the denominator on the right-hand side will be so large that the right-hand side is less than $\varepsilon/2$. Hence,

$$\frac{f(t_2)}{g(t_2)} - L < \frac{\varepsilon}{2} + \frac{\varepsilon}{2} = \varepsilon. \qquad (**)$$

By a parallel line of development from the left-hand inequality in equation (*), we can obtain (Exercise 5.66)

$$-\varepsilon < \frac{f(t_2)}{g(t_2)} - L$$

(verify!). This result, together with equation (**), implies that

$$\lim_{t_2 \to b^-} \frac{f(t_2)}{g(t_2)} = L.$$

■

■ Example 5.21

Find $\lim\limits_{x \to (\pi/2)^-} \frac{\sec x - 3}{\tan x + 2}$.

The numerator and denominator approach ∞ as $x \to (\pi/2)^-$. Designating $f(x) = \sec x - 3$ and $g(x) = \tan x + 2$, we see that $f'(x) = \sec x \tan x$ and $g'(x) = \sec^2 x$ are both continuous on $(0, \pi/2)$, that $\sec^2 x \neq 0$ on $(0, \pi/2)$, and that

$$\lim_{x \to (\pi/2)^-} \frac{f'(x)}{g'(x)} = \lim_{x \to (\pi/2)^-} \frac{\sec x \,\tan x}{\sec^2 x} = \lim_{x \to (\pi/2)^-} \frac{\tan x}{\sec x}$$
$$= \lim_{x \to (\pi/2)^-} \sin x$$

is equal to $L = 1$. By Theorem 5.16, we conclude that $\lim_{x\to(\pi/2)^-} [f(x)/g(x)] = 1$ also. ■

It is possible to formulate an analog of l'Hôpital's Rule that applies to pairs of numerical sequences rather than pairs of continuous functions. The proof is practically the same, except that arithmetic operations must be substituted for any differentiations and integrations (Boas, Jr., 1969; Huang, 1988). We state the result without proof.

Theorem 5.17 (Discrete l'Hôpital's Rule). *Suppose that $\{f_n\}_{n=1}^{\infty}, \{g_n\}_{n=1}^{\infty}$ are sequences in $\mathbf{R}^1$, that $\lim_{n\to\infty} f_n = \lim_{n\to\infty} g_n = 0$ or ∞, that $\Delta g_n = g_{n+1} - g_n$ does not change sign for all sufficiently large n, and that $\lim_{n\to\infty} \Delta f_n/\Delta g_n = L, L \in \mathbf{R}^1$ or* **Re**. *Then we have*

$$\lim_{n\to\infty} \frac{f_n}{g_n} = L.$$

■ Example 5.22

Find $\lim_{n\to\infty} \left[\frac{1^2+3^2+5^2+\cdots+(2n-1)^2}{2^2+4^2+6^2+\cdots+(2n)^2}\right]$.

The numerator and denominator approach ∞ as $n \to \infty$.

Designating $f_n = 1^2 + 3^2 + 5^2 + \cdots + (2n-1)^2$ and $g_n = 2^2 + 4^2 + 6^2 + \cdots + (2n)^2$, we find that $\Delta g_n = (2n+2)^2 > 0$ for all $n \geq 1$ and

$$\frac{\Delta f_n}{\Delta g_n} = \frac{(2n+1)^2}{(2n+2)^2} = \frac{1 + (1/n) + (1/4n^2)}{1 + (2/n) + (1/n^2)}.$$

Then $\lim_{n\to\infty} (\Delta f_n/\Delta g_n) = 1$, so by Theorem 5.17 $\lim_{n\to\infty} (f_n/g_n) = 1$ also. ■

5.10 DIFFERENTIABILITY OF FUNCTIONS $\mathbf{R}^n \to \mathbf{R}^1$

Suppose that $f: \mathrm{S} \to \mathbf{R}^1, \mathrm{S} \subseteq \mathbf{R}^2$, is a function. The **Cauchy quotient** is, formally,

$$\frac{f(\mathbf{p} \oplus \mathbf{h}) - f(\mathbf{p})}{\mathbf{h}},$$

where $\mathbf{p} = (x_1, x_2), \mathbf{h} = (h_1, h_2)$, and $\mathbf{p}, \mathbf{p} \oplus \mathbf{h} \in \mathrm{S}$. Division by a vector is not defined in $\mathbf{R}^2$, however, so Cauchy is of no help in trying to frame a definition of an acceptable definition of $f'(\mathbf{p})$. We could set up and possibly evaluate the following limit of an acceptable Cauchy quotient:

$$(D_1 f)(\mathbf{p}) = \lim_{h_1\to 0} \frac{f(x_1 + h_1, x_2) - f(\mathbf{p})}{h_1}.$$

If the limit exists (in $\mathbf{R}^1$), then we call $(D_1 f)(\mathbf{p})$ the **first-order partial derivative of f with respect to the first independent variable at the point p**. Similarly, if this limit exists:

$$(D_2 f)(\mathbf{p}) = \lim_{h_2 \to 0} \frac{f(x_1, x_2 + h_2) - f(\mathbf{p})}{h_2},$$

then we call $(D_2 f)$ $(\mathbf{p})$ the **first-order partial derivative of f with respect to the second independent variable at the point p**.

Neither of the two partial derivatives of f, however, can serve as "the" derivative of f at $\mathbf{p}$ for at least two reasons:

1. Each partial derivative measures the rate of change of f in a definite direction. There is no compelling reason for choosing any particular direction as the basis of a definition of the derivative.
2. It is not hard to concoct a function f in $\mathbf{R}^2$ for which both partial derivatives exist at some $\mathbf{p}$, but for which the function is discontinuous at $\mathbf{p}$ (Exercise 5.69). In the effort to generalize from functions in $\mathbf{R}^1$, it is desirable to preserve in higher dimensions the very basic Theorem 5.1.

A different approach from Cauchy's is needed; let us reexamine the one-dimensional case. Suppose that $f(x)$ has a derivative at $x = a$. The equation of the line tangent at this point is $y = f'(a)(x - a) + f(a)$ and, as Figure 5.10 suggests, in the close vicinity of $x = a$ the function f behaves nearly linearly *because* f has a derivative at $x = a$. If we define $e(h) = \frac{f(a+h)-f(a)}{h} - f'(a)$ and

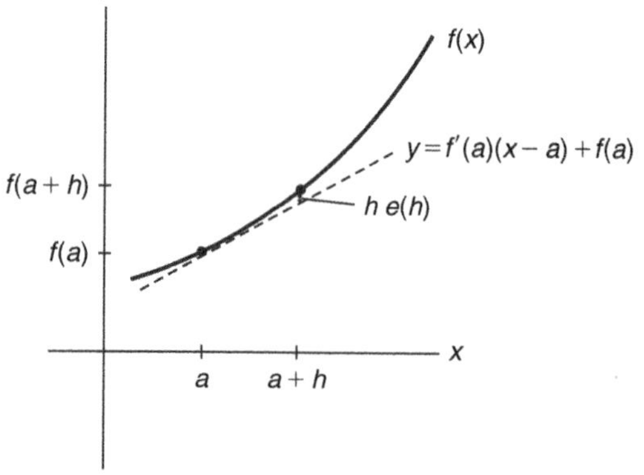

FIGURE 5.10

Linearization of the function f near $x = a$.

then set $e(h)h = E(h)|h|$, we obtain

$$\begin{aligned} \Delta f &= f(a+h) - f(a) \\ &= f'(a)h + E(h)|h| \\ &\approx f'(a)h, \quad (\lim_{h\to 0} E(h) = 0). \end{aligned}$$

The key observation here is that $f'(a)h$ is the value of a *linear* function. That is, we may think of the derivative of f at a, not as a mere number (which it is, of course) but rather as a linear mapping $\mathbf{L}$ that transforms any $h \in \mathbf{R}^1$ into $f'(a)\,h$:

$$\mathbf{L}(h) = f'(a)h.$$

Now let us move up to two dimensions. Let $f\colon \mathbf{S} \to \mathbf{R}^1, \mathbf{S} \subseteq \mathbf{R}^2$, be a function of two independent variables. Suppose that $\mathbf{p} = (x_1, x_2)$ is any interior point of $\mathbf{S}$ at which the two first-order partial derivatives of f exist and are continuous in a sufficiently small 2-ball about the point $\mathbf{p}$. Suppose also that the norm $||\mathbf{h}||$ of $\mathbf{h} = (h_1, h_2)$ is so small that $\mathbf{p} \oplus \mathbf{h} \in \mathbf{S}$. We have

$$\begin{aligned} \Delta f &= f(\mathbf{p} \oplus \mathbf{h}) - f(\mathbf{p}) \\ &= \left[f(x_1 + h_1, x_2 + h_2) - f(x_1, x_2 + h_2)\right] + \left[f(x_1, x_2 + h_2) - f(x_1, x_2)\right] \\ &= \left[(D_1 f)(x_1 + \theta_1 h_1, x_2 + h_2)\right] h_1 + \left[(D_2 f)(x_1, x_2 + \theta_2 h_2)\right] h_2, \end{aligned}$$

from the Mean-Value Theorem, and where $\theta_1, \theta_2 \in (0, 1)$ depend upon h_1, h_2.

Since $D_1 f, D_2 f$ are continuous at and near $\mathbf{p}$, we have

$$\begin{cases} (D_1 f)(x_1 + \theta_1 h_1, x_2 + h_2) = (D_1 f)(\mathbf{p}) + e_1(\mathbf{h}) \\ (D_2 f)(x_1, x_2 + \theta_2 h_2) = (D_2 f)(\mathbf{p}) + e_2(\mathbf{h}), \end{cases} \tag{*}$$

for h_1, h_2 near 0 and where $e_1(\mathbf{h}), e_2(\mathbf{h})$ are functions that tend to 0 as $\mathbf{h} \to \mathbf{0}$. In equations (*), let $e_1(\mathbf{h})h_1 = E_1(\mathbf{h})||\mathbf{h}||, e_2(\mathbf{h})h_2 = E_2(\mathbf{h})||\mathbf{h}||, E_1(\mathbf{h}) + E_2(\mathbf{h}) = E(\mathbf{h})$; then upon using row and column symbolism for vectors in $\mathbf{R}^2$,[8] we obtain for Δf

$$\begin{aligned} \Delta f &= f(\mathbf{p} \oplus \mathbf{h}) - f(\mathbf{p}) \\ &= \begin{pmatrix} (D_1 f)(\mathbf{p}) & (D_2 f)(\mathbf{p}) \end{pmatrix} \begin{pmatrix} h_1 \\ h_2 \end{pmatrix} + E(\mathbf{h})||\mathbf{h}||. \end{aligned}$$

We see, similar to the one-dimensional case, that the first term on the right-hand side is the value of a *linear mapping*, this time at the argument $\mathbf{h}$. This is

---

[8] Can you conjecture why we might do this? Resist peeking ahead to Exercise 5.75.

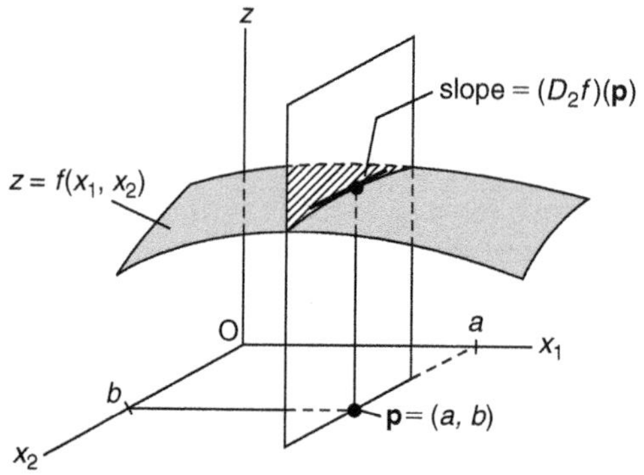

FIGURE 5.11

Geometric interpretation of $(D_2 f)(\mathbf{p})$ *as the slope of a line in the* $x_2 z$*-plane.*

true because partial derivatives are themselves linear mappings; thus, for any $\mathbf{h}, \mathbf{k} \in \mathbf{R}^2$, we have ($*$ denotes the inner product in $\mathbf{R}^2$)

$$\begin{aligned}\Big((D_1 f)(\mathbf{p}) \quad (D_2 f)(\mathbf{p})\Big) * (\mathbf{h} \oplus \mathbf{k}) &= \Big((D_1 f)(\mathbf{p}) \quad (D_2 f)(\mathbf{p})\Big) * \mathbf{h} \\ &\quad + \Big((D_1 f)(\mathbf{p}) \quad (D_2 f)(\mathbf{p})\Big) * \mathbf{k}.\end{aligned}$$

Still other aspects of the two-dimensional case are analogous to the one-dimensional case. For example, the partial derivatives have a geometric interpretation (Figure 5.11) similar to that of the one-dimensional derivative in Figure 5.10. Additionally, in the one-dimensional case, the symbol dx can be used in place of Δx or h for any increment of the independent variable x. The dx is called a **differential** of x. The **differential of a function** f at a point $a + h$ near a fixed point a is then defined to be

$$df = f'(a)\, dx = f'(a)h.$$

It is a function of the two independent, real variables a, h. The differential df is, as seen earlier, an approximation to $f(a + h) - f(a)$ when h is close to 0.

In the two-dimensional case the analogous definition of the differential of a function f at a point $\mathbf{p} \oplus \mathbf{h}$ is

$$df = \Big((D_1 f)(\mathbf{p}) \quad (D_2 f)(\mathbf{p})\Big) * \mathbf{h}.$$

It is a function of the four independent, real variables x_1, x_2, h_1, h_2. The df here is an approximation to $f(\mathbf{p} \oplus \mathbf{h}) - f(\mathbf{p})$ when $||\mathbf{h}||$ is close to zero (Figure 5.12). The vector $\Big((D_1 f)(p) \quad (D_2 f)(p)\Big)$ is known as the **gradient** of f at $\mathbf{p}$, and is commonly denoted by grad f, ∇f, or better yet, $\nabla f(\mathbf{p})$.

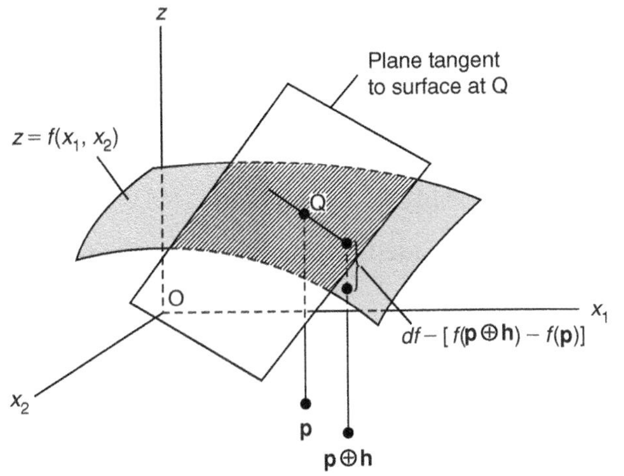

FIGURE 5.12
The difference $df - [f(\mathbf{p} \oplus \mathbf{h}) - f(\mathbf{p})]$ is small when $\mathbf{p} \oplus \mathbf{h}$ is close to $\mathbf{p}$, and is given by the vertical separation at $\mathbf{p} \oplus \mathbf{h}$ between the surface $z = f(x_1, x_2)$ and its tangent plane at Q.

■ Example 5.23

Estimate $f(x_1, x_2) = \sqrt[3]{1 + x_1}\,\sin(2x_2)$ at $(1, 1)$; assume values of $\pi \approx 3.141593$, $\sqrt{3} \approx 1.732051$.

Let $\mathbf{p} \oplus \mathbf{h} = (1, 1), \mathbf{h} = \left(2 - \left(\frac{53}{42}\right)^3, 1 - \frac{\pi}{3}\right)$. Then $\mathbf{p} = \left(\left(\frac{53}{42}\right)^3 - 1, \frac{\pi}{3}\right)$ and $f(\mathbf{p} \oplus \mathbf{h}) \approx f(\mathbf{p}) + \nabla f(\mathbf{p}) * \mathbf{h}$

$$= \frac{53}{42} \sin \frac{2\pi}{3} + \frac{\sin(2x_2)}{3\sqrt[3]{(1 + x_1)^2}}\bigg|_{\mathbf{p}} \left[2 - \left(\frac{53}{42}\right)^3\right]$$

$$+ 2\sqrt[3]{1 + x_1}\cos(2x_2)\Big|_{\mathbf{p}} \left[1 - \frac{\pi}{3}\right]$$

$$= \frac{53\sqrt{3}}{84} + \frac{\sqrt{3}/2}{3(53/42)^2}\left[2 - \left(\frac{53}{42}\right)^3\right] + \frac{106}{42}\left(\frac{-1}{2}\right)\left[1 - \frac{\pi}{3}\right]$$

$$\approx 1.150685.$$

The calculator value of $f(1, 1)$ is 1.145643; our error is 0.44%. ■

We are now ready to generalize and give meaning to differentiability and to the derivative of real-valued functions defined on $\mathbf{R}^n$, $n > 1$.

Definition. *A function $f\colon \mathbf{S} \to \mathbf{R}^1, \mathbf{S} \subseteq \mathbf{R}^n$, is said to be* **differentiable** *at $\mathbf{p} \in \mathbf{S}$ iff there exists a linear mapping $\mathbf{L}$ and a function $E(\mathbf{h}) : \mathbf{R}^n \to \mathbf{R}^1$, defined for $\mathbf{h} \in \mathbf{R}^n$*

of sufficiently small norm, such that

$$f(\mathbf{p} \oplus \mathbf{h}) = f(\mathbf{p}) + \mathbf{L} * \mathbf{h} + E(\mathbf{h})||\mathbf{h}||$$

and $\lim_{\mathbf{h}\to 0} E(\mathbf{h}) = 0$. *The linear mapping* $\mathbf{L}$ *is called the* **derivative** *of* f *at* $\mathbf{p}$. *A function* f *is differentiable on a set* $\mathbf{S}$ *iff it is differentiable at each* $\mathbf{p} \in \mathbf{S}$.

Some elementary deductions can be made from the preceding definition if we know that a certain function $f: \mathbf{S} \to \mathbf{R}^1, \mathbf{S} \subseteq \mathbf{R}^n$, is differentiable at a point $\mathbf{p} \in \mathbf{S}$. The following theorem is important.

Theorem 5.18. *If* $f: \mathbf{S} \to \mathbf{R}^1, \mathbf{S} \subseteq \mathbf{R}^n$, *is differentiable at* $\mathbf{p} \in \mathbf{S}$, *then*

(a) *the linear mapping* $\mathbf{L}$ *is unique for a given* $\mathbf{p} \in \mathbf{S}$;

(b) *the derivative of* f *at* $\mathbf{p}$ *is* $\nabla f(\mathbf{p})$;

(c) f *is continuous at* $\mathbf{p}$.

Proof. Work with the definitions of the partial derivatives and of differentiability at a point. The proof is left to you. ■

■ Example 5.24

Consider the function $f: \mathbf{R}^2 \to \mathbf{R}^1$, where $f: \mathbf{R}^2 \to \mathbf{R}^1$, where

$$f(\mathbf{p}) = \begin{cases} x_1x_2/(x_1^2 + x_2^2) & \mathbf{p} = (x_1, x_2) \neq 0 \\ 0 & \mathbf{p} = \mathbf{0}. \end{cases}$$

When $x_1 \neq 0$, then $f(x_1, 0) = 0$ and

$$\begin{aligned}(D_1f)(\mathbf{0}) &= \lim_{h_1\to 0} \frac{f(h_1, 0) - f(\mathbf{0})}{h_1} \\ &= \lim_{h_1\to 0} \frac{0}{h_1} \\ &= 0.\end{aligned}$$

Similarly, we have $(D_2f)(\mathbf{0}) = 0$, so both partial derivatives are defined at $\mathbf{0}$. However, along the line $x_2 = 2x_1$ we have $f(x_1, x_2) = 2x_1^2/5x_1^2$ and

$$\lim_{(x_1,x_2)\to(0,0)} f(x_1, x_2) = \lim_{x_1\to 0} \left[2x_1^2/5x_1^2\right] = 2/5 \neq f(\mathbf{0}).$$

Thus, f is not continuous at $\mathbf{p} = \mathbf{0}$, so it is not differentiable there either. ■

Although we have now the definition of differentiability for functions $f: \mathbf{R}^n \to \mathbf{R}^1$, it is better suited for theoretical than practical purposes. A criterion for determining differentiability would be useful. The Mean-Value Theorem proves helpful, once again. For simplicity, we prove the basic result for functions defined on $\mathbf{S} \subseteq \mathbf{R}^2$, but the result holds for $\mathbf{R}^n$ in general.

Theorem 5.19. *Let f be defined on some open set $\mathbf{S} \subseteq \mathbf{R}^2$ and assume that its two first-order partial derivatives exist everywhere on $\mathbf{S}$ and are continuous there. Then f is differentiable on $\mathbf{S}$.*

Proof. Let $\mathbf{p} = (x_1, x_2)$ be an arbitrary point in $\mathbf{S}$, and let $\mathbf{h} = (h_1, h_2) \neq (0, 0)$ be of small enough norm that $\mathbf{p} \oplus \mathbf{h}$ is also in $\mathbf{S}$. Then, as outlined previously, the Mean-Value Theorem leads to the existence of numbers $\theta_1, \theta_2 \in (0, 1)$ such that

$$\begin{cases} \left[(D_1 f)\left(x_1 + \theta_1 h_1, x_2 + h_2\right) - (D_1 f)(\mathbf{p})\right] = e_1(\mathbf{h}) \\ \left[(D_2 f)\left(x_1, x_2 + \theta_2 h_2\right) - (D_2 f)(\mathbf{p})\right] = e_2(\mathbf{h}). \end{cases}$$

Because $D_1 f, D_2 f$ are continuous on $\mathbf{S}$, then $\lim\limits_{\mathbf{h} \to \mathbf{0}} e_1(\mathbf{h}) = \lim\limits_{\mathbf{h} \to \mathbf{0}} e_2(\mathbf{h}) = 0$.

After defining $E(\mathbf{h})$ by

$$e_1(\mathbf{h}) h_1 + e_2(\mathbf{h}) h_2 = E(\mathbf{h}) ||\mathbf{h}||,$$

we obtain for $f(\mathbf{p} \oplus \mathbf{h})$

$$f(\mathbf{p} \oplus \mathbf{h}) = f(\mathbf{p}) + [\nabla f(\mathbf{p})] * \mathbf{h} + E(\mathbf{h}) ||\mathbf{h}||.$$

All that remains for the definition of differentiability of f at $\mathbf{p}$ is to show that $\lim_{\mathbf{h} \to \mathbf{0}} E(\mathbf{h}) = 0$. But for any nonzero $\mathbf{h}$ we automatically have $-1 \leq h_1 / ||\mathbf{h}|| \leq 1$, and similarly for h_2. Thus,

$$-|e_1(\mathbf{h})| \leq h_1 e_1(\mathbf{h}) / ||\mathbf{h}|| \leq |e_1(\mathbf{h})|,$$

and the Squeeze Theorem then gives

$$\lim_{\mathbf{h} \to \mathbf{0}} h_1 e_1(\mathbf{h}) / ||\mathbf{h}|| = 0,$$

and similarly for $h_2 e_2(\mathbf{h}) / ||\mathbf{h}||$. It follows that

$$\lim_{\mathbf{h} \to \mathbf{0}} E(\mathbf{h}) = \lim_{\mathbf{h} \to \mathbf{0}} \frac{h_1 e_1(\mathbf{h}) + h_2 e_2(\mathbf{h})}{||\mathbf{h}||} = 0,$$

so f is differentiable at $\mathbf{p}$. Since $\mathbf{p}$ was arbitrary, then f is differentiable on $\mathbf{S}$. ■

The theorem is a *sufficiency* condition for differentiability. There are functions $f : \mathbf{R}^2 \to \mathbf{R}^1$ that are differentiable at some point $\mathbf{p}$, but $D_1 f$ and/or $D_2 f$ fail to be continuous at that point. Of course, D_1f and D_2f certainly have to *exist* at $\mathbf{p}$. In the great *majority* of cases, if you are able to find formulas for D_1f and D_2f that are valid on some open set $\mathbf{S} \subseteq \mathbf{R}^2$, then f will be differentiable there.

■ Example 5.25

In Example 5.24 the function f is differentiable on any bounded, open set $\mathbf{S} \subseteq \mathbf{R}^2$ that excludes the origin. To see this, we first compute for any point $\mathbf{p} = (x_1, x_2)$ in $\mathbf{S}$

$$(D_1f)(\mathbf{p}) = \frac{x_2^3 - x_1^2 x_2}{\left(x_1^2 + x_2^2\right)^2}.$$

Each term in the numerator is continuous anywhere in $\mathbf{S}$, as are the terms in the denominator; these statements follow from Theorem 4.3. Further, the denominator is never 0 on $\mathbf{S}$, so by Theorem 4.3 again the quotient is continuous anywhere in $\mathbf{S}$. A similar conclusion holds for $(D_2f)(\mathbf{p})$. Theorem 5.19 now applies. ■

It can be surmised from the definition of differentiability in this section that the whole theory of differentiation could be developed along the lines begun in Section 5.2. A step toward this is the extension of Carathéodory's vision of the derivative to functions of several independent variables (Botsko and Gosser, 1985).

Theorem 5.20. *If* $\mathbf{T} \subseteq \mathbf{R}^n$ *is open, then the function* $f : \mathbf{T} \to \mathbf{R}^1$ *is differentiable at* $\mathbf{p} \in \mathbf{T}$ *iff there is a function* $g : \mathbf{T} \to \mathbf{R}^n$ *that is continuous at* $\mathbf{p}$ *and satisfies*

$$f(\mathbf{y}) = f(\mathbf{p}) + g(\mathbf{y}) * (\mathbf{y} - \mathbf{p})$$

for all $\mathbf{y} \in \mathbf{T}$.

Proof. $(\rightarrow)$ Suppose that f is differentiable at $\mathbf{p}$, so that

$$f(\mathbf{y}) = f(\mathbf{p}) + \mathbf{L} * (\mathbf{y} - \mathbf{p}) + E(\mathbf{y} - \mathbf{p})\|\mathbf{y} - \mathbf{p}\|$$

holds, where $\mathbf{L}$ is a linear operator and the function $E : \mathbf{R}^n \to \mathbf{R}^1$ satisfies $\lim_{\mathbf{y}\to\mathbf{p}} E(\mathbf{y} - \mathbf{p}) = 0$. From Theorem 5.18(b) we know that $\mathbf{L} = \left((D_1f)(\mathbf{p}) \quad (D_2f)(\mathbf{p}) \cdots (D_nf)(\mathbf{p})\right)$. Then

$$f(\mathbf{y}) = f(\mathbf{p}) + \sum_{i=1}^{n} \left[(D_if)(\mathbf{p})\right](y_i - x_i) + \frac{E(\mathbf{y} - \mathbf{p})}{\|\mathbf{y} - \mathbf{p}\|} \sum_{i=1}^{n} (y_i - x_i)^2, \qquad (*)$$

where x_i, y_i are the ith components of $\mathbf{p}$, $\mathbf{y}$, respectively. Now define for each $i = 1, 2, 3, \ldots, n$

$$g_i(\mathbf{y}) = \begin{cases} (D_if)(\mathbf{p}) + E(\mathbf{y} - \mathbf{p})\frac{y_i - x_i}{\|\mathbf{y} - \mathbf{p}\|} & \mathbf{y} \neq \mathbf{p} \\ (D_if)(\mathbf{p}) & \mathbf{y} = \mathbf{p}. \end{cases}$$

Equation (*) then simplifies when $\mathbf{y} \neq \mathbf{p}$ to

$$f(\mathbf{y}) = f(\mathbf{p}) + g(\mathbf{y})^*(\mathbf{y} - \mathbf{x}) \qquad (**)$$

where $g(\mathbf{y}) = \left(g_1(\mathbf{y}) \quad g_2(\mathbf{y}) \cdots g_n(\mathbf{y})\right)$. The function g is continuous at $\mathbf{p}$ since each component is continuous there. Equality in equation (**) then clearly holds when $\mathbf{y} = \mathbf{p}$, so equation (**) is true for all $\mathbf{y} \in \mathbf{T}$.

($\leftarrow$) Suppose, conversely, that there is a function $g : \mathbf{T} \to \mathbf{R}^n$ that is continuous at $\mathbf{p}$ and satisfies $f(\mathbf{y}) = f(\mathbf{p}) + g(\mathbf{y}) * (\mathbf{y} - \mathbf{p})$. For $i = 1, 2, \ldots, n$, define $e_i(\mathbf{y} - \mathbf{p})$: $\mathbf{T} \to \mathbf{R}^1$ by

$$e_i(\mathbf{y} - \mathbf{p}) = g_i(\mathbf{y}) - g_i(\mathbf{p}),$$

and then define $e(\mathbf{y}) : \mathbf{T} \to \mathbf{R}^n$ by

$$e(\mathbf{y}) = \left(e_1(\mathbf{y} - \mathbf{p}) \quad e_2(\mathbf{y} - \mathbf{p}) \cdots e_n(\mathbf{y} - \mathbf{p})\right).$$

It follows that

$$f(\mathbf{y}) = f(\mathbf{p}) + g(\mathbf{y}) * (\mathbf{y} - \mathbf{p}) = f(\mathbf{p}) + [g(\mathbf{p}) \oplus e(\mathbf{y})] * (\mathbf{y} - \mathbf{p}).$$

For any $\mathbf{p} \in \mathbf{T}, g(\mathbf{p})$ is just a $(1 \times n)$-dimensional array of constants and is, thus, a linear mapping $\mathbf{T} \to \mathbf{R}^n$. Set $\mathbf{L} = g(\mathbf{p})$ and define the function E by

$$E(\mathbf{y} - \mathbf{p}) = \begin{cases} \dfrac{e(\mathbf{y}) * (\mathbf{y} - \mathbf{p})}{\|\mathbf{y} - \mathbf{p}\|} & \mathbf{y} \neq \mathbf{p} \\ 0 & \mathbf{y} = \mathbf{p}. \end{cases}$$

For all $\mathbf{y} \in \mathbf{T}$ we then have

$$\begin{aligned} f(\mathbf{y}) &= f(\mathbf{p}) + \mathbf{L} * (\mathbf{y} - \mathbf{p}) + e(\mathbf{y}) * (\mathbf{y} - \mathbf{p}) \\ &= f(\mathbf{p}) + \mathbf{L} * (\mathbf{y} - \mathbf{p}) + E(\mathbf{y} - \mathbf{p})\|\mathbf{y} - \mathbf{p}\|. \end{aligned}$$

Finally, for each $i = 1, 2, \ldots, n$, $\lim_{\mathbf{y} \to \mathbf{p}} e_i(\mathbf{y} - \mathbf{p}) = \lim_{\mathbf{y} \to \mathbf{p}} [g_i(\mathbf{y}) - g_i(\mathbf{p})] = \mathbf{0}$ because g is continuous at $\mathbf{p}$. Thus, $\lim_{\mathbf{y} \to \mathbf{p}} e(\mathbf{y}) = \mathbf{0}$ and $\lim_{\mathbf{y} \to \mathbf{p}} E(\mathbf{y} - \mathbf{p}) = 0 = E(\mathbf{0})$; hence, from the definition, we have shown the differentiability of f at $\mathbf{p}$. ■

Incidentally, the first half of the theorem shows that if f is differentiable at $\mathbf{p}$, so that

$$f(\mathbf{y}) = f(\mathbf{p}) + g(\mathbf{y}) * (\mathbf{y} - \mathbf{p})$$

holds, then $g(\mathbf{p})$ is the derivative of f at $\mathbf{p}$; that is, $g(\mathbf{p}) = (\nabla f)(\mathbf{p})$. The form of this equation corresponds (in $\mathbf{R}^n$) to the definition in Section 5.2 and to Theorem 5.4. Continuing the development, we could use Theorem 5.20 to obtain chain rules for differentiation that would be extensions of Theorem 5.3.

EXERCISES

Section 5.1

5.1. **(a)** Use Cauchy's definition expressed in limit language to show that $f\colon \mathbf{R}^1 \to \mathbf{R}^1, f(x) = x^{1/3}$, has a derivative at any $x \neq 0$.

(b) Reexpress Cauchy's definition in δ, ε-language and use it to show that $f\colon [-1, \infty) \to \mathbf{R}^1, f(x) = \sqrt{1+x}$, has a derivative at $x = a = 3$.

5.2. Produce a function that illustrates the falsity of the converse of Theorem 5.1. What are the "sided" derivatives of f at a, if they exist?

5.3. Determine for which values of x the following functions have a derivative: (a) $f\colon \mathbf{R}^1 \to \mathbf{R}^1, f(x) = x|x|$, (b) $g\colon \mathbf{R}^1 \to \mathbf{R}^1, g(x) = |\cos(x)|$, (c) $h\colon [-1, 2] \to \mathbf{R}^1, h(x) = |x| - |x-1|$.

5.4. The **Dirichlet function** $f\colon \mathbf{R}^1 \to \mathbf{R}^1$ is $f(x) = \begin{cases} 1 & x \in \mathbf{Q} \\ 0 & x \notin \mathbf{Q}. \end{cases}$ Assume that you are not in possession of Example 4.6. Prove that f does not have a derivative anywhere.

5.5. Prove parts (a) and (c) of Theorem 5.2.

5.6. Assume that you are not yet in possession of the Chain Rule, but that you do have Theorem 5.2 and Example 5.4. How could you work out the derivative of $f\colon (-\pi/4, \pi/4) \to \mathbf{R}^1, f(\theta) = \sqrt{\cos(2\theta)}$?

5.7. Let $\mathbf{S}$ be a set of elements on which are defined two binary operations, $\oplus$ and $\otimes$, called (generically) **addition** and **multiplication**; that is, if $a, b \in \mathbf{S}$, then $a \oplus b \in \mathbf{S}$ and $a \otimes b \in \mathbf{S}$. Then the algebraic structure $< \mathbf{S}, \oplus, \otimes >$ is called a **ring** if the following hold: (1) $< \mathbf{S}, \oplus >$ is a group (refer to Exercise 4.19); (2) $\otimes$ obeys the associative law: if $a, b, c \in \mathbf{S}$, then $a \otimes (b \otimes c) = (a \otimes b) \otimes c$; (3) the two distributive laws are obeyed: if $a, b, c \in \mathbf{S}$, then $a \otimes (b \oplus c) = (a \otimes b) \oplus (a \otimes c)$ and $(a \oplus b) \otimes c = (a \otimes c) \oplus (b \otimes c)$. Let $\mathbf{S}$ be the set of all real-valued functions defined on $0 \leq x \leq 2$ that have a derivative at $x = 1$, and let $\oplus, \otimes$ be ordinary, pointwise addition, multiplication. Prove that $< \mathbf{S}, \oplus, \otimes >$ is a ring.

Section 5.2

5.8. Prove Theorem 5.4. Apply the theorem to the evaluation of the derivative of $f\colon [0, \infty) \to \mathbf{R}^1, f(x) = \sqrt{x+1}$, at $a = 2$.

5.9. A student argues that "Carathéodory's definition is not really so useful because it does not give an explicit formula for the derivative of a function $f(x)$." How would you respond?

5.10. Suppose that in Example 5.5 the function $f\colon \mathbf{R}^1 \to \mathbf{R}^1$ had been $f(x) = \cos x$. What, then, would be the function $\phi_1(x)$?

5.11. Use Carathéodory's definition to reprove Theorem 5.2(a).

5.12. **INVERSE FUNCTION THEOREM**

> Suppose that f is continuous and strictly monotonic on an open interval **I** that contains a and that $f'(a) \neq 0$. Then $g = f^{-1}$ has a derivative at $b = f(a)$ that is given by $g'(b) = 1/[f'(a)]$.

Begin with Carathéodory's statement that f has a derivative at $x = a$; let **U** be the open interval that is the domain of f^{-1}. Now rewrite Carathéodory's statement in terms of the variable $y = f(x)$, and make use of Theorems 4.3 and 4.4 and Exercise 4.52(d).

(a) Complete the proof of the Inverse Function Theorem.

(b) Apply the Theorem to $y = f(x) = x^3 + 1$ and compute the derivative of f^{-1} at $y = b = 2$.

5.13. **(a)** Suppose that $n \in \mathbf{N}$ is even and that $f : [0, \infty) \to \mathbf{R}^1$ is defined by $f(x) = x^n$. Use the Inverse Function Theorem (IFT) to show that the derivative of $g = f^{-1}$ is $g'(y) = n^{-1}y^{n^{-1}-1}, y > 0$.

(b) In this case suppose that $n \in \mathbf{N}, n > 1$ is odd and that $f : \mathbf{R}^1 \to \mathbf{R}^1$ is again $f(x) = x^n$. Apply the IFT again to obtain $g'(y)$.

5.14. Let $\alpha = m/n > 0$, where $m, n \in \mathbf{N}$, and let $F : [0, \infty) \to \mathbf{R}^1$ be defined as $F(x) = x^\alpha$. Think of F as $f[g]$, where $f(x) = x^m, g(x) = x^{1/n}$. Show that for $x > 0$ we have $F'(x) = \alpha x^{\alpha - 1}$. What is $F'(0)$ if $\alpha > 1$? This Exercise thus extends Example 5.1 to exponents $c \in \mathbf{Q}$.

5.15. Consider the function $\text{Tan}^{-1}x, 0 \leq \text{Tan}^{-1}x < \pi/2$, the inverse of the restriction to quadrant 1 of the tangent function. Use the IFT to derive the formula for the derivative of $\text{Tan}^{-1}x$. Is there a constant b such that the ellipse $x^2 + (y^2/b^2) = 1$ and $\text{Tan}^{-1}x$ are perpendicular in quadrant 1? If so, estimate it.

5.16. Confirm the last line of Example 5.7.

Section 5.3

5.17. Write out the proof of the analog of Theorem 5.5 for functions with a relative maximum at c. Produce a nonconstant function f and a point $c \in \mathbf{I}$ such that $f'(c) = 0$ but c is not the location of a relative extremum.

5.18. A function f is defined by $f(x) = \begin{cases} -x^2 & x > 1 \\ -2x^3 + x & x \leq 1. \end{cases}$ Students A and B proceed as follows: A differentiates $-x^2$, sets the derivative equal to 0, and concludes from Theorem 5.5 that f has a relative extremum at $x = 0$. B differentiates $-2x^3 + x$, sets $-6x^2 + 1 = 0$, and concludes from Theorem 5.5 that $\pm\sqrt{6}/6$ are locations of relative extrema. How would you grade these two student papers? Has there been a violation of Theorem 5.5? Discuss.

5.19. Show by examples that Rolle's Theorem can fail if

(a) f is not continuous everywhere on $[a, b]$.

(b) f is continuous on $[a, b]$ but f' does not exist everywhere on (a, b).

(c) Will Theorem 5.6 hold if, instead of f being continuous on $[a, b]$, we have that

$$\lim_{x \to a^+} f(x) = \lim_{x \to b^-} f(x) = k \in \mathbf{R}^1?$$

5.20. The function $f(x) = \begin{cases} -x^2 + x + 2 & x \le 1 \\ \frac{1}{2}x - 1 & x > 1 \end{cases}$ satisfies $f(-1) = f(2) = 0$, but the function clearly does not have a derivative everywhere on $(-1, 2)$. May we conclude from Rolle's Theorem that there is no $c \in (-1, 2)$ at which $f'(c) = 0$? Discuss.

5.21. How many real roots does $x^5 + x^3 - 2x - 1 = 0$ have? Do this analytically; do not use the computer.

5.22. Refer to Figure 5.13. Let f be continuous on $[a, b]$ and have a derivative everywhere on (a, b). Also let g be the function whose values are the differences between the y-coordinate of a point on the secant AB and the corresponding point below (or above) on the curve.

(a) Write a formula for $g(x)$.

(b) Apply Rolle's Theorem to g.

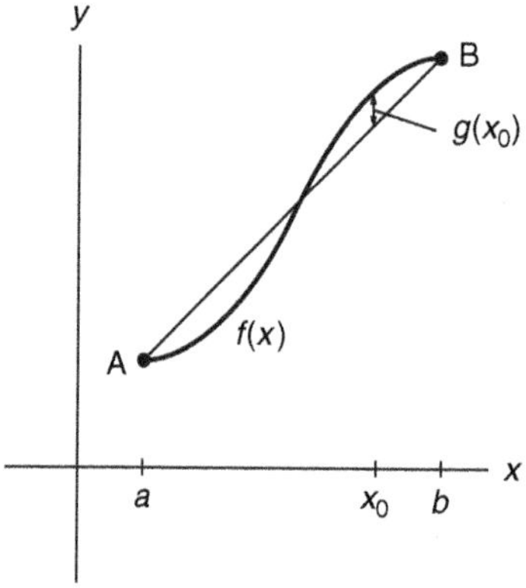

FIGURE 5.13

Diagram for Exercise 5.22.

Section 5.4

5.23. **(a)** The equation $f(x) = f(a) + f'(c)(x - a), x < c < a$, is satisfied if $f(x) = \sqrt[3]{x}$ and we take $a > 0, x = -a, c = a/(3\sqrt{3})$. But f certainly does not have a derivative everywhere in $(-a, a)$. Do we have a violation of Theorem 5.7? Discuss.

(b) The equation $f(x) = f(a) + f'(c)(x - a)$ for some $c \in (x, a)$ is not true if $f(x) = x^{-1}$ and $a > 0 > x$. This follows because $f(x) - f(a) < 0$ but $f'(c) = -c^2 < 0$ and $x - a < 0$, so $f'(c)(x - a) > 0$, a contradiction. Is this a violation of Theorem 5.7? Discuss.

5.24. **(a)** Prove that if $0 < \theta < \pi/4$ holds, then $\tan\theta < 4\theta/\pi$. (Missouri MAA Examination, 2003)

(b) If $a < b < c, f'(x)$ is strictly increasing on (a, c), and $f(x)$ is continuous on $[a, c]$, then prove that

$$(b - a)f(c) + (c - b)f(a) > (c - a)f(b).$$

(Missouri MAA Examination, 2005)

5.25. **(a)** Show how to deduce the derivative of $\sin\theta$ for $-\pi/2 < \theta < \pi/2$ from knowledge of Examples 5.2 and 5.4.

(b) If $0 < \theta < \pi/2$ holds, show that $\sin\theta < \theta$.

(c) Deduce an analogous inequality if $-\pi/2 < \theta < 0$.

(d) Use parts (b) and (c) to establish $\lim_{\theta\to 0} \frac{\sin\theta}{\theta} = 1$.

(e) Show that an argument with the Mean-Value Theorem yields $\cos\,\theta > 1 - \theta^2$, but that integration plus a geometric argument sharpens this to $\cos\,\theta > 1 - (\theta^2/2)$.

5.26. Show, as requested, in connection with Figure 5.5, that the areas K of the parallelograms there are given by the indicated determinant. Also show that if $f(t), g(t)$ have derivatives at $t = c$, then $K'(c) = 0$ implies

$$\begin{vmatrix} f(a) & g(a) & 1 \\ f'(c) & g'(c) & 0 \\ f(b) & g(b) & 1 \end{vmatrix} = 0.$$

5.27. Prove Corollary 5.7.1.

5.28. We borrow from Section 5.7 (or Table 5.1) the fact that the derivative of the logarithmic function $\ln x, x > 0$, is x^{-1}; also, we need $\ln 1 = 0$.

(a) Show that $0 < a < b$ implies $\frac{b-a}{b} < \ln\left(\frac{b}{a}\right) < \frac{b-a}{a}$.

(b) Obtain, as needed for Corollary 5.7.2, that $\ln x \le x - 1$ for any $x > 0$, with equality only when $x = 1$.

5.29. Consider $f : [-1/2, \infty) \to \mathbf{R}^1, f(x) = \sqrt{1+2x}$. In Theorem 5.7, let $a = 3/2$ and let b be variable. Complete the following table by computing c that corresponds to each b. Make a conjecture on the behavior of c.

b	$f'(c)$	c
2		
7/4		
25/16		
49/32		

5.30. A slightly weakened form of the Mean-Value Theorem can be proved without the use of any sophisticated theorems on the real number system (Halperin, 1954). Read this short article and write up the proof in your own words. For complementary reading, look at Boas, Jr. (1981).

5.31. The **Hölder Inequality**, in a common formulation, states that if $\{a_1, a_2, \ldots, a_n\}$, $\{b_1, b_2, \ldots, b_n\}$ are two sets of nonnegative numbers, then

$$\left[\sum_{i=1}^{n} (a_i)^p\right]^{1/p} \left[\sum_{i=1}^{n} (b_i)^q\right]^{1/q} \ge \sum_{i=1}^{n} a_i b_i,$$

where $p, q > 0$ satisfy $p^{-1} + q^{-1} = 1$. Prove this from Corollary 5.7.2.

5.32. Let $\mathbf{S} \cong [1,\ \infty) \cap \mathbf{Q}$; prove that $3(6r-1)^r > 2(6r+1)^r$ then holds for any $r \in \mathbf{S}$.

Section 5.5

5.33. Formulate and prove the analog of Theorem 5.9 for relative maxima. Then use a form of the First Derivative Test to identify relative extrema, if any, of these functions on the indicated intervals:

(a) $f(x) = x\cos x - \sin x, \quad -4 \le x \le 4;$

(b) $g(x) = x^3 + (3/x), \quad (-2, -1/2) \cup (1/2, 2);$

(c) $h(x) = x|x^2 - 27|, \quad -\sqrt{27} \le x < \sqrt{27}.$

5.34. An alternative proof of the First Derivative Test to the one presented in Theorem 5.9 can be given. Use proof by contradiction. Assume in $[a, b]$ that there is a point $x_1 \neq c$ such that $f(x_1) < f(c)$. Now apply the Intermediate-Value Theorem and reach a contradiction. Write out the proof in your own words.

5.35. This exercise and the next give some practice with higher-order derivatives. In order to more efficiently handle some kinds of differentiation problems, **Leibniz's Rule** is useful. This theorem, stated in a letter of 1695 to Johann Bernoulli, states that the nth derivative of the product of two functions f, g that have nth-order derivatives is given by the "binomial-like" expansion

$$(fg)^{(n)} = \sum_{j=0}^{n} \binom{n}{j} f^{(n-j)} g^{(j)},$$

where $g^{(j)}$ means the jth derivative of g if $j > 0$ and $g^{(0)}$ means $g(x)$. Prove Leibniz's Rule.

5.36. An application of Leibniz's Rule is to the sequence of functions $\{L_n(x)\}_{n=0}^{\infty}$, defined by

$$L_n(x) = e^x f^{(n)}(x), \quad f(x) = x^n e^{-x},$$

and known as the **Laguerre polynomials**.[9]

(a) Work out $L_6(x)$ (don't use brute force!).

(b) For arbitrary n, determine rigorously $L_n(0)$.

(c) The Laguerre polynomials satisfy **Laguerre's differential equation**:

$$xy'' + (1 - x)y' + ny = 0.$$

Determine rigorously, for arbitrary n, the value of $L_n'(0)$.

5.37. **(a)** Complete the proof of Theorem 5.10.

(b) Formulate and prove the analog of Theorem 5.10 for relative maxima. Then use a form of the Second Derivative Test to identify relative extrema, if any, of these functions on the indicated intervals. If the Test is inconclusive anywhere, state this fact.

(i) $f(x) = 3x^4 - 4x^3 - 12x^2 + 5, \quad -3 < x < 3;$

---

[9] After the French mathematician **Edmond Laguerre** (1834–1886). The polynomials are important in the quantum mechanics of the hydrogen atom. **Schrödinger** solved the hydrogen atom in the first of his four marvelous papers of 1926.

(ii) $g(x) = x^{2/3}(4 - 2x), \quad 0 < x < 3;$
(iii) $h(x) = 2\sin x + \cos^2 x, \quad -2 < x < 2;$
(iv) $k(x) = 5x^5 - 20x^4 + 66, \quad -4 < x < 4.$

5.38. Reprove the inequality in Exercise 5.32 by using a first and a second derivative; assume that $r \geq 1$. (Missouri MAA Examination, 2004)

5.39. Consult (Creighton, 1975) on an alternative Second Derivative Test. Write up a summary of this short paper in your own words.

5.40. **(a)** Refer to Figure 5.7. Find the value of $a > 0$ that minimizes the area of the parabolic segment POQ. Verify that you have, indeed, found a relative minimum.
(b) Refer to Figure 5.7. Find the value of a that minimizes the length of the parabolic arc that connects P and Q. Verify that you have, indeed, found a relative minimum.

Section 5.6

5.41. Use mathematical induction to arrive at the statement of Taylor's Theorem that is given as Theorem 5.11.

5.42. Verify entry 1 in Table 3.2 as to magnitude and compute an upper bound to the error in this value of sin 1 by the use of Lagrange's form of the remainder.

5.43. Prove, as requested in Example 5.14, that if $n \in \mathbf{N}$, then $[(n+1)/3]^{n+1} < (n+1)!$.

5.44. In Theorem 5.11 let $a = 0$. Work out $T_9(\theta)$ for the function $f(\theta) = \tan\theta$. As it is difficult to work with the remainder $R_n(\theta)$ for this function as of yet, let us assume at this point that $T_9(\theta)$ could serve as an approximation to $f(\theta)$ when $\theta = 1/2$, for example. Estimate $\tan(1/2)$, and compare with the calculator value of 0.5463025.

5.45. It is known that when $a = 0$ in Theorem 5.11, all the nonzero coefficients in the Taylor series for $f(\theta) = \tan\theta$ are positive. See if you can prove this. A good strategy is to first look for patterns in the successive derivatives of f.

5.46. Prove Corollary 5.11.2. How does the hint supplied for the proof of this corollary suggest a briefer, alternative proof of the result in Example 5.14?

5.47. A function f is stipulated to have the following properties: (1) $y = f(x)$ satisfies the differential equation $xy' = 1$ on some set of nonzero real numbers (2) $f(1) = 0$.
(a) Work out the Taylor series for f at $a = 1$.
(b) Does the Taylor series certainly represent f if $1 \leq x < 2$?

5.48. Suppose that f' is continuous on $[a, x]$ and that f'' exists everywhere on (a, x). We attempt to approximate f on $[a, x]$ by a second-degree polynomial $P(x)$ that agrees in value with f at a and x, and such that $f'(a) = P'(a)$:

$$P(t) = f(a) + f'(a)(t - a) + k(t - a)^2.$$

Now define the function g by $g(t) = f(t) - P(t)$.
(a) Apply Rolle's Theorem twice and deduce k.
(b) What is the connection between part (a) and Taylor's Theorem?

5.49. If x_1 is close to x, a zero of f, then $0 = f(x) \approx f(x_1) + f'(x_1)(x - x_1)$, and so $x \approx x_1 - [f(x_1)/f'(x_1)]$, which is the start of the algorithm for the well-known **Newton-Raphson Method**. Let us extend the method by beginning, instead, with

$$f(x) \approx f(x_1) + f'(x_1)(x - x_1) + \frac{f''(x_1)(x - x_1)^2}{2}.$$

(a) Obtain a revised algorithm from this (Parker, 1959).
(b) Use the new algorithm to find all real roots of $x^4 - x^3 - 16x^2 + 18x - 36 = 0$.

Section 5.7

5.50. Show, as requested, that if $x \neq 0$ and the natural number k exceeds $|x|$, then we have

$$\left[1 + \frac{x}{k+1}\right]^{k+1} > \left[1 + \frac{x}{k}\right]^k.$$

5.51. Refer to the nine properties of $\exp(x)$ given in the text.
(a) Explain briefly properties (3), (4), and the second part of (7).
(b) Do the inequalities in property (5) hold if $|h| > 1$? Explain how you know that $\exp(h)$ is continuous at $h = 0$.
(c) Explain briefly the steps in property (9).
(d) Let $a > 0$ be arbitrary. We define $a^x, x \in \mathbf{R}^1$, to mean $e^{x\ln(a)}$. What, then, is the derivative of a^x?

5.52. Refer to the five properties of $\ln x$ given in the text.
(a) Account briefly for property (1).
(b) Obtain the functional equation given in property (3).
(c) Prove both parts of property (4).
(d) Let $F(y) = 1 + y - e^{y-2y^2}$; work out $F'(y)$ and $F''(y)$ and show that $1 + y > e^{y-2y^2}$ for any $y \in (-0.20, 0) \cup (0, 0.20)$.
(e) **(Bernoulli's Inequality)** Here, Corollary 5.7.3 is extended to irrational $\alpha > 1$. Use Exercise 5.51(d) to establish the derivative of $(1 + x)^\alpha$, and then go on to prove Bernoulli's Inequality.

5.53. Prove that for all $x > 3/2$, we have $\left(\frac{2}{3}\right)^x < \frac{1}{2x-3}$.

5.54. Consider the series $\sum_{i=1}^{\infty} \frac{1}{p_i}$ of reciprocals of the primes. Intuition is no guide in deciding if this series converges. An elegant solution can be obtained by appealing to the Taylor series for e^x. Read this proof, as well as two others that are cited in Dence and Dence (1999), and write up a summary of your reading.

5.55. If $|x| < 1$, then the Taylor series $1 + \sum_{k=1}^{\infty} \binom{\alpha}{k} x^k$ converges to some function $f(x)$. Assume that term-by-term differentiation is permitted.[10]
(a) Do this, and show that $\alpha f(x) = (1 + x)f'(x)$.
(b) Deduce the function f from the result in part (a).

---

[10] This topic is considered in Section 7.5.

5.56. If you did not do Exercise 2.11, please do it now and then go on to show that $\{x_n\}_{n=1}^{\infty}$ converges.[11]

Section 5.8

5.57. Consider the three power series: (A) $\sum_{n=0}^{\infty}(x-1)^n$, (B) $\sum_{n=1}^{\infty}\frac{(x-1)^n}{n}$, (C) $\sum_{n=1}^{\infty}\frac{(x-1)^n}{n^3}$. Show that they have the same radius of convergence, but that the behavior at the endpoints is different in each case.

5.58. Finish the proof of Theorem 5.13. Deduce $R = 1/\rho$ in Example 5.17 by first showing that for all sufficiently large n, $0 < \ln n < \sqrt{n}$ holds.

5.59. Complete the proof of Theorem 5.14. Next, in this theorem let $a = 0, f(x) = (1+x)^{1/2}$. Compute an upper bound to $|R_6(3/4)|$ according to the theorem. Compare this to the actual value obtained by evaluating $|f(x) - T_6(x)|$ at $x = \frac{3}{4}$. What lesson can you draw from this exercise?

5.60. Assume that $f(x) = \frac{7x-4}{x^2+x-6}$ is represented by its Taylor series $\sum_{k=0}^{\infty} c_k x^k$ on the appropriate interval **I**. What is the value of the sum $\sum_{k=0}^{\infty} c_k$? As a complement, compute the partial sum $s_8 = \sum_{k=0}^{8} c_k$.

5.61. **(Binomial Theorem)** A very important function to which Taylor's Theorem can be applied is the **binomial function** f, defined by $f(x) = (1+x)^{\alpha}$, $\alpha \in \mathbf{R}^1 \backslash [\mathbf{N} \cup \{0\}]$.

(a) Show that the Taylor series for f about $a = 0$ is $1 + \sum_{n=1}^{\infty}\binom{\alpha}{n}x^n$, where the **generalized binomial coefficient** $\binom{\alpha}{n}$ was defined in Exercise 2.10.

(b) Show that the radius of convergence of the series is $R = 1$.

(c) Use Lagrange's form of the remainder to show that if $0 \le x < 1$, then $\lim_{n\to\infty} R_n(x) = 0$.

(d) Use Theorem 5.14 (Cauchy's form of the remainder) to show that if $-1 < x < 0$, then for some c that satisfies $-1 < x < c < 0$ we have

$$|R_n(x)| = \left[\frac{\left(1-\frac{c}{x}\right)}{(1+c)}\right]^n [1+c]^{\alpha-1} \left|(\alpha-n)\binom{\alpha}{n}x^{n+1}\right|.$$

(e) In part (d), show that the quantity inside the first brackets on the right is bounded from above by 1, that the second bracketed factor is bounded, and that the quantity between the absolute value bars (if regarded as the $(n+1)$st term of an infinite series) tends to 0 as $n \to \infty$.

(f) Conclude that for any $\alpha \in \mathbf{R}^1 \backslash [\mathbf{N} \cup \{0\}]$, the function f is represented by its Taylor series if $-1 < x < 1$.

(g) Estimate $\sqrt[3]{13}$.

---

[11] The limit, **Euler's Constant** γ, is fundamental in analysis and has been estimated in a variety of ways (Johnsonbaugh, 1981; Knuth, 1962).

Section 5.9

5.62. Evaluate:

(a) $\lim_{x\to 0} \frac{\cos(x^2)-1}{x^3 \sin(x)}$;

(b) $\lim_{x\to 0+} \frac{x^2}{\ln x - \ln(\sin x)}$;

(c) $\lim_{n\to\infty} P(n)$ in Example 5.2;

(d) $\lim_{x\to\infty} \frac{\ln\left(1+3e^{-\sqrt{x}}\right)}{\sqrt{2/x}}$;

(e) $\lim_{x\to\infty} \left[\frac{x+\ln 2}{x-\ln 2}\right]^x$.

5.63. In connection with Theorem 5.15, prove that the quotient $f(x)/g(x)$ is real for all $x \in (a,\ b)$.

5.64. The proof of Theorem 5.15 for the extended case where the quotient of derivatives is unbounded follows along lines similar to those of the theorem itself. Begin by letting an arbitrary, large $M > 0$ be given.

(a) Complete the proof that if $\lim_{x\to b-} [f'(x)/g'(x)] = \infty \in \mathbf{Re}$, then $\lim_{x\to b-} [f(x)/g(x)] = \infty$, also.

(b) Determine $\lim_{x\to 0-} \frac{(1-x)^{-1/2}-(1-2x)^{-1/2}}{x^2}$.

5.65. Corollary 5.15.1, where $x \to \infty$, represents another variant of Theorem 5.15. Assume that f, g are differentiable on (a, ∞), that $f(x)$ and $g(x)$ both approach 0 as $x \to \infty$, and that $g(x) \neq 0, g'(x) \neq 0$ on (a, ∞). Begin by defining the new variable $z = x^{-1}$ and the new function

$$F(z) = \begin{cases} f\left(x^{-1}\right) & 0 < z < a^{-1} \\ 0 & z = 0, \end{cases}$$ and similarly for $g(x^{-1})$.

(a) Complete the proof that if $\lim_{x\to\infty} [f'(x)/g'(x)] = L \in \mathbf{R}^1$, then $\lim_{x\to\infty} [f(x)/g(x)] = L$.

(b) Determine $\lim_{x\to -\infty} (1-2^{1/x})^{\sin(1/x)}$.

5.66. In the proof of Theorem 5.16, give the details that lead up to $-\varepsilon < \frac{f(t_2)}{g(t_2)} - L$. Now apply the theorem to this limit: $\lim_{x\to 0-} \frac{1-\ln(-2x)}{1-\ln(-\sin 3x)}$.

5.67. Determine $\lim_{n\to\infty} \left[\sum_{k=1}^{n} k^k / \sum_{k=0}^{n} (k+1)^k\right]$.

5.68. In his books in the 1820s, Cauchy introduced the following function, now known as the **Cauchy function**:

$$f(x) = \begin{cases} e^{-x^2} + e^{-1/x^2} & x \neq 0 \\ e^{-x^2} & x = 0. \end{cases}$$

(a) Let $g(x) = e^{-1/x^2}$. Show that for any $n \in \mathbf{N}$, $\lim_{x\to 0} \frac{g(x)}{x^n} = 0$.

(b) Build on the result of part (a) and determine, for any $n \in \mathbf{N}$, the value of $g^{(n)}(0)$.

(c) Prove that the Taylor series for f is $\sum_{n=0}^{\infty} (-1)^n \frac{\left(x^2\right)^n}{n!}$.

(d) Determine the radius of convergence R of the Taylor series.

(e) For which $x \in (-R, R)$ does the remainder $R_n(x)$ not approach 0 as $n \to \infty$, or equivalently, for which x does the Taylor series represent f?

Section 5.10

5.69. Let $\mathbf{p} = (x_1, x_2) \in \mathbf{R}^2$ and define the function

$$f(\mathbf{p}) = \begin{cases} \dfrac{-2x_1x_2}{x_1^2 + x_2^2} & \mathbf{p} \neq \mathbf{0} \\ 0 & \mathbf{p} = \mathbf{0}. \end{cases}$$

Show that although f is discontinuous at $\mathbf{p} = \mathbf{0}$, both first-order partial derivatives of f exist there.

5.70. Let $f(x) = \sin x$ and $a = \pi/2$. What, explicitly, is the function $E(h)$? Show that $\lim_{h \to 0} E(h) = 0$. Let $f(x_1, x_2) = x_1 + x_2^2 + 2x_1x_2$ and $\mathbf{p} = (1, 2)$. What, explicitly, is the function $E(\mathbf{h})$? Prove that $\lim_{\mathbf{h} \to \mathbf{0}} E(\mathbf{h}) = 0$.

5.71. Estimate $f(x_1, x_2) = x_1^{x_2}$ at $(e, \pi/2)$; assume, as in Example 5.23, values of $e = 2.718282, \pi = 3.141593, \sqrt{11} = 3.316625, \ln(11/4) = 1.011601$.

5.72. It is useful to think of $\nabla f(\mathbf{p})$ as the result of a *vector operator* $\nabla \cong (D_1 \quad D_2)$ operating on a **scalar function**; that is, a function $f\colon \mathbf{S} \to \mathbf{R}^1, \mathbf{S} \subseteq \mathbf{R}^2$, and similarly for functions defined in $\mathbf{R}^n, n > 2$. Suppose that f, g are two scalar functions defined in $\mathbf{R}^2$ that have partial derivatives at $\mathbf{p}$.

(a) Show that $[\nabla(f+g)](\mathbf{p}) = (\nabla f)(\mathbf{p}) \oplus (\nabla g)(\mathbf{p})$.

(b) Show that $[\nabla(fg)](\mathbf{p}) = f(\nabla g)(\mathbf{p}) \oplus g(\nabla f)(\mathbf{p})$.

5.73. **(a)** Let $F\colon \mathbf{S} \to \mathbf{R}^1, \mathbf{S} \subseteq \mathbf{R}^3$, be defined by $F(\mathbf{p}) = \ln\left(\frac{1}{\|\mathbf{p}\|}\right), \mathbf{p} = (x_1, x_2, x_3)$. Work out $(\nabla F)(\mathbf{p})$ for $\mathbf{p} \neq \mathbf{0}$.

(b) The inner product of the vector operator ∇ and a vector ∇f could be notated as $\nabla * (\nabla f)$, or more simply, as $\nabla^2 f$. Work out $\nabla^2 F$ for the function in part (a).

5.74. Prove all parts of Theorem 5.18. How, then, do you know that the function f in Exercise 5.69 is not differentiable at $\mathbf{p} = \mathbf{0}$?

5.75. The definition of differentiability carries over to functions $f : \mathbf{S} \to \mathbf{R}^m, \mathbf{S} \subseteq \mathbf{R}^n, m > 1$. In this case, what would the representation of $\mathbf{L}$ look like?
The linear mapping $\mathbf{L}$, for all cases $m \geq 1$, $n > 1$ is known as the **Fréchet derivative**.[12]

5.76. Let $f\colon \mathbf{S} \to \mathbf{R}^1, \mathbf{S} \subseteq \mathbf{R}^2$, be defined by $f(\mathbf{p}) = x_1^{x_2}, \mathbf{p} = (x_1, x_2)$, and $\mathbf{S}$ is the square $0 < x_1 \leq 1, 0 \leq x_2 \leq 1$. Prove that f is differentiable on $\mathbf{S}$.

5.77. **Theorem 5.21 (Equality of Mixed Second Partials).** Let $f\colon \mathbf{S} \to \mathbf{R}^1, \mathbf{S} \subseteq \mathbf{R}^2$, be a function such that $f, D_1f, D_2f, D_{2,1}f = D_2(D_1f)$ are continuous on the open set $\mathbf{S}$. Then $D_{1,2}f = D_1(D_2f)$ exists on $\mathbf{S}$ and at each $\mathbf{p} \in \mathbf{S}$ we have $(D_{1,2}f)(\mathbf{p}) = (D_{2,1}f)(\mathbf{p})$.

---

[12] After the French analyst and topologist **Maurice Fréchet** (1878–1973). His definition of the derivative appeared around 1914.

(a) Let $\mathbf{p} = (a, b) \in \mathbf{S}$ be arbitrary, and define the function $F: \mathbf{S} \to \mathbf{R}^1$, by $F(x_1) = f(x_1, b + h_2) - f(x_1, b)$. Show that $F(a + h_1) - F(a) = h_1[(D_1 f)(a + \theta_1 h_1, b + h_2) - (D_1 f)(a + \theta_1 h_1, b)]$, where $\theta_1 \in (0,\ 1)$ and $\mathbf{h} = (h_1, h_2)$ is so small that $\mathbf{p} \oplus h \in \mathbf{S}$.

(b) Use the result in (a) to establish that

$$[f(a + h_1, b + h_2) - f(a + h_1, b)] - [f(a, b + h_2) - f(a, b)]$$
$$= h_1 h_2 (D_{2,1} f)(a + \theta_1 h_1, b + \theta_2 h_2),$$

where $\theta_2 \in (0,\ 1)$.

(c) Finally, use the result in (b) to prove that $(D_{1,2} f)(\mathbf{p}) = (D_{2,1} f)(\mathbf{p})$.

5.78. Suppose that there is a $\delta > 0$ such that the real-valued functions f, g are defined on $\mathbf{B}_n(\mathbf{a}; \delta)$. Further, suppose that f is differentiable at $\mathbf{a}$, $f(\mathbf{a}) = 0$, and g is continuous at $\mathbf{a}$. Use Theorem 5.20 to prove that fg is differentiable at $\mathbf{a}$.

REFERENCES

Cited Literature

Barrett, L.C. and Jacobson, R.A., "Extended Laws of the Mean," *Amer. Math. Monthly*, **67**, 1005–1007 (1960). Our geometric route to the Mean-Value Theorem was drawn from this paper.

Blumenthal, L.M., "Concerning the Remainder Term in Taylor's Formula," *Amer. Math. Monthly*, **33**, 424–426 (1926). The author's "determinantal" approach to Taylor's formula leads to expressions for the remainder due to Schlömilch, Lagrange, and Cauchy.

Boas, R.P., Jr., "l'Hôpital's Rule Without Mean-Value Theorems," *Amer. Math. Monthly*, **76**, 1051–1053 (1969). Author proves the Rule using only one simple fact from continuity and one simple fact regarding integration. The source of our Theorem 5.16.

Boas, R.P., Jr., "Who Needs Those Mean-Value Theorems, Anyway?" *Two-Year Coll. Math. J.*, **12**, 178–181 (1981). Author advocates replacing use of the Mean-Value Theorem in calculus with the Mean-Value Inequality: $(b - a) \min f'(x) \leq f(b) - f(a) \leq (b - a) \max f'(x)$.

Borden, R.S., *A Course in Advanced Calculus*, Dover Publications, New York, 1998, pp. 48–55. A nice discussion of several classic inequalities. The indicated pages would make a good outside reading assignment.

Botsko, M.W. and Gosser, R.A., "On the Differentiability of Functions of Several Variables," *Amer. Math. Monthly*, **92**, 663–665 (1985). Our Theorem 5.20 is drawn from this important paper.

Creighton, J.H.C., "A Strong Second Derivative Test," *Amer. Math. Monthly*, **82**, 287–289 (1975). Interesting paper; this author's Second Derivative Test assumes continuity of $f''(x)$ on a neighborhood about x_0 and that x_0 is not a limit point of the zeros of $f''(x)$.

Dunham, W., "Nondifferentiability of the Ruler Function," *Math. Mag.*, **76**, 140–142 (2003). Very nice paper; recommended.

Goodner, D.B., "An Extension of Cauchy's Generalized Law of the Mean," *Amer. Math. Monthly*, **69**, 907–909 (1962). The extension is to pairs of functions f, g that have bounded derivatives at each point of an open interval (a, b).

Grabiner, J.V., *The Origins of Cauchy's Rigorous Calculus*, MIT Press, Cambridge, MA, 1981, pp. 114–116, 167–170. This fine work should be required reading of every serious mathematics student. The first citation is to Cauchy's definition of the derivative, wherein the δ, ε–notation appears for the first time in mathematics. The second citation is to Cauchy's derivation of the Mean-Value Theorem and the associated Mean-Value Inequality.

Grabiner, J.V., "Who Gave You the Epsilon? Cauchy and the Origins of Rigorous Calculus," *Amer. Math. Monthly*, **90**, 185–194 (1983). A nice historical appraisal of Cauchy's achievements in real analysis in the context of the algebra of inequalities and ever-changing attitudes toward rigor.

Hall, L.M., "A Dozen Minima for a Parabola," *Coll. Math. J.*, **34**, 139–141 (2003). A nice collection of minimization problems, all dealing with mathematical properties of the parabola $y = x^2$.

Halperin, I., "A Fundamental Theorem of the Calculus," *Amer. Math. Monthly*, **61**, 122–123 (1954). This is a clever proof of the Mean-Value Inequality that is quite elementary. Nice.

Hardy, G.H., *A Course of Pure Mathematics*, 10th ed., Cambridge University Press, Cambridge, 1967, pp. 217–218, 316–317, 434–435. This book is always good for clarity. The first citation is to Hardy's note on older "proofs" of the Chain Rule. The second and third citations are part of his development of the circular functions and lead up to an analytical proof that $d(\sin x)/dx = \cos x$.

Hartig, D., "l'Hôpital's Rule Via Integration," *Amer. Math. Monthly*, **98**, 156–157 (1991). An approach to l'Hôpital's Rule that does not use the Mean-Value Theorem; similar in spirit to Boas, Jr. (1969).

Hildebrandt, T.H., "A Simple Continuous Function with a Finite Derivative at No Point," *Amer. Math. Monthly*, **40**, 547–548 (1933). You may read this, but a good exercise, however, would be to write a short program to print out the graphs of the first few partial sums of the series representation for the author's $\phi(x)$.

Huang, X.-C., "A Discrete l'Hôpital's Rule," *Coll. Math. J.*, **19**, 321–329 (1988). Interesting extensions of l'Hôpital's Rule to ratios of differences.

Johnsonbaugh, R., "The Trapezoidal Rule, Stirling's Formula, and Euler's Constant," *Amer. Math. Monthly*, **88**, 696–698 (1981). A very interesting paper that ties together the three topics in the title; includes an elementary method for the estimation of γ. Recommended.

Kalman, D., "Rediscovering Taylor's Theorem," *Coll. Math. J.*, **16**, 103–107 (1985). Author presents a proof of Taylor's Theorem that emerges as the solution of an interpolation problem. Of interest to budding numerical analysts.

Kline, M., *Mathematics: The Loss of Certainity*, Oxford University Press, New York, 1980, pp. 160–162. Brief description of the logical morass into which analysis had fallen in the early 1800s.

Knuth, D.E., "Euler's Constant to 1271 Places," *Math. Comp.*, **16**, 275–281 (1962). The technique used the Euler-MacLaurin Summation Formula. By now, a great many more decimal places of γ are known, but 1271 places were impressive in 1962.

Kountourogiannis, D. and Loya, P., "A Derivation of Taylor's Formula with Integral Remainder," *Math. Mag.*, **76**, 217–219 (2003). The method of proof here involves multiple integration of a derivative, $f^{(n+1)}$.

Kowalewski, G., "Über Bolzanos nichtdifferenzierbare stetige Funktion," *Acta Math.*, **44**, 315–319 (1923). Plunge in if you're prepared to tackle the German.

Kuhn, S., "The Derivative a la Carathéodory," *Amer. Math. Monthly*, **98**, 40–44 (1991). Very interesting paper; the source of our discussion in Section 5.2.

Maz'ya, V. and Shaposhnikova, T., *Jacques Hadamard: A Universal Mathematician*, American Mathematical Society, Providence, RI, 1998. A fascinating read about one of the most influential mathematicians of the nineteenth/twentieth-century interface.

McCarthy, J., "An Everywhere Continuous Nowhere Differentiable Function," *Amer. Math. Monthly*, **60**, 709 (1953). The example in question is similar in nature to that in Hildebrandt (1933).

Parker, F.D., "Taylor's Theorem and Newton's Method," *Amer. Math. Monthly*, **66**, 51 (1959). This slight extension of the Newton-Raphson method is seldom mentioned in texts; the source of our Exercise 5.49.

Reich, S., "On Mean Value Theorems," *Amer. Math. Monthly*, **76**, 70–73 (1969). A reading of this will give you the flavor of the idea that mean-value theorems did not end with Cauchy.

Sanderson, D.E., "A Versatile Vector Mean Value Theorem," *Amer. Math. Monthly*, **79**, 381–383 (1972). This extension of the Mean-Value Theorem is for functions with domains in $\mathbf{R}^1$ and ranges in $\mathbf{R}^n$. The proof requires only multiple application of Rolle's Theorem. For further extension to functions $f : \mathbf{R}^m \to \mathbf{R}^n$ see Furi, M. and Martelli, M., "A Multidimensional Version of Rolle's Theorem," *Amer. Math. Monthly*, **102**, 243–249 (1995).

Schaumberger, N., "More Applications of the Mean-Value Theorem," *Coll. Math. J.*, **16**, 397–398 (1985). The use of the Mean-Value Theorem to derive $\ln x \leq x - 1 (x > 0)$ is quite easy; that the AM–GM Inequality can be obtained from this is much more interesting. See also the related paper Schaumberger, N., *ibid*, **20**, 320 (1989).

Spiegel, M.R., "On the Derivatives of Trigonometric Functions," *Amer. Math. Monthly*, **63**, 118–120 (1956). The author begins by first finding the derivative of $\text{Sin}^{-1}x$, for which he needs the formulas for the area of a triangle and of a sector of a circle, however. The latter is the rub in a development of the theory of the circular functions.

Swann, H., "Commentary on Rethinking Rigor in Calculus: The Role of the Mean Value Theorem," *Amer. Math. Monthly*, **104**, 241–245 (1997). Calculus instructors should read this article. We like the author's comment: "The mean value theorem is actually a friendly theorem.... It provides one more test of the effectiveness of Weierstrass' definition of limit and continuity ...".

Truesdell, C., "The New Bernoulli Edition," *Isis*, **49**, 54–62 (1958). This article summarizes background and plans by authorities in Switzerland a half-century ago to publish the letters, diaries, unpublished and published works of several of the Bernoullis. The project is still under way. The first volume published (1955) contains, among other items, the correspondence between Johann Bernoulli and the Marquis de l'Hôpital.

Wen, Liu, "A Nowhere Differentiable Continuous Function," *Amer. Math. Monthly*, **107**, 450–453 (2000). The example described here is certainly not appropriate for a first calculus course (as the author asserts), but is appropriate for the second.

Additional Literature

Anderson, G., Vamanamurty, M. and Vuorinen, M., "Monotonicity Rules in Calculus," *Amer. Math. Monthly*, **113**, 805–816 (2006).

Anselone, P.M. and Lee, J.W., "Differentiability of Exponential Functions," *Coll. Math. J.*, **36**, 388–393 (2005).

Azpeitia, A.G., "On the Lagrange Remainder of the Taylor Formula," *Amer. Math. Monthly*, **89**, 311–312 (1982).

Boas, Jr., R.P., "Intermediate Forms Revisited," *Math. Mag.*, **63**, 155–159 (1990) (l'Hôpital's Rule).

Boerner, H., "Constantin Carathéodory," in Gillispie, C.C. (ed.), *Dictionary of Scientific Biography*, Vol. III, Charles Scribner's Sons, NY, pp. 62–63.

Georgiadou, M., *Constantin Carathéodory: Mathematics and Politics in Turbulent Times*, Springer-Verlag, Berlin, 2004.

Griffel, D.H., *Applied Functional Analysis*, Dover Publications, Mineola, NY, 2002, pp. 308–312 (the Fréchet derivative).

Komornik, V., "Another Short Proof of Descartes's Rule of Signs," *Amer. Math. Monthly*, **113**, 829–830 (2006) (application of Rolle's Theorem).

Meyerson, M.D., "Every Power Series is a Taylor Series," *Amer. Math. Monthly*, **88**, 51–52 (1981).

Moser, J.M., "A Note on the Mean Value Theorem for Derivatives," *Math. Comp. Educ.*, **18** (1), 36–37 (1984).

Rosenholtz, I. and Smylie, L., " 'The Only Critical Point in Town' Test," *Math. Mag.*, **58**, 149–150 (1985).

Subramanian, P.K., "Successive Differentiation and Leibniz's Theorem," *Coll. Math. J.*, **35**, 274–282 (2004).

Taylor, A.E., "Derivatives in the Calculus," *Amer. Math. Monthly*, **49**, 631–642 (1942) (relative extrema, concavity, tangents, and convexity).

Taylor, A.E., "L'Hospital's Rule," *Amer. Math. Monthly*, **59**, 20–24 (1952).

Trombetta, M., "Tangent Lines and the Inverse Function Differentiation Rule," *Coll. Math. J.*, **35**, 258–261 (2004).

CHAPTER 6

Integration

"Therefore, first of all: What is one to understand by $\int_a^b f(x)dx$?"
Georg F.B. Riemann

CONTENTS

Reviewed in this chapter	Elementary properties of integrals; Simpson's Rule; the Integral Test.
New in this chapter	The Riemann integral; the Darboux integral; Fundamental Theorem of Calculus; improper integrals; integration in $\mathbf{R}^n$.

6.1 INTRODUCTION; INTEGRATION ACCORDING TO RIEMANN

For **Isaac Newton** (1642–1727), the English father of calculus, differentiation was the primary process in calculus. Integration (as it was called later) was to be done by using a derivative[1] in reverse (Boyer, 1959). Thus, if f is continuous on $[a, b]$, then integration consisted in finding a function F such that for each $x \in [a, b]$

$$F'(x) = f(x).$$

Much later, after Newton, the function F was referred to as an **antiderivative**[2] by some writers, and as a **primitive** by others. Newton's use of derivatives in reverse became known, suggestively, as **antidifferentiation**.

Newton, and others, used familiar techniques such as substitution, integration by parts, and partial fractions, together with tables of selected antiderivatives, just as we do today. Newton did not conceive of defining integration independently of differentiation. And if he had, he might not have realized that there are some derivatives that cannot be integrated (in the later, Riemann sense, of course).

The German polymath **Gottfried Wilhelm von Leibniz** (1646–1716) (Hofmann, 1973) did see the integral differently from Newton. For Leibniz, integration meant a "summation of many quantities" (but, exactly of what, he could not say) and the integral was to be interpreted as the area under a curve (all of this is imprecise). The following modernized exposition captures the flavor of his thinking.

In Figure 6.1 the point a is fixed and x is variable. The area A under the positive-valued, continuous curve $y = f(t)$ clearly depends upon both a and x. It is plausible that there should be a function F such that $F(x) - F(a)$ gives this area. This is certainly the case if $y = f(t)$ is the equation of a straight line,

---

[1] Newton called it a **fluxion**, from the Latin noun *fluxus*, meaning "a flow," and derived from the Latin verb *fluere*, "to flow." By 1599, the word "fluxion" had taken on the technical meaning of a continuous, progressive change.

[2] Newton called this a **fluent**, from the present participle *fluens* (of the verb *fluere*), which means "flowing." By 1705 the term "fluent" could refer to a stream or to more abstract objects. We are indebted to Prof. Frank K. Flinn for assistance in these matters. Newton's adopted words "fluxion" and "fluent" actually had been used (in their Latin equivalents) in the writing of the medieval technical scholar **Richard Suiseth** (14th century), who was known in his day as "Calculator."

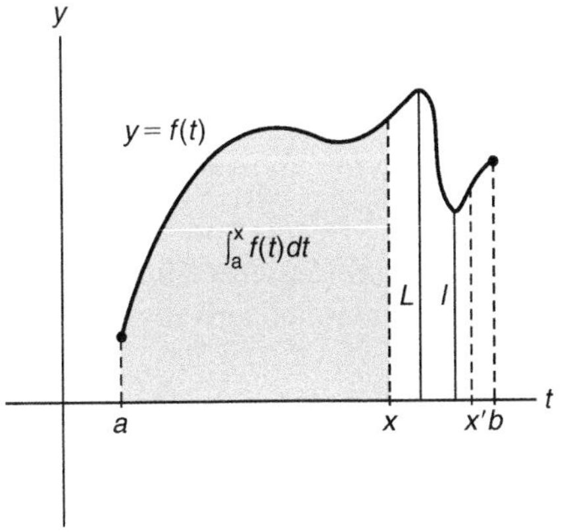

FIGURE 6.1
The integral as an area under a curve.

$y = mt + B, m, B > 0$, for then we have

$$A = \left[\frac{f(a) + f(x)}{2}\right](x - a)$$
$$= \left[\frac{(ma + B) + (mx + B)}{2}\right](x - a)$$
$$= \left[\frac{mx^2}{2} + Bx\right] - \left[\frac{ma^2}{2} + Ba\right]$$

for the area of the resulting trapezoid. Leibniz also considered the area to be given by an integral, namely

$$A = F(x) - F(a) = \int_a^x f(t)dt.$$

Now let x' be near x; for the interval $[x, x']$, we denote the minimum and maximum values of f by l, L, respectively. These numbers exist (Theorem 4.10); Leibniz, of course, did not have this theorem. The area from $t = x$ to $t = x'$ is $F(x') - F(x)$. Approximations by rectangles give

$$l(x' - x) \leq F(x') - F(x) \leq L(x' - x),$$

or equivalently,

$$l \leq \frac{F(x') - F(x)}{x' - x} \leq L.$$

Since f is continuous on $[a, b]$, then $x' \to x$ implies $l \to f(x)$ and $L \to f(x)$. At the same time, the indicated quotient approaches $f(x)$ (Squeeze Theorem). By definition, $f(x)$ must also be $F'(x)$. As $F(a)$ is a constant, its derivative is 0 and consequently,

$$\frac{d}{dx}[F(x) - F(a)] = F'(x) = \frac{d}{dx}\int_a^x f(t)dt.$$

The final result, therefore, is that

$$\frac{d}{dx}\int_a^x f(t)dt = f(x), \qquad (*)$$

and we see that Leibniz's conception of the integral is consistent with that of Newton, who *could have* written equation (*) as $D[D^{-1}f] = f$, where D is a symbol for differentiation and D^{-1} is a symbol for antidifferentiation.

Equation (*), suitably stated, is one half of the celebrated **Fundamental Theorem of the Calculus** (see Section 6.5). Both Newton and Leibniz recognized its extreme importance. It was developed a bit more rigorously in Colin MacLaurin's book *Treatise of Fluxions* (1742) (Grabiner, 1997). Although Leibniz did not call it so, the quantity $\int_a^x f(t)dt$ in equation (*) became known as an **indefinite integral** of f (if x is not specified) or as a **definite integral** of f (if x is specified).

Although Leibniz had the germ of the idea for how to define integration apart from differentiation, he was not yet able to execute on it. Whenever he needed to evaluate an integral, he resorted to antidifferentiation as the key step. An integral as elementary as $\int_0^x \cos t\, dt$ would have been challenging to work out by some sort of "summation of many quantities" (Exercise 6.4).

It is easy to see that both Newton's and Leibniz's approaches to integration were rather restricted. We might wish to integrate functions whose domains are unbounded, or functions that are discontinuous in some places or whose ranges are unbounded. Newton, for example, could only integrate derivatives that are defined and bounded everywhere on a finite interval, and Leibniz would have had trouble with the many different kinds of "summations of many quantities" that would arise. Two ingredients were missing in 1685: a better understanding of area and a correct theory of limits.

The function f defined by

$$f(x) = \begin{cases} 2 & 0 \le x < 1/2, 1/2 < x \le 1 \\ 1 & x = 1/2 \end{cases}$$

could not have been integrated on [0, 1] by Newton. This, or functions like it, led Cauchy and, later, Riemann, to improve upon Newtonian integration.[3] Their starting points were the ideas of Leibniz. Cauchy correctly pointed out that the integral should be formulated as a *limit* of a summation of terms rather than as merely a "summation of many quantities."

In the remainder of this section and in the next two we shall proceed along lines outlined by and motivated by Riemann's brief work on the integral. The objective is to probe the theory of Riemann integration, the first serious theory of integration that a student should encounter. Several theorems will be presented, and you will be requested to supply some of the details. Chapters 1 through 5 contain all that you will need. Except for Section 6.6*, not much attention will be paid to the mechanics of the evaluation of integrals.

Georg F. B. Riemann (1826–1866), born in the village of Breselenz in Lower Saxony in northern Germany, was one of the most original and brilliant mathematicians of all time (Freudenthal, 1975). His development of the integral, which appeared in a manuscript of 1854,[4] applies to functions f that are defined and bounded on a closed, bounded interval $[a, b]$. On such an interval let $P = \{a_k\}_{k=0}^{n}$ be a sequence of $n + 1$ points that satisfy

$$a = a_0 < a_1 < a_2 < \cdots < a_n = b;$$

P is called a **partition** of the interval.[5] The points may, but need not be, evenly spaced. For a given partition P, the norm $\mathfrak{N}$ is the length of the longest subinterval:

$$\mathfrak{N} = \max_k \left\{\Delta a_k\right\}, \Delta a_k = a_k - a_{k-1} > 0.$$

A partition P is said to be **tagged** if a point x_k, $k = 1, 2, \ldots, n$, called the **tag**, is chosen *arbitrarily* from each subinterval $[a_{k-1}, a_k]$ (Figure 6.2): $a_{k-1} \le x_k \le a_k$.

[3] It has been said that Riemann's definition represented an essential advance over the older Cauchy definition of the integral (Grabiner, 1981). In fact, it can be shown that the two definitions are not essentially different (Kristensen, Poulsen, and Reich, 1962).
[4] Riemann's exposition was a small section of a much larger manuscript (not published until shortly after his death) on the representation of functions by Fourier series (Weber, 1953).
[5] Commonly used synonyms for "partition" are **mesh**, **net**, **grid**, and **decomposition**. The terms "net" and "grid" are geometrically appealing, but "partition" is firmly entrenched.

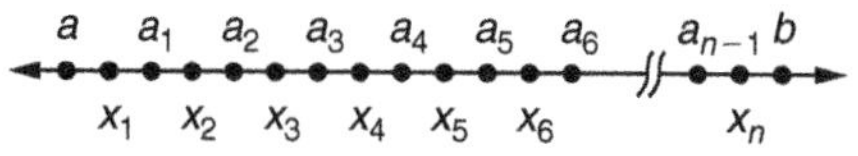

FIGURE 6.2
A tagged partition of $[a, b]$.

We denote a tagged partition by an ordered pair of two sequences:

$$\overline{P}\left(\{a_k\}_{k=0}^{n}, \{x_k\}_{k=1}^{n}\right),$$

or as just $\overline{P}$, for short. We now give Riemann's definition of the integral:

Definition. *A function f is **Riemann-integrable** on $[a, b]$ iff there is a number $I \in \mathbf{R}^1$ for which, given any $\varepsilon > 0$, there is a $\delta > 0$ such that for every tagged partition $\overline{P}$ of norm $\mathfrak{N} < \delta$, the inequality*

$$\left|\sum_{k=1}^{n} f(x_k)\Delta a_k - I\right| < \varepsilon$$

*holds. The number I is called the **Riemann integral** of f over $[a, b]$, and we write*

$$I = \int_a^b f(x)dx,$$

*or sometimes as just $\int_a^b f$. We shall also say that $\int_a^b f$ **exists**.*

Any such sum of the form $S = \sum_{k=1}^{n} f(x_k)\Delta a_k$ is called a **Riemann sum**. Its limit is the desired "sum of many quantities" that Leibniz was seeking, but never found.

The following comments on Riemann's definition of the integral provide additional clarification:

1. A Riemann sum is a numerical approximation to the value of a Riemann integral and thus can be used to estimate areas of simple, planar figures (Exercises 6.6, 6.62).

■ Example 6.1

Let $n = 16$ and $f(x) = x^{-1}$. In Figure 6.2 let us choose equally spaced subdivisions of $[1, 2]$ and choose each $x_k = (a_{k-1} + a_k)/2$. Then the Riemann sum S is (verify!)

$$S = \sum_{k=1}^{16} \frac{32}{31 + 2k} \frac{1}{16}$$

$$= 2\left[\frac{1}{33} + \frac{1}{35} + \cdots + \frac{1}{63}\right]$$

$$\approx 0.693025.$$

This compares favorably with $\int_1^2 x^{-1} dx = \ln 2 \approx 0.693147$. ■

2. In limit language, the definition of the Riemann integral becomes

$$I = \lim_{\mathfrak{N} \to 0} \sum_{k=1}^{n} f(x_k)\Delta a_k,$$

provided that the limit exists. For given values of a, b, n, the minimum value of $\mathfrak{N}$ is $(b - a)/n$, corresponding to that partition $\overline{P}$ where all the subintervals are of equal length. Hence, $\mathfrak{N} \to 0$ implies $n \to \infty$. Nevertheless, the limit is formed with respect to $\mathfrak{N}$, not n, as the sole independent variable. There do exist in the literature, however, ordinary sequence definitions of the Riemann integral (Sklar, 1960) (Exercise 6.7).

3. For each choice of $\mathfrak{N}$, there are infinitely many tagged partitions of $[a, b]$.

These will, in general, yield different values for the corresponding Riemann sums, unless f is a constant function. The limit, when it exists, is still unique, however (Burk, Goel, and Rodríguez, 1986). Suppose that there were two limits, I_1 and I_2. Assume that $I_2 > I_1$; let $\varepsilon = (I_2 - I_1)/2$. By hypothesis, there is then a $\delta > 0$ such that for all tagged partitions of $[a, b]$ with $\mathfrak{N} < \delta$ we have

$$I_2 - \varepsilon < \sum_{k=1}^{n} f(x_k)\Delta a_k < I_1 + \varepsilon.$$

The two ends of the inequalities lead to $(I_2 - I_1)/2 < \varepsilon$, a contradiction; hence, $I_1 = I_2$.

6.2 NECESSARY CONDITIONS FOR RIEMANN-INTEGRABILITY

There are several conditions that can be called necessary conditions for a function f to be Riemann-integrable. Of course, one such condition is contained in the definition of Riemann-integrability; others now follow.

Theorem 6.1. *If f is Riemann-integrable on $[a, b]$, then f is bounded there.*

Proof. Let P be an arbitrary partition of $[a, b]$ and assume that f is unbounded in the subinterval $[a_{j-1}, a_j]$. If $M > 0$ is given, then a point $x_j \in [a_{j-1}, a_j]$ can be chosen so that $\sum_{k=1}^{n} f(x_k)\Delta a_k > M$. As M is arbitrary, this shows that the Riemann sum has no limit in $\mathbf{R}^1$, so f is not Riemann-integrable on $[a, b]$. ■

Theorem 6.2. *If $f(x) \le g(x)$ for all $x \in [a, b], b > a$, and both functions are Riemann-integrable on $[a, b]$, then*

$$\int_a^b f \le \int_a^b g.$$

Proof. Assume the contrary and let $\varepsilon = \frac{1}{2}\left[\int_a^b f - \int_a^b g\right]$; reach a contradiction.

The proof is left to you. ■

■ Example 6.2

For any real t, we have $\cos t \le 1$. Choose an arbitrary $x \ge 0$; then by Theorem 6.2, if we assume integrability, we will have $\int_0^x \cos t\, dt \le \int_0^x dt$, or $\sin x \le x$. Replace x by t in this and integrate again from $t = 0$ to $t = x$: $-\cos x + 1 \le x^2/2$, or more completely, $-1 \le -\cos x \le x^2/2 - 1$. Again, replace x by t, integrate, and repeat this cycle three more times. We obtain, for any $x \ge 0$ (verify!),

$$x - \frac{x^3}{6} \le \sin x \le x - \frac{x^3}{6} + \frac{x^5}{120}.$$

Thus, we have bracketed $\sin x$ by two successive Taylor polynomials *without* direct appeal to Taylor's Theorem (Leonard and Duemmel, 1985). In particular, we find at

$$x = 1: \quad 0.83333 \le 0.84147 = \sin 1 \le 0.84167$$

$$x = 2: \quad 0.6666 \le 0.9093 = \sin 2 \le 0.9333.$$

■

■ Example 6.3

The Taylor polynomial $T_6(x)$ for e^{-x^2} is $1 - x^2 + (x^4/2) - (x^6/6)$. Let $F(x) = T_6(x) - e^{-x^2}$; then $F'(x) = 2x\left(-1 + x^2 - \frac{1}{2}x^4 + e^{-x^2}\right)$. We can establish that $F'(x) \le 0$ on $[0, 1]$ (verify!), so F decreases there. Hence, $F(0) = 0$ implies

$$1 - x^2 + \frac{x^4}{2} - \frac{x^6}{6} \le e^{-x^2}, \quad 0 \le x \le 1.$$

Application of Theorem 6.2 by integration of both sides of the inequality over $[0, 1]$ yields

$$\int_0^1 e^{-x^2}\, dx \ge \left|\left(x - \frac{x^3}{3} + \frac{x^5}{10} - \frac{x^7}{42}\right)\right|_0^1 \approx 0.74286.$$

In fact, the integral on the left-hand side has the approximate value of 0.74683, but this requires some additional work (Exercise 6.43). We have, of course, assumed the Riemann-integrability of e^{-x^2}; that this is so will become apparent later. ■

It is reasonable that there should be connections between the antiderivative of Newton and the Riemann integral. The most important of these connections is given here; we have already made implicit use of it in Examples 6.2 and 6.3.

Theorem 6.3. *If an antiderivative of f on $[a, b]$ is $D^{-1}f = F$ and if f is Riemann-integrable over $[a, b]$ to I, then $F(b) - F(a) = I$.*

Proof. By definition, $F'(x) = f(x)$ at each $x \in [a, b]$. Let $\overline{P}$ denote any tagged partition of $[a, b]$, as in Figure 6.2. For any n, the finite series $\sum_{k=1}^{n}[F(a_k) - F(a_{k-1})]$ is a telescoping series whose sum is $F(a_n) - F(a_0) = F(b) - F(a)$. On each subinterval $[a_{k-1}, a_k]$ we apply the Mean-Value Theorem (Theorem 5.7)

$$F(a_k) - F(a_{k-1}) = f(x_k)(a_k - a_{k-1}),$$

for some $x_k \in (a_{k-1}, a_k)$. Summing over k, we obtain

$$F(b) - F(a) = \sum_{k=1}^{n} f(x_k)\Delta a_k.$$

The right-hand side is a specific Riemann sum. We now choose $\overline{P}$ to be any partition in which the division points a_k are equally spaced, so that $\mathfrak{N} = (b - a)/n$. In this case, $n \to \infty$ implies $\mathfrak{N} \to 0$. But f is Riemann-integrable, so the Riemann sum converges to I as $n \to \infty$. ■

In common with the Newtonian antidifferentiation operator D^{-1}, the Riemann integral enjoys the property of linearity (see later). Although Theorem 6.4 was not really necessary for Examples 6.2 and 6.3 because Theorem 6.3 was sufficient for that purpose (assuming, of course, integrability), Theorem 6.4 is used all the time and it remains valid even when we are unable to obtain antiderivatives of some of the integrands. This, in fact, is the case most of the time!

Theorem 6.4. *If f, g are Riemann-integrable on $[a, b]$ and $c \in \mathbf{R}^1$, then*

(i) $\int_a^b cf = c\int_a^b f$;

(ii) $\int_a^b (f + g) = \int_a^b f + \int_a^b g$.

Proof.

(i) The equality is trivial if $c = 0$. Let $c \neq 0$ and let any $\varepsilon > 0$ be given. Then there is a $\delta > 0$ such that for every tagged partition $\overline{P}$ of norm

$\mathfrak{N} < \delta$, the inequality

$$\left|\sum_{k=1}^{n} f(x_k)\Delta a_k - I\right| < \frac{\varepsilon}{|c|}$$

holds. Multiplication by $|c|$ then gives

$$|c|\left|\sum_{k=1}^{n} f(x_k)\Delta a_k - I\right| = \left|\sum_{k=1}^{n} [cf(x_k)]\Delta a_k - cI\right| < \varepsilon,$$

which is equivalent to $\int_a^b cf = c\int_a^b f = cI$.

(ii) Proof of this part is left to you. ■

Theorem 6.5 (Finite Additivity). *Let f be defined on $[a,b]$ and Riemann-integrable on the subintervals $[a,c]$ and $[c,b]$, $a < c < b$. Then f is integrable on $[a,b]$ and*

$$\int_a^b f = \int_a^c f + \int_c^b f.$$

Proof. By Theorem 6.1, f is bounded separately on $[a,c]$ and $[c,b]$, so it is bounded on $[a,c] \cup [c,b] = [a,b]$; let $|f(x)| < K$ for $a \le x \le b$. Let I_1, I_2 denote the subintegrals $I_1 = \int_a^c f(x)dx$, $I_2 = \int_c^b f(x)dx$. Now let $\varepsilon > 0$ be given. There exist numbers $\delta_1, \delta_2 > 0$ such that for any tagged partition $\overline{P}_1$ on $[a,c]$ of norm $\mathfrak{N}_1 < \delta_1$ and for any tagged partition $\overline{P}_2$ on $[c,b]$ of norm $\mathfrak{N}_2 < \delta_2$ we have

$$\underbrace{\left|\sum_{i=1}^{r} f(x_i)\Delta a_i - I_1\right| < \frac{\varepsilon}{4}}_{[a,c]}, \quad \underbrace{\left|\sum_{j=1}^{s} f(x_j)\Delta a_j - I_2\right| < \frac{\varepsilon}{4}}_{[c,b]}.$$

If $\delta_3 = \min\{\delta_1, \delta_2\}$, then both inequalities hold simultaneously for all tagged partitions of norm $\mathfrak{N} < \delta_3$.

Suppose that the point c lies in the jth subinterval: $a_{j-1} < c \le a_j$. Regardless of whether c is a partition point or not, a Riemann sum for f on $[a,b]$ and with a $\overline{P}$ of norm $\mathfrak{N} < \delta_3$ can be written as

$$S_1 = \sum_{k=1}^{j-1} f(x_k)\Delta a_k + f(x_j)\Delta a_j + \sum_{k=j+1}^{n} f(x_k)\Delta a_k.$$

If c is now regarded as a new partition point, then the following is also a Riemann sum for f on $[a, b]$:

$$S_2 = \left[\sum_{k=1}^{j-1} f(x_k)\Delta a_k + f(c)(c - a_{j-1})\right] + \left[f(c)(a_j - c) + \sum_{k=j+1}^{n} f(x_k)\Delta a_k\right].$$

The sum in each pair of brackets is, in magnitude, less than $\frac{\varepsilon}{4}$ because they approximate I_1, I_2, respectively. Hence, $|S_2 - (I_1 + I_2)| < \varepsilon/2$.

Finally, we let $\delta = \min\{\delta_3, \varepsilon/(4K)\}$. All inequalities so far hold simultaneously. Additionally, upon subtraction we find $|S_1 - S_2| = |f(x_j) - f(c)|\Delta a_j < (K + K)\left(\frac{\varepsilon}{4K}\right) = \varepsilon/2$. Hence, by the Triangle Inequality

$$\begin{aligned} \left|S_1 - (I_1 + I_2)\right| &= \left|S_1 - S_2 + [S_2 - (I_1 + I_2)]\right| \\ &\leq \left|S_1 - S_2\right| + \left|S_2 - (I_1 + I_2)\right| < \varepsilon. \end{aligned}$$

This says that

$$\lim_{\mathfrak{N}\to 0} S_1 = \int_a^c f + \int_c^b f = \int_a^b f.$$

■

Definition. *If $b > a$ and f is Riemann-integrable on $[a, b]$, then we define $\int_b^a f$ to be $-\int_a^b f$.*

Corollary 6.5.1. *The conclusion of Theorem 6.5 remains valid if c is exterior to $[a, b]$, provided that $\int_a^c f$ and $\int_b^c f$ exist.*

The Riemann integral has one elementary property that the Newtonian antiderivative does not possess. From this property (see next) we get our first glimpse at the somewhat wider applicability of the Riemann integral. The wider applicability of the Riemann integral has many benefits; for example, it allows us to develop a good theory of the **area** of sets in $\mathbf{R}^2$, a nice treatment of which can be found in Knopp (1969).

Theorem 6.6. *If f is Riemann-integrable on $[a, b]$ to I and the value of f at one point is changed, then f is still Riemann-integrable on $[a, b]$ and the value of the integral is still I.*

Proof. Let $x = c$ be the exceptional point and define the function g by

$$g(x) = \begin{cases} f(x) & x \neq c \\ C & x = c, \end{cases}$$

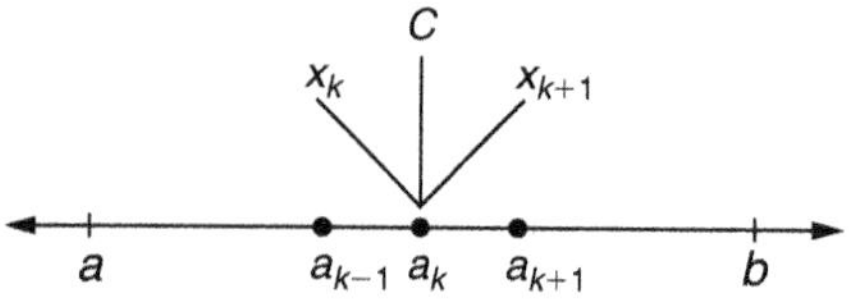

FIGURE 6.3
Supercoincidence of the exceptional point c.

where $C \neq f(c)$. The point $x = c$ might not coincide with any of the tags in a partition $\overline{P}$ of $[a, b]$. At most, however, it could coincide with two tags, say x_j and x_{j+1}, if these coincide with the partition point a_j (Figure 6.3). The exceptional point will change, therefore, at most two terms in any Riemann sum for f.

Thus, for a tagged partition $\overline{P}$ of norm $\mathfrak{N}$, we have

$$\left|\sum_{k=1}^{n} f(x_k)\Delta a_k - \sum_{k=1}^{n} g(x_k)\Delta a_k\right| \leq \left|f(x_k) - g(c)\right| \Delta a_k + \left|f(x_{k+1}) - g(c)\right| \Delta a_{k+1}$$
$$\leq 2\left[|f(c)| + |C|\right]\mathfrak{N}$$

(why?). If $\varepsilon > 0$ is given, show how to choose $\delta > 0$ so that tagged partitions of norm $\mathfrak{N} < \delta$ lead to $\left|\sum_{k=1}^{n} g(x_k)\Delta a_k - I\right| < \varepsilon$.

The completion of the proof is left to you. ■

Corollary 6.6.1. *Theorem 6.6 extends to the general case where any finite number of points of a Riemann-integrable function are changed.*

■ Example 6.4

Let f be the function back in Section 6.1 that would have bothered Newton:

$$f(x) = \begin{cases} 2 & 0 \leq x < 1/2, 1/2 < x \leq 1 \\ 1 & x = 1/2. \end{cases}$$

We then define the new function

$$g(x) = \begin{cases} f(x) & 0 \leq x < 1/2,\ 1/2 < x \leq 1 \\ 2 & x = 1/2, \end{cases}$$

and from Theorem 6.6 it follows (assuming integrability) that

$$\int_0^1 f(x)dx = \int_0^1 g(x)dx = 2x\Big|_0^1 = 2.$$

■

■ **Example 6.5**

Let the function g be defined by

$$g(x) = \begin{cases} x^2 & x \in \left\{[-1, -1/2) \bigcup_{k=2}^{6} (-1/k, -1/(k+1))\right\} \cup (-1/7, 1] \\ 10 & x = -1/k, k = 2, 3, \ldots, 7. \end{cases}$$

By Corollary 6.6.1 (and assuming integrability), we find that the Riemann integral is $\int_{-1}^{1} g = 2/3$. ■

The Newtonian antiderivative of the function in Example 6.5 does not exist because g is not the derivative of any function, and Newtonian antidifferentiation can handle only derivatives. Thus, the Riemann integral has a capability not possessed by the Newtonian antiderivative. In fact, the definition of the Riemann integral can even be extended to functions f that are *undefined* at a finite number of points because Theorem 6.6 and its corollary show that the values that might be assigned to f at these points are irrelevant.

6.3 SUFFICIENCY CONDITIONS FOR RIEMANN-INTEGRABILITY

Theorems 6.1 through 6.6 have all had Riemann-integrability as part of their hypotheses. We turn now to sufficiency conditions for this; we pursue two lines of development, here and in the next section.

Definition. *A* ***step-function*** $\sigma(x)$ *is a function defined on* $[a, b]$ *that has a constant (but, not necessarily, identical) value on each open subinterval* (a_{k-1}, a_k) *of some partition of* $[a, b]$.

Step-functions are very simple functions (Figure 6.4); we should expect them to have nice properties. If $P = \{a_k\}_{k=0}^{n}$ is a partition of the closed, bounded interval $[a, b]$, we let σ_k denote the value of $\sigma(x)$ when $x \in (a_{k-1}, a_k), k = 1, 2, \ldots, n$.

Theorem 6.7. *If* σ *is a step-function on* $[a, b]$*, then*

$$\int_a^b \sigma = \sum_{k=1}^{n} \sigma_k \Delta a_k.$$

Proof. Let $P_1 = \{a_{k1}\}_{k=0}^{n}$ be an arbitrary partition of $[a_0, a_1]$, and for each $k \in \{1, 2, \ldots, n\}$ let x_{k1} be a tag of $[a_{k-1,1}, a_{k1}]$. Then if σ is initially restricted

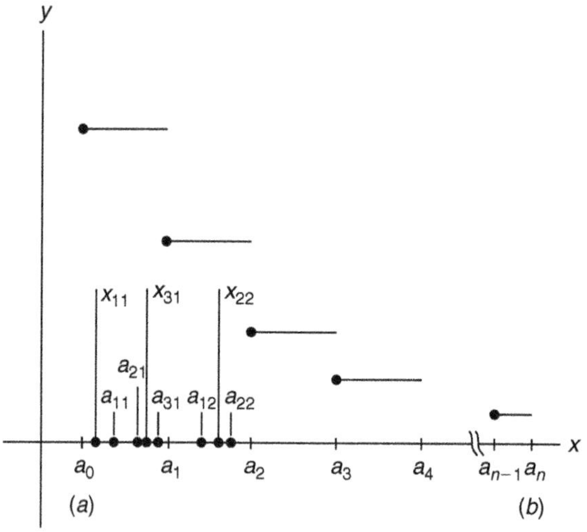

FIGURE 6.4
A step-function on $[a, b]$.

to $[a_0, a_1]$ and is redefined at its endpoints (if necessary), we have

$$\int_{a_0}^{a_1} \sigma(x)dx = \lim_{\mathfrak{N}\to 0} \sum_{k=1}^{n} \sigma(x_{k1})\Delta a_{k1} = \lim_{\mathfrak{N}\to 0} \sum_{k=1}^{n} \sigma_1 \Delta a_{k1}$$

$$= \sigma_1 \lim_{\mathfrak{N}\to 0} \sum_{k=1}^{n} \Delta a_{k1} = \sigma_1 \lim_{\mathfrak{N}\to 0} (a_1 - a_0) = \sigma_1 \Delta a_1.$$

Hence, the restriction of σ to $[a_0, a_1]$ is integrable and mathematical induction, together with Theorem 6.5, then extends this to all of $[a, b]$. ■

The reason for introducing step-functions is so that we may use them to bracket other functions. The result is a kind of "Squeeze Theorem" that can be applied to integrals.

Theorem 6.8. *Let f be defined on $[a, b], b > a$. Suppose that for any $\varepsilon > 0$ and for all $x \in [a, b]$ there are step-functions $\rho(x), \sigma(x)$ such that $\rho(x) \leq f(x) \leq \sigma(x)$ and $0 \leq \int_a^b [\sigma(x) - \rho(x)]dx < \varepsilon$. Then f is Riemann-integrable on $[a, b]$.*

Proof. Let $\varepsilon > 0$ be given and assume that there are step-functions $\rho(x), \sigma(x)$ such that for $x \in [a, b]$ we have $\rho(x) \leq f(x) \leq \sigma(x)$ and $0 \leq \int_a^b [\sigma(x) - \rho(x)]dx < \varepsilon$.

From Theorems 6.2 and 6.7 we have $\int_a^b \rho(x)dx \leq \int_a^b \sigma(x)dx$. Hence, if $\sigma(x)$ is momentarily fixed, then by the Completeness Axiom

$$\sup_{\rho} \int_a^b \rho(x)dx = I_1$$

exists for all such integrals where $\rho(x) \leq f(x)$. Similarly,

$$\inf_{\sigma} \int_a^b \sigma(x)dx = I_2$$

exists for all such integrals where $\sigma(x) \geq f(x)$. We cannot have $I_2 < I_1$, since then we would have $I_2 < \int_a^b \sigma(x)dx$ for some step-function $\sigma(x)$, a contradiction. Thus, $I_1 \leq I_2$ holds. But $0 \leq \int_a^b [\sigma(x) - \rho(x)]dx < \varepsilon$ is true for any $\varepsilon > 0$, so we conclude that it is true in the limit of $\varepsilon = 0$, that is, $I_1 = I_2 = I$.

Accordingly, we can choose step-functions $\rho(x), \sigma(x)$ such that

$$\int_a^b \rho(x)dx > I - \frac{\varepsilon}{2}, \quad \int_a^b \sigma(x)dx < I + \frac{\varepsilon}{2}.$$

Then choose $\delta > 0$ so small that for any partition with $\mathfrak{N} < \delta$ we have for the Riemann sums

$$\sum_{k=1}^{n} \rho_k \Delta a_k > \int_a^b \rho(x)dx - \frac{\varepsilon}{2}, \sum_{k=1}^{n} \sigma_k \Delta a_k < \int_a^b \sigma(x)dx + \frac{\varepsilon}{2}.$$

Combination of the four preceding inequalities then gives (verify!)

$$I - \varepsilon < \sum_{k=1}^{n} \rho_k \Delta a_k \leq \sum_{k=1}^{n} f(x_k) \Delta a_k \leq \sum_{k=1}^{n} \sigma_k \Delta a_k < I + \varepsilon.$$

It follows that $\lim_{\mathfrak{N} \to 0} \sum_{k=1}^{n} f(x_k)\Delta a_k = I$, so f is Riemann-integrable on $[a, b]$. ■

■ Example 6.6

Let $f : [0, 2] \to \mathbf{R}^1$ be defined by $f(x) = -x^2 + 2x$. Suppose that the $2n + 1$ partition points a_k are equidistant; also, for any $k = 1, 2, \ldots, 2n$ and all $x \in (a_{k-1}, a_k)$, let

$$\rho_k(x) = \begin{cases} f(a_{k-1}) & k = 1, 2, \ldots, n \\ f(a_k) & k = n+1, n+2, \ldots, 2n \end{cases}$$

$$\sigma_k(x) = \begin{cases} f(a_k) & k = 1, 2, \ldots, n \\ f(a_{k-1}) & k = n+1, n+2, \ldots, 2n. \end{cases}$$

Since f is symmetrical about the line $x = 1$ and has a maximum at $x = 1$, then these definitions ensure that for all $x \in [0, 2]$ we have $\rho(x) \le f(x) \le \sigma(x)$. The step-functions are assumed to have been redefined at all endpoints so as to give them "sided" continuity there. At this point, drawing a sketch would be extremely helpful.

We can now establish that (Exercise 6.17)

$$\int_a^b [\sigma(x) - \rho(x)]dx = \int_0^2 [\sigma(x) - \rho(x)]dx$$

$$= \frac{2}{2n} \sum_{k=1}^{2n} [\sigma_k(x) - \rho_k(x)] = \frac{2}{n}.$$

Hence, if $\varepsilon > 0$ is given, then n is chosen so that $n > \left\lceil \frac{2}{\varepsilon} \right\rceil$. It then follows that

$$\int_a^b [\sigma(x) - \rho(x)]dx < \varepsilon,$$

and by Theorem 6.8, $f(x) = -x^2 + 2x$ is Riemann-integrable on $[0, 2]$. ■

Theorem 6.9. *If f is defined and monotonic on $[a, b]$, $b > a$, then f is Riemann-integrable on $[a, b]$.*

Proof. Assume that f is increasing on $[a, b]$. For any partition $P = \{a_k\}_{k=1}^n$ of norm $\mathfrak{N}$, define the step-functions

$$\rho(x) = f(a_{k-1}), \quad a_{k-1} \le x < a_k \quad (k = 1, 2, \ldots, n)$$

$$\sigma(x) = f(a_k), \quad a_{k-1} \le x < a_k \quad (k = 1, 2, \ldots, n).$$

Show that $0 \leq \int_a^b [\sigma(x) - \rho(x)]dx \leq \mathfrak{M}[f(b) - f(a)]$ and indicate how to choose $\delta > 0$ so that Theorem 6.8 can be applied. The completion of the proof is left to you. ■

A function f defined on $[a, b]$ is **finitely piecewise monotonic** there iff $[a, b]$ can be partitioned into a finite number of closed subintervals, on each of which f is monotonic. For example, $f(x) = \sin x$ is finitely piecewise monotonic on $[0, 3\pi/2]$.

Corollary 6.9.1. *If f is defined and finitely piecewise monotonic on $[a, b]$, then it is Riemann-integrable there.*

Proof. The proof is left to you. ■

■ Example 6.7

The following integrals exist:

(a) $\int_1^2 f(x)dx$ (Example 6.1)

(b) $\int_0^x \cos t\, dt$ (Example 6.2)

$(x \in \mathbf{R})$

(c) $\int_0^1 e^{-x^2} dx$ (Example 6.3)

(d) $\int_1^4 \frac{\ln(1+y)}{2+y} dy$. ■

6.4 THE CONTRIBUTION OF DARBOUX

Riemann gave a second necessary and sufficient condition in his 1854 paper for the existence of the integral, but supplied no details. This was done in 1875 by the French geometer **Jean-Gaston Darboux** (1842–1917).

Let f be defined and bounded on $[a, b], b > a$. We define $M = \sup_x f(x)$ and $m = \inf_x f(x), x \in [a, b]$. Similarly, for each closed subinterval $[a_{k-1}, a_k]$ of a partition P of $[a, b]$ we define $M_k = \sup_x f(x), m_k = \inf_x f(x)$, where $x \in [a_{k-1}, a_k]$. These numbers exist by the Completeness Axiom. The **upper Darboux sum** that corresponds to f and to P is

$$U(P, f) = \sum_{k=1}^{n} M_k \Delta a_k,$$

and the **lower Darboux sum** is

$$L(P, f) = \sum_{k=1}^{n} m_k \Delta a_k.$$

It is apparent that since for each k we have $m \le m_k \le M_k \le M$, then $L(P,f)$ is bounded from above by $\sum_{k=1}^{n} M\Delta a_k = M(b-a)$ and a $U(P,f)$ is bounded from below by $\sum_{k=1}^{n} m\Delta a_k = m(b-a)$, independently of the choice of partition P. It is also clear that $L(P,f) \le U(P,f)$, since all Δa_k's are positive numbers. By the Completeness Axiom, the following real numbers then exist:

$$\overline{\int_a^b} f = \inf_P U(P,f), \quad \underline{\int_a^b} f = \sup_P L(P,f),$$

where inf, sup mean with respect to all possible partitions P of $[a,b]$. The two real numbers are called, respectively, the **upper Darboux integral** and the **lower Darboux integral**.

■ Example 6.8

Let $f(x) = 16x^3 - 24x^2 + 9x + 1$ and $n = 10$. The following data are obtained for subintervals $[a_{k-1}, a_k]$ of $[0, 1]$ and of uniform length 0.1:

k	m_k	M_k	k	m_k	M_k
1	1	1.676	6	1.216	3/2
2	1.676	1.968	7	1.028	1.216
3	1.968	2	8	1	1.032
4	1.784	1.972	9	1.032	1.324
5	3/2	1.784	10	1.324	2

Hence, $L(P,f) = (1 + 1.676 + 1.968 + \cdots + 1.324)(0.1) = 1.353$, and $U(P,f) = (1.676 + 1.968 + 2 + \cdots + 2)(0.1) = 1.647$. Consequently, $\overline{\int_0^1} f \le 1.647$ and $\underline{\int_0^1} f \ge 1.353$. ■

Darboux then made the following definition:

Definition. *A function f is **integrable** on $[a,b]$, $b > a$, iff $\overline{\int_a^b} f = \underline{\int_a^b} f$, and we write (provisionally) for the common value (D) $\int_a^b f$.*

The prefatory (D) indicates that we do not know at this point if Darboux's integral is the same as Riemann's. Darboux proved that it is (and so shall we). His formulation was a model for still later theories of integration. The following new concept is needed in Darboux's formulation:

Definition. *A partition P^* of $[a,b]$ is a **refinement** of partition P of $[a,b]$ iff the division points a_k^* in P^* are all of the division points a_k in P plus at least one new point.*

In what follows, a sequence of three lemmas followed by a sequence of four theorems will take us up to the very important connection between continuity and Riemann-integrability (Theorem 6.13).

Lemma 6.4.1. *If f is a bounded function on $[a,b]$ and P^* is a refinement of P, then* $L(P,f) \leq L(P^*,f) \leq U(P^*,f) \leq U(P,f)$.

Proof. The middle inequality is true because it holds for any partition of $[a,b]$. Suppose now that P^* contains just one more point than does P; denote this extra point c and suppose that $c \in (a_{j-1}, a_j)$. We define

$$\begin{cases} m_k^* = m_k = \inf_x f(x) & x \in [a_{k-1}, a_k], k \neq j \\ m_c^* = \inf_x f(x) & x \in [a_{j-1}, c] \\ m_j^* = \inf_x f(x) & x \in [c, a_j]. \end{cases}$$

Figure 6.5 shows possible geometric situations for subinterval $[a_{j-1}, a_j]$; there are only three allowed pairs of relations for m_j, m_c^*, m_j^*.

Thus, we always have $m_j \leq m_c^*$ and $m_j \leq m_j^*$. In view of this, we obtain for $L(P,f)$

$$\begin{aligned} L(P,f) &= \sum_{k=1}^{j-1} m_k \Delta a_k + m_j(a_j - c + c - a_{j-1}) + \sum_{k=j+1}^{n} m_k \Delta a_k \\ &= \sum_{k=1}^{j-1} m_k \Delta a_k + m_j(c - a_{j-1}) + m_j(a_j - c) + \sum_{k=j+1}^{n} m_k \Delta a_k \\ &\leq \sum_{k=1}^{j-1} m_k^* \Delta a_k + m_c^*(c - a_{j-1}) + m_j^*(a_j - c) + \sum_{k=j+1}^{n} m_k^* \Delta a_k \\ &= L(P^*,f). \end{aligned}$$

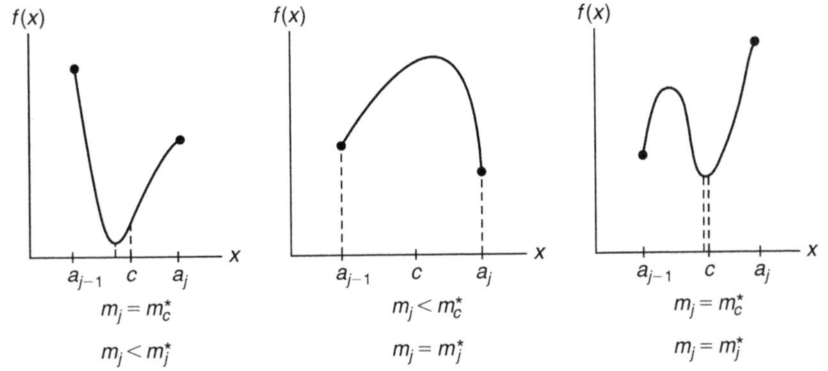

FIGURE 6.5
Inequalities in a partition refinement.

The case where P^* contains many more points than P is handled by mathematical induction. The details of it, and of the analogous inequality $U(P^*,f) \leq U(P,f)$, are left to you. ■

Lemma 6.4.2. *If f is a bounded function on $[a,b]$ and P_1, P_2 are any two partitions of $[a,b]$ whatsoever, then $L(P_1,f) \leq U(P_2,f)$.*

Proof. Let $P^* = P_1 \cup P_2$ denote the partition whose division points are those of P_1 *and* those of P_2. Then P^* is a refinement of P_1 and a refinement of P_2. It follows from Lemma 6.4.1 that

$$L(P_1,f) \leq L(P^*,f) \leq U(P^*,f) \leq U(P_2,f).$$ ■

Lemma 6.4.3. *If f is a bounded function on $[a,b]$, then $\underline{\int_a^b} f \leq \overline{\int_a^b} f$.*

Proof. Momentarily fix the partition P_1; Lemma 6.4.2 shows that $L(P_1,f)$ is a lower bound for the set of all upper Darboux sums $U(P,f)$ and, in particular, to the infimum of this set:

$$L(P_1,f) \leq \overline{\int_a^b} f.$$

Letting P_1 now be variable, we see that $\overline{\int_a^b} f$ is an upper bound to the set of all lower Darboux sums $L(P,f)$ and, in particular, to the supremum of this set:

$$\underline{\int_a^b} f \leq \overline{\int_a^b} f.$$ ■

Theorem 6.10. *Let f be a bounded function on $[a,b]$. Then f is Darboux-integrable on $[a,b]$ if for every $\varepsilon > 0$ there is a $\delta > 0$ such that for all partitions P of $[a,b]$ of norm $\mathfrak{N} < \delta$ we have*

$$U(P,f) - L(P,f) < \varepsilon.$$

Proof. Suppose that for each $\varepsilon > 0$ there is a $\delta > 0$ such that $U(P,f) - L(P,f) < \varepsilon$ holds for all partitions P of norm $\mathfrak{N} < \delta$. Then, by definition,

$$\overline{\int_a^b} f \leq U(P,f)$$

$$= [U(P,f) - L(P,f)] + L(P,f)$$
$$< \varepsilon + L(P,f)$$
$$\leq \varepsilon + \underline{\int_a^b} f.$$

As this is true for all $\varepsilon > 0$, it is true in the limit of $\varepsilon = 0$; that is, $\overline{\int_a^b} f \leq \underline{\int_a^b} f$. But by Lemma 6.4.3 we have $\overline{\int_a^b} f \geq \underline{\int_a^b} f$; the two inequalities imply $\overline{\int_a^b} f = \underline{\int_a^b} f$. ■

Theorem 6.10 may be considered the sufficiency part of a "Cauchy-like" criterion for Darboux-integrability. The obvious necessary part of the criterion is also valid.

Theorem 6.11. *Let f be a bounded, Darboux-integrable function on $[a, b]$. Then for any $\varepsilon > 0$ there is a $\delta > 0$ such that for all partitions P of $[a, b]$ of norm $\mathfrak{N} < \delta$, we have $U(P,f) - L(P,f) < \varepsilon$.*

Proof. Let $\varepsilon > 0$ be given; then, by definition of infimum and supremum, there are partitions P_1, P_2 such that

$$L(P_1,f) > \underline{\int_a^b} f - \frac{\varepsilon}{4}, \quad U(P_2,f) < \overline{\int_a^b} f + \frac{\varepsilon}{4}.$$

The new partition $Q = P_1 \cup P_2$ is a refinement of both P_1 and P_2, so by Lemma 6.4.1 we have

$$L(P_1,f) \leq L(Q,f) \leq U(Q,f) \leq U(P_2,f),$$

or equivalently,

$$U(Q,f) - L(Q,f) \leq U(P_2,f) - L(P_1,f)$$
$$< \left(\overline{\int_a^b} f + \frac{\varepsilon}{4}\right) - \left(\underline{\int_a^b} f - \frac{\varepsilon}{4}\right)$$
$$= \left(\overline{\int_a^b} f - \underline{\int_a^b} f\right) + \frac{\varepsilon}{2} = \varepsilon/2. \qquad (*)$$

Since f is bounded, then $|f(x)| \le K, x \in [a,b]$. Also, suppose that Q has $n > 1$ subintervals; set $\delta = \frac{\varepsilon}{8(n-1)K}$. Let P be any partition of $[a,b]$ with norm $\mathfrak{N} < \delta$. We define the refinement $P^* = P \cup Q$. Suppose P^* were to have only one point that is not in P. Call this point c, and suppose that it falls in subinterval $[a_{j-1}, a_j]$ of P. All terms of $L(P^*, f), L(P, f)$ would coincide except for those involving $[a_{j-1}, a_j]$. We obtain upon subtraction and use of the Triangle Inequality,

$$\begin{aligned} L(P^*,f) - L(P,f) &= m_c^*(c - a_{j-1}) + m_j^*(a_j - c) - m_j(a_j - a_{j-1}) \\ &< |m_c^*|(c - a_{j-1}) + |m_j^*|(a_j - c) + |m_j|(a_j - a_{j-1}) \\ &< K(c - a_{j-1} + a_j - c) + K(a_j - a_{j-1}) \\ &< 2K\mathfrak{N}. \end{aligned}$$

But, in actuality, P^* has at most $n-1$ points that are not in P (Figure 6.6), so if all such points are included, we obtain

$$\begin{aligned} L(P^*,f) - L(P,f) &< 2(n-1)K\mathfrak{N} \\ &< \frac{\varepsilon}{4}. \end{aligned}$$

From Lemma 6.4.1 again, since P^* is a refinement of Q, then $L(Q,f) \le L(P^*,f)$, so

$$L(Q,f) - L(P,f) < \frac{\varepsilon}{4}.$$

Similar reasoning to the previous steps produces for the analogous upper Darboux sums (verify!)

$$U(P,f) - U(Q,f) < \frac{\varepsilon}{4}.$$

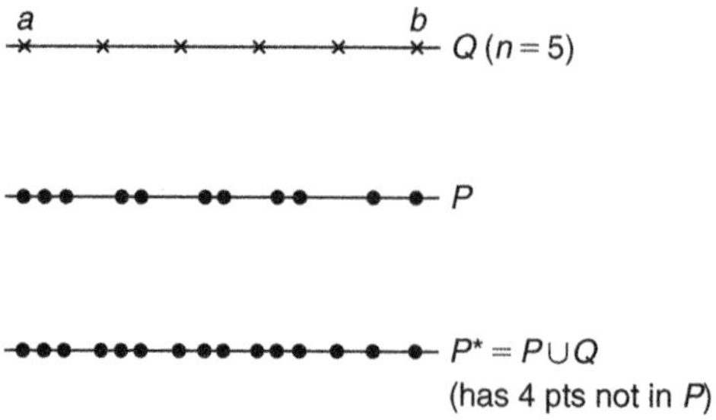

FIGURE 6.6

Tallying points in a refinement.

Combination of the two inequalities gives

$$U(P,f) - L(P,f) < [U(Q,f) - L(Q,f)] + \frac{\varepsilon}{2} < \varepsilon,$$

from equation (*). Thus, if $\varepsilon > 0$, then any partition P with norm $\mathfrak{N} < \delta = \frac{\varepsilon}{8(n-1)K}$ satisfies $U(P,f) - L(P,f) < \varepsilon$. ■

We are now ready to establish the connection between the Darboux integral and the Riemann integral. This will allow us to use facts about either formulation to prove theorems valid for both integrals.

Theorem 6.12. *Let f be a bounded function on $[a,b]$. Then f is Riemann-integrable on $[a,b]$ iff f is Darboux-integrable there, and the two integrals are equal.*

Proof. ($\rightarrow$) Suppose that f is Riemann-integrable:

$$\int_a^b f = I.$$

Let $\varepsilon > 0$ be given; then there is a $\delta > 0$ such that for all tagged partitions $\overline{P}$ with norm $\mathfrak{N} < \delta$ we have

$$\left| \sum_{k=1}^{n} f(x_k)\Delta a_k - I \right| < \varepsilon. \qquad (*)$$

We now make a specific choice for each of the tags; for each subinterval $[a_{k-1}, a_k]$ choose x_k to be any point in $[a_{k-1}, a_k]$ such that $f(x_k) \leq m_k + \varepsilon$. The corresponding Riemann sum is then

$$\begin{aligned} S = \sum_{k=1}^{n} f(x_k)\Delta a_k &\leq \sum_{k=1}^{n} (m_k + \varepsilon)\Delta a_k \\ &= \sum_{k=1}^{n} m_k \Delta a_k + \varepsilon(b-a) \\ &= L(P,f) + \varepsilon(b-a). \qquad (**) \end{aligned}$$

Hence, we obtain from equations (*) and (**)

$$I - \varepsilon - \varepsilon(b-a) < S - \varepsilon(b-a) \leq L(P,f) \leq \underline{\int_a^b} f,$$

and since this is true for every $\varepsilon > 0$, then in the limit of $\varepsilon = 0$ we have

$$I \leq \underline{\int_a^b} f.$$

Similar reasoning in which each tag x_k is any point in $[a_{k-1}, a_k]$ such that $f(x_k) \geq M_k - \varepsilon$ leads to the result (verify!)

$$\overline{\int_a^b} f \leq I.$$

Thus, we have, upon combination, $\overline{\int_a^b} f \leq \underline{\int_a^b} f$. Since Lemma 6.4.3 has given us $\underline{\int_a^b} f \leq \overline{\int_a^b} f$ (whether f is Darboux-integrable or not), we can conclude that $\underline{\int_a^b} f = \overline{\int_a^b} f$, so $I = (D) \int_a^b f$.

($\leftarrow$) Suppose that f is Darboux-integrable. Let $\varepsilon > 0$ be given; there is $\delta > 0$ such that for every partition P with norm $\mathfrak{N} < \delta$ we have from Theorem 6.11

$$U(P, f) - L(P, f) < \varepsilon. \tag{*}$$

Now suppose that tags are chosen for each subinterval of $[a, b]$; we have for $k = 1, 2, \ldots, n$ that $m_k \leq x_k \leq M_k$, so

$$L(P, f) = \sum_{k=1}^{n} m_k \Delta a_k \leq \sum_{k=1}^{n} f(x_k) \Delta a_k \leq \sum_{k=1}^{n} M_k \Delta a_k = U(P, f). \tag{**}$$

From equations (*) and (**) and the definitions of the upper and lower Darboux integrals, we obtain

$$\overline{\int_a^b} f \leq U(P, f) < L(P, f) + \varepsilon \leq S + \varepsilon$$

and

$$\underline{\int_a^b} f \geq L(P, f) > U(P, f) - \varepsilon \geq S - \varepsilon.$$

Finally, from Darboux's definition of the integral we obtain

$$-\varepsilon < S - (D)\int_a^b f < \varepsilon.$$

As $\varepsilon > 0$ is arbitrary, this shows that $(D)\int_a^b f = \int_a^b f$. ■

■ Example 6.9

Let f be the Dirichlet function (see Example 4.6), and let P be any partition of $[0, 1]$. We have

$$\overline{\int_0^1} f = \inf_P \sum_{k=1}^{n} M_k \Delta a_k = 1$$

$$\underline{\int_0^1} f = \sup_P \sum_{k=1}^{n} m_k \Delta a_k = 0.$$

Since $0 \neq 1$, it follows from Theorem 6.12 that f is not Riemann-integrable. ■

Theorem 6.13. *If f is continuous on $[a, b]$, then f is Riemann-integrable on $[a, b], b > a$.*

Proof. By Theorem 4.16, f is uniformly continuous on the compact set $[a, b]$. Hence, if $\varepsilon > 0$ is given, then there is a $\delta > 0$ such that for any $x_1, x_2 \in [a, b]$ that satisfy $|x_1 - x_2| < \delta$, we have $|f(x_1) - f(x_2)| < \frac{\varepsilon}{b-a}$. Now let P be any partition of $[a, b]$ with norm $\mathfrak{N} < \delta$. Then on each subinterval $[a_{k-1}, a_k]$ it follows that $0 \leq M_k - m_k < \frac{\varepsilon}{b-a}$, so

$$\begin{aligned} U(P, f) - L(P, f) &= \sum_{k=1}^{n} (M_k - m_k)\Delta a_k \\ &< \frac{\varepsilon}{b-a} \sum_{k=1}^{n} \Delta a_k = \varepsilon. \end{aligned}$$

By Theorems 6.10 and 6.12 we conclude that f is (Riemann)-integrable on $[a, b]$. ■

■ Example 6.10

The following integrals exist, by virtue of Theorem 6.13:

(a) all the integrals in Example 6.7;

(b) $\int_0^{2\pi} \sin(\sqrt{x})dx$;

(c) $\int_0^{-1} \frac{2x^2-3x-5}{x^3+2x-2}dx$;

(d) $I = \int_0^{\pi/4} \frac{dz}{\sqrt{1-k^2\sin^2 z}}$, $(\sin^{-1} k = \pi/3)$.[6] ■

Corollary 6.13.1 (Mean-Value Theorem for Integrals). *If f is continuous on $[a,b]$, then there is a point $c \in [a,b]$ at which*

$$\int_a^b f(x)dx = f(c)(b-a).$$

Proof. By Theorem 6.13, the integral exists. Let $M = \sup_x f(x)$ and $m = \inf_x f(x), x \in [a,b]$; use Theorems 6.2, 4.10, and 4.14. The completion of the proof is left to you. ■

■ Example 6.11

Since $f(t) = \frac{1}{1+t^2}$ is continuous on $\mathbf{R}^1$, then $F(x) = \int_0^x \frac{dt}{1+t^2}$ exists for any $x \in \mathbf{R}^1$. Suppose $x_2 > x_1$; then by Theorem 6.5 and Corollary 6.5.1, we have

$$F(x_2) - F(x_1) = \int_0^{x_2} \frac{dt}{1+t^2} - \int_0^{x_1} \frac{dt}{1+t^2} = \int_{x_1}^{x_2} \frac{dt}{1+t^2}.$$

From the Mean-Value Theorem for Integrals, there is a $c \in [x_1, x_2]$ such that $\int_{x_1}^{x_2} \frac{dt}{1+t^2} = \frac{1}{1+c^2}(x_2 - x_1) > 0$. Hence, $F(x)$ is an increasing function. Further, $F(0) = 0$ implies that $F(x) > 0$ for $x > 0$. The function F is an entry point into a formal development of the trigonometric functions (Exercises 6.38, 6.50, 6.51). ■

We defined earlier a finitely piecewise monotonic function. Rather analogously, a **finitely piecewise continuous** function on $[a,b]$ is one for which $[a,b]$ can be partitioned into a finite number of closed subintervals, on the interiors of which f is continuous and for which both one-sided limits at the endpoints are finite. Step-functions are simple examples of finitely piecewise continuous functions.

---

[6] This is an **elliptic integral of the first kind**; more on elliptic integrals in Exercises 6.15, 6.16, and 6.46.

Theorem 6.14. *If f is finitely piecewise continuous on $[a, b]$, then f is integrable on $[a, b]$.*

Proof. Let P be the partition $a = a_0 < a_1 < a_2 < \cdots < a_n = b$, and suppose that f is continuous on each open interval $(a_{k-1}, a_k), k = 1, 2, \ldots, n$. Assume also that on each open interval the two limits $\lim_{x \to a_{k-1}^+} f(x)$, $\lim_{x \to a_k^-} f(x)$ exist in $\mathbf{R}^1$.

For each k we define the function

$$f_k(x) = \begin{cases} f(x) & a_{k-1} < x < a_k \\ \lim_{x \to a_{k-1}^+} f(x) & x = a_{k-1} \\ \lim_{x \to a_k^-} f(x) & x = a_k. \end{cases}$$

Then f_k is continuous on $[a_{k-1}, a_k]$ and by Theorem 6.13 each integral $\int_{a_{k-1}}^{a_k} f_k(x)dx$ exists. But by Corollary 6.6.1, the values of f_k at the two points a_{k-1}, a_k are irrelevant, so if these values are changed back to $f(a_{k-1})$ and $f(a_k)$, then the integrand can be rewritten as f and we have

$$\int_{a_{k-1}}^{a_k} f(x)dx = \int_{a_{k-1}}^{a_k} f_k(x)dx.$$

Finally, Theorem 6.5 extends this procedure to all the subintervals in P and f is integrable on all of $[a, b]$. ■

■ Example 6.12

The proviso of the assumption of integrability can now be removed from Example 6.5. As another example, suppose that $\mathbf{S} = \{n^{-1} : n = 1, 2, \ldots, 100\}$. Then if g is defined by $g(x) = \begin{cases} x^3 & x \notin \mathbf{S} \\ 2 & x \in \mathbf{S} \end{cases}$, we have $\int_0^2 g(x)dx = 4$. ■

6.5 THE FUNDAMENTAL THEOREM OF THE CALCULUS

The celebrated Fundamental Theorem of the Calculus was presented by Newton and Leibniz in their work. Newton's predecessor, **Isaac Barrow** (1630–1677), was aware of the relationship between finding the tangent to a curve and the calculation of an area under that curve, but he probably did not appreciate its significance (Kline, 1972).

It has become standard to state the modern version of the Fundamental Theorem as a pair of independent statements, one involving integration of derivatives and the other involving differentiation of indefinite integrals. Our exposition is based on Cunningham (1965).

Theorem 6.15 (Fundamental Theorem of the Calculus—A). *If f is continuous on $[a, b]$ and differentiable on (a, b) and f' is integrable on $[a, b]$, then*

$$\int_a^b f' = f(b) - f(a).$$

Proof. Let $P = \{a_k\}_{k=0}^n$ denote a partition of $[a, b]$ with norm $\mathfrak{N}$. The difference $f(b) - f(a)$ is given by a telescoping sum

$$f(b) - f(a) = \sum_{k=1}^n [f(a_k) - f(a_{k-1})].$$

On each subinterval $[a_{k-1}, a_k]$ we can apply the Mean-Value Theorem because f is continuous on $[a, b]$ and differentiable on (a, b). The completion of the proof is left to you. ■

As stated, Theorem 6.15 applies to integrals whose integrands are derivatives that are defined on (a, b) and are bounded on $[a, b]$. There are, of course, other functions that are not derivatives but that are still Riemann-integrable, and to these integrals Theorem 6.15 (as stated) does not apply. However, by a suitable modification of the wording, the Fundamental Theorem of the Calculus—A can be made to apply to all Riemann integrable functions (Botsko, 1991).

On the other hand, not every derivative is Riemann-integrable. The condition of integrability of f' is needed in the hypotheses of Theorem 6.15. A standard example is the function

$$f(x) = \begin{cases} x^2 \sin(1/x^2) & 0 < x \le 1 \\ 0 & x = 0. \end{cases}$$

Its derivative is (verify!)

$$f'(x) = \begin{cases} 2x \sin(1/x^2) - 2x^{-1} \cos(1/x^2) & 0 < x \le 1 \\ 0 & x = 0. \end{cases}$$

But $f'(x)$ is unbounded in any neighborhood of the origin, so the integral $\int_0^1 f'$ does not exist (Theorem 6.1). There are even functions f such that f' is defined and bounded on $[a, b]$, and yet, f' is not integrable (Chatterji, 1988). Finally, a criterion exists that tells us just when a derivative f' is Riemann-integrable over $[a, b]$ (van de Lune, 1975).

Integration formulas, no matter how complicated and no matter how they are deduced (Exercise 6.35), depend upon the Fundamental Theorem of the Calculus—A. The importance of the theorem, therefore, cannot be overestimated.

■ Example 6.13

The integrand in $\int_{2a}^{4a} \frac{\sqrt{x^2-a^2}}{x} dx$, $a > 0$, is the derivative of the function

$$f(x) = \sqrt{x^2 - a^2} - a\sec^{-1}(x/a)$$
$$(0 \le \mathrm{Sec}^{-1} u < \pi/2 \text{ if } u \ge 1).$$

This function is differentiable (and continuous) at all $x \ge a$, as is the integrand. By Theorem 6.13 the integral exists, and Theorem 6.15 gives

$$\int_{2a}^{4a} \frac{\sqrt{x^2 - a^2}}{x} dx = f(4a) - f(2a)$$
$$= a\left[\left(\sqrt{15} - \sqrt{3}\right) + \left(\mathrm{Sec}^{-1}2 - \mathrm{Sec}^{-1}4\right)\right].$$ ■

Theorem 6.16 (Fundamental Theorem of the Calculus—B). *If f is continuous on $[a, b]$ and $F(x) = \int_a^x f$ for each $x \in [a, b]$, then*

$$F'(x) = \frac{d}{dx}\int_a^x f(t)dt = f(x).$$

Proof. Let $|h|$ be nonzero but sufficiently small so that $x + h \in [a, b]$ for all $x \in [a, b]$. At the left (right) endpoints, h must be positive (negative). Then

$$F(x+h) - F(x) = \int_a^{x+h} f - \int_a^x f$$
$$= \int_x^{x+h} f \quad \text{from Theorem 6.5.}$$

This holds because f is continuous on $[a, b]$ and on any subinterval of it, so f is integrable there. From the Mean-Value Theorem for Integrals (Corollary 6.13.1), we can write

$$\int_{x}^{x+h} f = hf(c_h),$$

for some c_h satisfying $x < c_h < x+h$ (if $h > 0$), or $x+h < c_h < x$ (if $h < 0$). Hence, we obtain

$$\frac{F(x+h)-F(x)}{h} = f(c_h). \qquad (*)$$

As $h \to 0$, $c_h \to x$, and since f is continuous on $[a,b]$, then $f(c_h) \to f(x)$. The left-hand side of equation $(*)$ must also approach a limit; by definition, this is $F'(x)$, and we arrive at $F'(x) = f(x)$. ■

Corollary 6.16.1. *If f is continuous on $[a,b]$, then*

$$F(x) = \int_{a}^{x} f(t)dt$$

is also continuous on $[a,b]$.

Proof. This follows immediately from Theorems 6.16 and 5.1. ■

■ Example 6.14

A major use of the Fundamental Theorem of the Calculus—B is in connection with the generation of transcendental functions. Recalling Approach 2 following Example 4.8, we see that $\ln x = \int_1^x \frac{dt}{t}$ is continuous at every $x > 0$, since $1/t$ is continuous at every $t > 0$. ■

■ Example 6.15

The **Fresnel integrals**, which occur in the physics of diffraction at a slit, are defined as

$$C(x) = \int_{0}^{x} \cos(\pi t^2/2)dt, \quad S(x) = \int_{0}^{x} \sin(\pi t^2/2)dt.$$

The integrals exist for any real x because the integrands are continuous on $[0,x]$. Hence, by the Fundamental Theorem of the Calculus—B, we then have

$$C'(x) = \cos(\pi x^2/2), \quad S'(x) = \sin(\pi x^2/2).$$

These functions are transcendental functions. It is possible to deduce that $C(x)$, $S(x)$ are themselves transcendental functions.[7] ■

Refer once again to Figure 6.1. Do you see that the figure provides a geometric interpretation of the Fundamental Theorem of the Calculus—B?

Theorems 6.15 and 6.16 have to be regarded as amazing. The processes of differentiation and integration are defined independently of each other, but the two theorems show that differentiation and integration are related in a simple and pretty way.

6.6* NUMERICAL INTEGRATION—SIMPSON'S RULE

Any procedure for the estimation of the value of a definite integral that does not make use of the Fundamental Theorem of the Calculus and does not express the value in exact, closed form (e.g., as $\pi^2/6$) may be called **numerical integration. Simpson's Rule**, named after the self-taught English mathematician **Thomas Simpson** (1710–1761) and published by him in 1743, is one such procedure.

Let f be continuous on $[a, b]$, and let $P = \{a_k\}_{k=0}^{n}$ be a partition of the interval into n subintervals of width $h = (b - a)/n$; we stipulate that n be even. Choose any odd k, $0 < k < n$, and define $u = x - a_k$. We seek a parabola $Q_k(u) = Au^2 + Bu + C$ that passes through the three points $(-h, f(a_{k-1}))$, $(0, f(a_k))$, $(h, f(a_{k+1}))$ (Figure 6.7). This leads to the system of equations

$$\begin{cases} Ah^2 - Bh + C &= f(a_{k-1}) \\ C &= f(a_k) \\ Ah^2 + Bh + C &= f(a_{k+1}). \end{cases}$$

The solution of this system is, from Cramer's Rule,

$$A = \begin{vmatrix} f(a_{k-1}) & -h & 1 \\ f(a_k) & 0 & 1 \\ f(a_{k+1}) & h & 1 \end{vmatrix} \Bigg/ \begin{vmatrix} h^2 & -h & 1 \\ 0 & 0 & 1 \\ h^2 & h & 1 \end{vmatrix}$$

---

[7] For this we need a little algebra, namely, that sums, differences, products, and quotients of algebraic functions are themselves algebraic (Herstein, 1975). The strategy now is to show that if f is algebraic, then f' is also algebraic (try to prove this last statement). The particular improper Fresnel integrals $C(\infty)$ and $S(\infty)$ have the common value ½ (Olds, 1968).

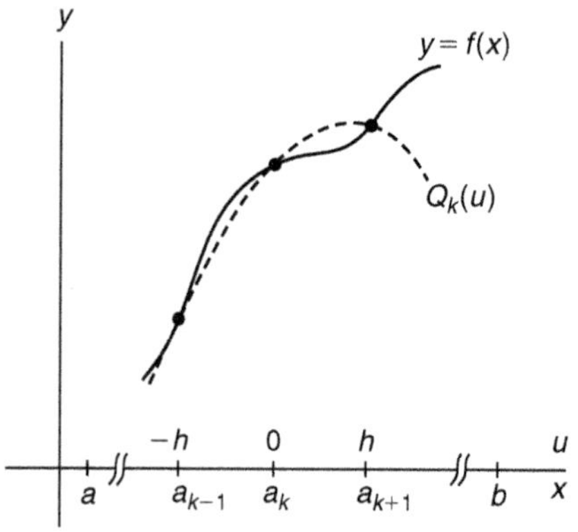

FIGURE 6.7
A parabola $Q_k(u)$ that passes through three points on a curve.

$$= \frac{1}{2h^2}[f(a_{k-1}) - 2f(a_k) + f(a_{k+1})]$$

$$B = \frac{-1}{2h}[f(a_{k-1}) - f(a_{k+1})]$$

$$C = f(a_k).$$

Then, upon evaluation of $\int_a^b Q_k(u)du$ in order to find the "area" enclosed by the parabolic arc, we find (verify!)

$$\int_a^b Q_k(u)du = \frac{h}{3}[f(a_{k-1}) + 4f(a_k) + f(a_{k+1})].$$

Finally, we apply this formula to successive cases $k = 1, 3, 5, \ldots, n-1$, and add the contributions due to each parabolic arc. The final result, $A_n(f)$, for the estimated "area" is as follows.

Theorem 6.17 (Simpson's Rule). *If f is continuous on $[a, b]$ and a partition $P = \{a_k\}_{k=0}^{n}$ is constructed as described earlier, then*

$$\int_a^b f(x)dx \approx \frac{h}{3}[f(a_0) + 4f(a_1) + 2f(a_2) + 4f(a_3) + 2f(a_4)$$
$$+ \cdots + 4f(a_{n-1}) + f(a_n)]$$
$$= A_n(f).$$

■ Example 6.16

The integral in Example 6.10(d) cannot be evaluated analytically, but must be evaluated numerically. Taking $n = 8$, we have from Theorem 6.17,

$$
\begin{aligned}
I &= F(k, \pi/4) \approx A_8(f) \\
&= \frac{\pi}{96}\left\{1 + 4\left[1 - \frac{3}{4}\sin^2(\pi/32)\right]^{-1/2} + 2\left[1 - \frac{3}{4}\sin^2(\pi/16)\right]^{-1/2}\right. \\
&\quad \left. + 4\left[1 - \frac{3}{4}\sin^2(3\pi/32)\right]^{-1/2} + \cdots + 4\left[1 - \frac{3}{4}\sin^2(7\pi/32)\right]^{-1/2} + \sqrt{\frac{8}{5}}\right\} \\
&\approx \frac{\pi}{96}\{1 + 4.0144894 + 2.0291712 + 4.1327230 + 2.1197992 \\
&\quad + 4.3817660 + 2.2814255 + 4.7872142 + 1.2649111\} \\
&= 0.8512243.
\end{aligned}
$$

The tabulated value is 0.85122375 (Abramowitz and Stegun, 1965). ■

The previous example is quite satisfying. It is possible to improve upon Theorem 6.17 by modifying the mechanics of the handling of the parabolic arcs (Richert, 1985; Velleman, 2004). We can also replace the parabolic arcs by cubic, quartic, or higher-degree splines in order to achieve greater accuracy.

Nevertheless, despite its relative simplicity, Simpson's Rule tends to work surprisingly well. How well does it work? We can quantify this if sufficient continuity conditions are imposed upon f.

Theorem 6.18. *If $f, f', f'', f^{(3)}, f^{(4)}$ are continuous on $[a, b]$, if $A_n(f)$ is the nth-order Simpson approximation to $\int_a^b f(x)dx$, and if $M = \sup_x |f^{(4)}(x)|$, then*

$$\left|\int_a^b f(x)dx - A_n(f)\right| \le (b-a)^5 M/(180n^4).$$

Proof. Corresponding to each odd $k, k = 1, 3, 5, \ldots, n-1$, we define the differentiable function

$$g_k(h) = \int_{a_k-h}^{a_k+h} f(x)dx - \frac{h}{3}[f(a_k - h) + 4f(a_k) + f(a_k + h)],$$

where $h = a_{k+1} - a_k$ is now regarded as a continuous variable (that is, the norm of the partition of $[a, b]$ is allowed to change continuously). The expression on the right can be interpreted as the error in the area made by approximating the integral shown by an "area" under one parabolic arc.

Four differentiations of $g_k(h)$ lead ultimately to (verify!)

$$g_k^{(4)}(h) = \frac{-h}{3}\left[f^{(4)}(a_k+h) + f^{(4)}(a_k-h)\right] - \frac{1}{3}\left[f^{(3)}(a_k+h) - f^{(3)}(a_k-h)\right].$$

It is easily verified that $g_k(0) = g_k'(0) = g_k''(0) = g_k^{(3)}(0) = g_k^{(4)}(0) = 0$. Also, in obtaining $g_k'(h)$ from $g_k(h)$, we have called upon Exercise 6.41.

Since $f^{(4)}$ is continuous on $[a, b]$, it assumes a maximum value somewhere there (Theorem 4.10), that is, there is a number $M > 0$ such that $-M \le f^{(4)}(x) \le M$ for all $x \in [a, b]$. It then follows that

$$\int_0^h -M\,dx \le \int_0^h f^{(4)}(a_k \pm x)dx \le \int_0^h M\,dx,$$

or $\left|f^{(3)}(a_k+h) - f^{(3)}(a_k-h)\right| \le 2Mh$. An application of the Triangle Inequality to $g_k^{(4)}(h)$ above now gives

$$\left|g_k^{(4)}(h)\right| \le \frac{h}{3}(M+M) + \frac{1}{3}(2Mh) = 4hM/3.$$

Writing this as $-4tM/3 \le g_k^{(4)}(t) \le 4tM/3$, where t is now a dummy variable, we integrate four times, using 0 as the lower limit on the integrals and h alternately as the upper limit and as a new dummy variable. We arrive at

$$-h^5M/90 \le g_k(h) \le h^5M/90.$$

For a given (even) n, there are $n/2$ odd h's, so addition of all the errors yields, finally,

$$\begin{aligned}\left|\int_a^b f(x)dx - A_n(f)\right| &\le \sum_k |g_k(h)| \\ &\le (n/2)(h^5M/90) \\ &= (b-a)^5M/\left(180n^4\right),\end{aligned}$$

since $h = (b-a)/n$. ■

■ Example 6.17

What order of Simpson approximation to $\int_0^{\pi/2} \sin(2x)dx$ will guarantee an error of magnitude less than 1×10^{-5}?

If $f(x) = \sin(2x)$, it is found that $f^{(4)}(x) = 16\sin(2x)$, so $|f^{(4)}(x)| \le M = 16$. Hence, we require from Theorem 6.18

$$\frac{(\pi/2)^5 \cdot 16}{180n^4} < 10^{-5},$$

or $n > 17$. Thus, the 18th-order Simpson approximation ($A_{18}(f)$) will certainly do. A lower-order approximation may also suffice. ■

■ Example 6.18

Consider $\int_0^1 \sqrt{8t^2+1}dt$; the integrand has an antiderivative, and we find that $\int_0^1 \sqrt{8t^2+1}dt \approx 1.81161262$. Use of Theorem 6.17, with $n = 6$, gives $A_6(f) \approx 1.8116324$, so $\left|\int_0^1 \sqrt{8t^2+1}dt - A_6(f)\right| \approx 1.98 \times 10^{-5}$. If we let $f(t) = (8t^2+1)^{1/2}$, then we can establish that $f^{(4)}(t) = 192[f(t)]^{-5}\{4 - 5[f(t)]^{-2}\}$, and that $f^{(4)}(0) = -192$ and $f^{(4)}(1) = 1984/729 \approx 2.72$. Potential local extrema of $f^{(4)}(t)$ occur where $f^{(5)}(t) = 0$; this leads to $t = \sqrt{3/32}$, and $f^{(4)}(\sqrt{3/32}) = 49{,}152\sqrt{7}/2401 \approx 54$. Hence, $|f^{(4)}(t)|$ is bounded above by $M = 192$, and we have from Theorem 6.18

$$\begin{aligned}\left|\int_0^1 \sqrt{8t^2+1}dt - A_6(f)\right| &\le (1-0)^5(192)/[180(6^4)] \\ &\approx 8.23 \times 10^{-4}.\end{aligned}$$

Note that the upper bound to the error that is supplied by Theorem 6.18 is not necessarily very sharp. ■

6.7 IMPROPER INTEGRALS

The symbols $\int_0^\infty e^{-t^2}dt$ and $\int_3^4 \frac{dt}{\sqrt{t-3}}$ are not, strictly speaking, the symbols for Riemann integrals. Yet they look like they could usefully represent meaningful quantities, and like the notation $\sum_{n=1}^\infty a_n$, they call for some extension of the relevant theory. Also, like the notation $\sum_{n=1}^\infty a_n$, the notation for improper integrals such as the preceding two requires some care in its definition, as neither integral can be viewed as the limit of a Riemann sum as the norm $\mathfrak{N} \to 0$. The first way in which an integral cannot be a limit of a Riemann sum is if the interval of integration is infinite.

Definition. *Let $a \in \mathbf{R}^1$ be fixed and suppose that for all real $x \ge a$ the function f is Riemann-integrable on $[a, x]$. If the upper limit of integration is allowed to be* *unbounded, the symbol $\int_a^\infty f(t)dt$ shall denote an* ***improper integral of the first kind.***[8]

■ Example 6.19

All the following are symbols for improper integrals of the first kind, either directly from the definition or from minor extensions of the definition:

(a) $\int_0^\infty e^{-t^2} dt$;

(b) $\int_\infty^1 (\ln t)dt$;

(c) $\int_{-\infty}^1 \frac{dt}{1+t^2}$;

(d) $\int_{-\infty}^\infty \frac{t^2 dt}{1+t^4}$. ■

The second way in which an integral cannot be a limit of a Riemann sum is if on a finite interval of integration $[a, b]$ there is a point at which f is unbounded (see Theorem 6.1).

Definition. *Suppose that there exists an $x \in [a, b]$, $-\infty < a < b < \infty$, at which f is unbounded or indeterminate, and suppose that for positive ε*

(a) *$\varepsilon < \min\{x - a, b - x\}$ implies that $\int_a^{x-\varepsilon} f(t)dt$ and $\int_{x+\varepsilon}^b f(t)dt$ are Riemann-integrable,* or

(b) *$\varepsilon < b - a$ implies that either $\int_{a+\varepsilon}^b f(t)dt$ or $\int_a^{b-\varepsilon} f(t)dt$ is Riemann-integrable.*

Then the symbol $\int_a^b f(t)dt$ shall denote an ***improper integral of the second kind.***

■ Example 6.20

All the following are symbols for improper integrals of the second kind, either directly from the definition or from minor extensions of the definition:

(a) $\int_{-1}^1 \frac{dt}{t^3}$;

(b) $\int_3^4 \frac{dt}{\sqrt{t-3}}$;

(c) $\int_{\pi/4}^{\pi/2} \sqrt{(\sin t)(\tan t)}dt$;

(d) $\int_{-1}^1 \frac{t\,dt}{\sqrt[3]{t^2-1}}$. ■

---

[8]The designations improper integrals of the "first kind" and "second kind" are employed by Shilov, but these terms are modifications of "infinite integral of the first (second) kind," which were introduced by Hardy. These latter terms suggest analogies with infinite series.

So far, improper integrals have been nothing more than symbols; this was the way in which the treatment of infinite series began. The theory of improper integrals, in fact, has a number of parallels to that of infinite series.[8] We proceed to add some content to the concept of an improper integral; for simplicity, we adopt the generic symbol $I(f)$ for an improper integral of the function f.

Property 1. (Scalar Multiplication) The product $(\cdot)$ of $c \in \mathbf{R}^1$ and the improper integral $I(f)$ is the new improper integral $I(cf)$.

Property 2. (Integral Addition) Addition $(\oplus)$ of the two improper integrals $I(f)$ and $I(g)$ on a common interval of integration is the new improper integral $I(f+g)$ on that same interval of integration.

By far, the most important definitional property of an improper integral is that of convergence (or not) of the integral.

Property 3. (Convergence) An improper integral of the first kind is said to **converge** to the number $I \in \mathbf{R}^1$ iff $\lim_{x\to\infty} \int_a^x f(t)dt = I$ (or, analogously, $\lim_{x\to-\infty} \int_x^b f(t)dt = I$). We then write $\int_a^\infty f(t)dt = I$ (or, analogously, $\int_{-\infty}^b f(t)dt = I$).

An improper integral $\int_a^b f(t)dt$ of the second kind converges to $I \in \mathbf{R}^1$ iff $\lim_{\varepsilon\to 0+} \int_a^{x-\varepsilon} f(t)dt = I_1 \in \mathbf{R}^1$, $\lim_{\varepsilon\to 0+} \int_{x+\varepsilon}^b f(t)dt = I_2 \in \mathbf{R}^1$, and $I_1 + I_2 = I$ (and, analogously, if the aberrant point x is a or b). We then write $\int_a^b f(t)dt = I$.

An improper integral that does not converge is said to **diverge** (or to be **divergent**).

As with infinite series, Properties 1 and 2 have nothing to do with convergence. Fortunately, also, these properties lead to desirable practical consequences.

Theorem 6.19. *Suppose that improper integrals $I(f)$ and $I(g)$, presumed to be defined on a common interval of integration, converge to A and B, respectively. Then if $c \in \mathbf{R}^1$, we have (a) $c \cdot I(f) = cA$, and (b) $I(f) \oplus I(g) = A + B$.*

Proof.

(a) We prove this part for an $I(f)$ of the first kind; the proof for an improper integral of the second kind is analogous. Suppose that $I(f) = \int_a^\infty f(t)dt$; let $F(x) = \int_a^x f(t)dt$, $x \geq a$. Then for any $c \in \mathbf{R}^1$, Theorem 6.4(i) gives

$$cF(x) = c\int_a^x f(t)dt = \int_a^x cf(t)dt.$$

By hypothesis and by Theorem 1.10(i), applied to the case where the domain of $F(x)$ is a subset of **Re**, we then have

$$\lim_{x\to\infty} cF(x) = c \lim_{x\to\infty} F(x) = cA = \lim_{x\to\infty} \int_a^x cf(t)dt = \int_a^\infty cf(t)dt.$$

(b) Proof of this part is left to you. ■

■ Example 6.21

In Example 6.19, only (b) diverges. To prove that (c) converges, for example, we write

$$F(x) = \int_x^1 \frac{dt}{1+t^2} = \operatorname{Tan}^{-1} t\Big|_x^1$$
$$= (\pi/4) - \operatorname{Tan}^{-1} x.$$

Then, by definition (and Exercise 6.50), we obtain

$$\int_{-\infty}^1 \frac{dt}{1+t^2} = \lim_{x\to-\infty} [(\pi/4) - \operatorname{Tan}^{-1} x]$$
$$= 3\pi/4.$$

The result is interpretable graphically in Figure 6.8.

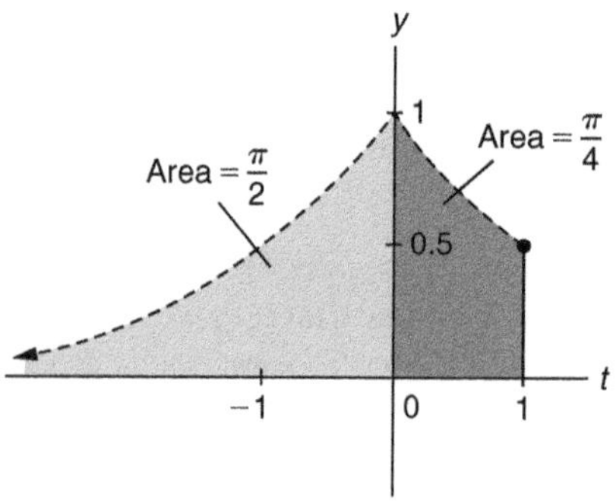

FIGURE 6.8
Geometric meaning of the improper integral $\int_{-\infty}^1 \frac{dt}{1+t^2}$.

■

■ Example 6.22

In Example 6.20, only (a) diverges. The integrand there is unbounded in any interval about $t = 0$. Then, by definition,

$$\int_{-1}^{1} \frac{dt}{t^3} = \lim_{\varepsilon\to 0+} \int_{-1}^{-\varepsilon} \frac{dt}{t^3} + \lim_{\varepsilon\to 0+} \int_{\varepsilon}^{1} \frac{dt}{t^3}$$

$$= \left[\lim_{\varepsilon\to 0+} \left(\frac{-1}{2\varepsilon^2} + \frac{1}{2}\right)\right] + \left[\lim_{\varepsilon\to 0+} \left(-\frac{1}{2} + \frac{1}{2\varepsilon^2}\right)\right].$$

The result is an indeterminate expression of the form $-\infty + \infty$; hence, the improper integral diverges. ■

Some comments are pertinent to Example 6.19(d) and Example 6.22. The following test, which is quite analogous to the Comparison Test in Theorem 3.4, would be useful in connection with Example 6.19(d).

Theorem 6.20 (Comparison Test for Improper Integrals). *Suppose that for all $t \geq a$ we have $0 \leq f(t) \leq g(t)$. If $\int_a^\infty g(t)dt$ converges and f is Riemann-integrable on any interval $[a, x]$, $x = a$, then $\int_a^\infty f(t)dt$ converges and*

$$\int_a^\infty f(t)dt \leq \int_a^\infty g(t)dt.$$

Proof. It follows from Theorem 6.2 that

$$0 \leq \int_a^x f(t)dt \leq \int_a^x g(t)dt \leq \int_a^\infty g(t)dt.$$

Since $F(x) = \int_a^x f(t)dt$ and $G(x) = \int_a^x g(t)dt$ are bounded above by $\int_a^\infty g(t)dt$, which converges by hypothesis, and since $F(x), G(x)$ are increasing functions, then from the Completeness Axiom both $F(x), G(x)$ have limits as $x \to \infty$. An adaptation of Theorem 2.8 then gives $\lim_{x\to\infty} F(x) \leq \lim_{x\to\infty} G(x)$. ■

Theorem 6.20 has a plausible geometric interpretation; we show this in Figure 6.9. Note that the theorem is stated for nonnegative-valued integrands.

■ Example 6.23

In Example 6.19(d) choose any $a \in (-\infty, \infty)$; a common, but arbitrary, choice is $a = 0$. Then for $c \geq 0$, define $F(x) = \int_0^c \frac{t^2dt}{1+t^4} + \int_c^x \frac{t^2dt}{1+t^4}$, $x > c$. The first integral exists because $t^2/(1 + t^4)$ is continuous on $[0, c]$. For the second

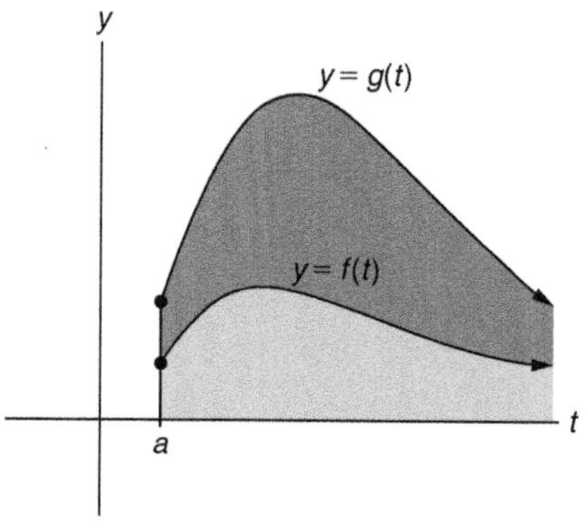

FIGURE 6.9
The geometric meaning of the Comparison Test for Improper Integrals of the First Kind.

integral we have $\frac{t^2}{1+t^4} < \frac{1}{t^2}$, so by Theorem 6.20 we obtain

$$\lim_{x\to\infty} \int_c^x \frac{t^2 dt}{1+t^4} \leq \lim_{x\to\infty} \int_c^x \frac{dt}{t^2} = \lim_{x\to\infty} \left(\frac{-1}{x} + \frac{1}{c} \right) = \frac{1}{c}.$$

It follows that $\int_0^\infty \frac{t^2 dt}{1+t^4}$ converges. Parallel reasoning shows that $\int_{-\infty}^0 \frac{t^2 dt}{1+t^4}$ also converges; hence, the integral in Example 6.19(d) converges. ■

A different issue arises in connection with Example 6.22. It might be argued that the terms $-1/(2\varepsilon^2)$ and $1/(2\varepsilon^2)$ cancel and, of course the $1/2$ and $-1/2$ also cancel, so the final limit ought to be 0. The definition of the value of an improper integral of the second kind in which the aberrant point is an interior point, however, requires that the two indicated limits be handled *separately*.

But just as it has been possible to construct alternative methods of summability of infinite series (see Section 3.2), so it has been possible to define alternative ways of attaching a value to certain improper integrals. Thus, let f be defined on $[a, b]$, except possibly at $c \in (a, b)$, where f is either undefined or indeterminate. Then if $\int_a^b f(t)dt$ diverges but the single limit

$$\lim_{\varepsilon\to 0^+} \left[\int_a^{c-\varepsilon} f(t)dt + \int_{c+\varepsilon}^b f(t)dt \right] = L$$

happens to exist (in $\mathbf{R}^1$), then we call this limit the **Cauchy principal value** of $\int_a^b f(t)dt$, and we write $(P)\int_a^b f(t)dt = L$.

■ Example 6.24

It follows that although $\int_{-1}^{1} \frac{dt}{t^3}$ diverges, we have $(P)\int_{-1}^{1} \frac{dt}{t^3} = 0$. ■

For some improper integrals that converge, it may still be possible to define a Cauchy principal value, obtained (as required, by definition) from the simultaneous symmetric evaluation of two limits. The following example shows that when $I(f)$ is convergent, its value I is still the same as its Cauchy principal value.

■ Example 6.25

The integral $I(f) = \int_0^\infty \frac{dt}{\sqrt{t}(1+t)}$ is improper at both limits.

Choose any $c \in (0, \infty)$; then

$$I = \lim_{x\to\infty} \int_c^x \frac{dt}{\sqrt{t}(1+t)} + \lim_{\varepsilon\to 0+} \int_\varepsilon^c \frac{dt}{\sqrt{t}(1+t)}.$$

An antiderivative is $2\,\mathrm{Tan}^{-1}\sqrt{t}$ (verify!), so from Theorem 6.15 we have

$$\begin{aligned} I &= \lim_{x\to\infty}\left[2\,\mathrm{Tan}^{-1}\sqrt{x} - 2\,\mathrm{Tan}^{-1}\sqrt{c}\right] + \lim_{\varepsilon\to 0^+}\left[2\,\mathrm{Tan}^{-1}\sqrt{c} - 2\,\mathrm{Tan}^{-1}\sqrt{\varepsilon}\right] \\ &= \left(\pi - 2\,\mathrm{Tan}^{-1}\sqrt{c}\right) + \left(2\,\mathrm{Tan}^{-1}\sqrt{c} - 0\right) \\ &= \pi. \end{aligned}$$

On the other hand, we can bypass the point c and write

$$\begin{aligned} (P)\int_0^\infty \frac{dt}{\sqrt{t}(1+t)} &= \lim_{\varepsilon\to 0+} \int_\varepsilon^{1/\varepsilon} \frac{dt}{\sqrt{t}(1+t)} \\ &= \lim_{\varepsilon\to 0+}\left[2\,\mathrm{Tan}^{-1}\sqrt{1/\varepsilon} - 2\,\mathrm{Tan}^{-1}\sqrt{\varepsilon}\right] \\ &= 2(\pi/2) - 2(0) \\ &= \pi. \end{aligned}$$

■

Theorem 6.20 was a first look at the parallelism between improper integrals and infinite series (and sequences). We continue this theme in Example 6.26 and Theorems 6.21 and 6.22.

■ Example 6.26

Does $\int_0^{\infty} \frac{\sin t}{2+\cos t} dt$ converge?

Here, we use a sequence definition of convergence: $I(f)$ converges to $I \in \mathbf{R}^1$ iff for *each* sequence $\{x_n\}_{n=1}^{\infty}$ such that $\lim_{n\to\infty} x_n = \infty$, we have $\lim_{n\to\infty} I(f)(x_n) = I$. If $\{x_n\}_{n=1}^{\infty}$ is defined by $x_n = 2\pi n$, then

$$\begin{aligned} I(f)(x_n) &= \int_0^{2\pi n} \frac{\sin t}{2+\cos t} dt \\ &= -\ln(2+\cos t)|_0^{2\pi n} \\ &= 0, \end{aligned}$$

so $\lim_{n\to\infty} I(f)(x_n) = 0$.

On the other hand, if $\{x_n^*\}_{n=1}^{\infty}$ is defined by $x_n^* = (4n+1)\pi/2$, then $\lim_{n\to\infty} x_n^*$ is still ∞, but now we find

$$\begin{aligned} I(f)(x_n^*) &= -\ln(2+\cos t)|_0^{(4n+1)\pi/2} \\ &= \ln(3/2), \end{aligned}$$

so $\lim_{n\to\infty} I(f)(x_n^*) = \ln(3/2) \neq \lim_{n\to\infty} I(f)(x_n)$. We conclude from earlier remarks that $I(f)$ does not converge. ■

Theorem 6.21 (Limit Comparison Test for Improper Integrals). *Let f, g be two positive functions defined on $[a, b)$ and Riemann-integrable on any closed subinterval $[a, x] \subset [a, b)$. If $0 < \lim_{t\to b^-} [f(t)/g(t)] < \infty$ holds, then the two improper integrals $I(f)$ and $I(g)$ converge or diverge together.*

Proof. Let $L = \lim_{t\to b^-} [f(t)/g(t)]$; by definition, this implies that for any $\varepsilon > 0$ there is a $t_n > 0$ such that

$$L - \varepsilon \leq \frac{f(t)}{g(t)} \leq L + \varepsilon$$

whenever $t_0 \leq t < b$ (if $b \in \mathbf{R}^1$) or whenever $t \geq t_0$ (if $b = \infty$). Since $g(t) > 0$ and $0 < L < \infty$, then

$$(L-\varepsilon)g(t) \leq f(t) \leq (L+\varepsilon)g(t),$$

and Theorem 6.20 gives for $t_0 \le x < b$

$$(L-\varepsilon)\int_{t_0}^{x} g(t)dt \le \int_{t_0}^{x} f(t)dt \le (L+\varepsilon)\int_{t_0}^{x} g(t)dt. \qquad (*)$$

Choose ε small enough so that $L > \varepsilon$. Then if $I(f)$ converges, so does $\int_{t_0}^{x} f(t)dt$, and so does $\int_{t_0}^{x} g(t)dt$ from the first inequality in equation (*).

Finally, we conclude that $I(g)$ converges. From the second inequality in equation (*), similarly, we have that convergence of $I(g)$ implies convergence of $I(f)$. Conclusions about divergence also follow from equation (*) and Theorem 6.2. ■

Corollary 6.21.1. *If f, g are as previously but $\lim_{t\to b^-}[f(t)/g(t)] = 0$, then the convergence of $I(g)$ implies the convergence of $I(f)$.*

■ Example 6.27

Let $I(f) = \int_0^{\infty} \sqrt{t}e^{-t^2}dt$; choose $g(t) = (t^2+1)^{-1}$. Then we have $I(g) = \lim_{x\to\infty}[\tan^{-1}x - \tan^{-1}0] = \pi/2$. Two applications of l'Hôpital's Rule give

$$\lim_{t\to\infty}\left[\sqrt{t}e^{-t^2}/(t^2+1)^{-1}\right] = 0.$$

It follows from Corollary 6.21.1 that $I(f)$ converges. ■

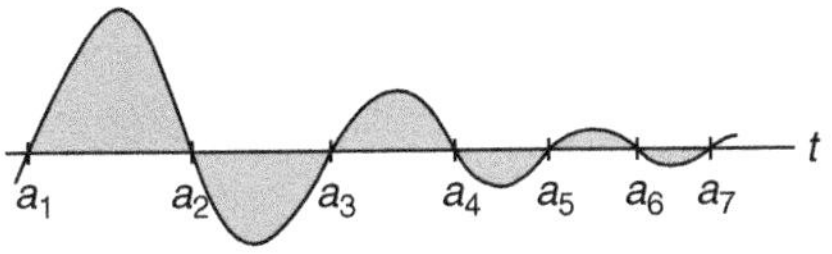

FIGURE 6.10
Diagram for Leibniz's Test.

We are not always so lucky to have improper integrals in which the integrand is of one sign throughout. An analog for improper integrals of the Alternating Series Test for infinite series (Theorem 3.9) is sometimes referred to as Leibniz's Test for Improper Integrals. Refer to Figure 6.10 in connection with the statement of the theorem.

Theorem 6.22 (Leibniz's Test for Improper Integrals). *Let f be defined on $[a,\infty)$ and have infinitely many zeros there at $a_1, a_2, a_3, \ldots$, where $a \le a_1 < a_2 < a_3 < \cdots$ and $\lim_{n\to\infty} a_n = \infty$. Suppose f alternates in sign regularly, $f(t) > 0$ if*

$a_{2n-1} < t < a_{2n}$ and $f(t) < 0$ if $a_{2n} < t < a_{2n+1}$. For each interval the quantity

$$F_n = \int_{a_n}^{a_{n+1}} f(t)dt$$

is presumed to exist. Also, suppose that $|F_{n+1}| \le |F_n|$ for all $n \ge N$ and that $\lim_{n\to\infty} |F_n| = 0$. Then the improper integral $I(f)$ converges.

Proof. For any $x > a$ let n be such that $a_n \le x < a_{n+1}$. By Theorem 6.5, applied repeatedly, we have

$$\int_a^x f(t)dt = \int_a^{a_1} f(t)dt + [F_1 + F_2 + \cdots + F_{n-1}] + \int_{a_n}^x f(t)dt.$$

The first integral on the right is a constant. The quantity in the brackets converges as $n \to \infty$ because the terms there satisfy the hypotheses of Theorem 3.9. Finally, the last integral on the right is not larger than $\int_{a_n}^{a_{n+1}} |f(t)|dt = |F_n|$ in magnitude and by hypothesis $|F_n|$ approaches 0 as $n \to \infty$. It follows, by summation, that $\int_a^x f(t)dt$ approaches a finite value as $x \to \infty$. ■

■ Example 6.28

Consider $\int_1^\infty \frac{\sin t}{\sqrt{t}} dt$. The zeros of the integrand occur at $t = \pi, 2\pi, 3\pi, \ldots$. Furthermore, for any $t \ge 1$ we have

$$\frac{|\sin t|}{\sqrt{t}} > \frac{|\sin(t+\pi)|}{\sqrt{t+\pi}},$$

so

$$\begin{aligned} |F_n| &= \int_{n\pi}^{(n+1)\pi} \frac{1}{\sqrt{t}} |\sin t| dt \\ &> \int_{n\pi}^{(n+1)\pi} \frac{1}{\sqrt{t+\pi}} |\sin(t+\pi)| dt \\ &= \int_{(n+1)\pi}^{(n+2)\pi} \frac{1}{\sqrt{x}} |\sin x| dx \quad (x = t + \pi) \\ &= |F_{n+1}|. \end{aligned}$$

Also, $$|F_n| \le \int_{n\pi}^{(n+1)\pi} \left\{ \left(\frac{1}{\sqrt{t}}\right)_{\max} |\sin t|_{\max} \right\} dt$$

$$= \frac{1}{\sqrt{n\pi}} \int_{n\pi}^{(n+1)\pi} dt$$

$$= \sqrt{\pi/n}.$$

This latter approaches 0 as $n \to \infty$, so $\lim_{n\to\infty} |F_n| = 0$. The hypotheses of Leibniz's Test are all satisfied, so the given integral converges. ■

In view of the close parallelism between the theory of infinite series and that of improper integrals, it is reasonable that there should exist a convergence test for infinite series that uses concepts from improper integrals. Our exposition is drawn from (Hardy, 1967).

Theorem 6.23 (Integral Test).[9] *If $f(t)$ is a nonconstant function that is positive, decreasing, and continuous for all real $t \ge n_0$, $n_0 \in \mathbf{N}$, then the series $\sum_{k=n_0}^{\infty} f(k)$ converges iff the improper integral $I(f) = \int_{n_0}^{\infty} f(t)dt$ converges. When $I(f) = L \in \mathbf{R}^1$, then $\sum_{k=n_0}^{\infty} f(k)$ has a sum strictly less than $L + f(n_0)$.*

Proof. For any natural number $k \ge n_0$, we have $f(k) \ge f(t) \ge f(k+1)$, when $k \le t \le k+1$. For each such k, Theorem 6.2 gives

$$0 \le \int_k^{k+1} [f(k) - f(t)]\, dt \le \int_k^{k+1} [f(k) - f(k+1)]\, dt = f(k) - f(k+1).$$

Hence, if for each k we define F_{k-n_0} by

$$F_{k-n_0} = f(k) - \int_k^{k+1} f(t)dt,$$

then $$0 \le F_{k-n_0} = \int_k^{k+1} [f(k) - f(t)]\, dt \le f(k) - f(k+1). \qquad (*)$$

[9] The Test was discovered by and presented by MacLaurin in his 1742 book *Treatise of Fluxions*, and was rediscovered in the next century by Cauchy.

Summing both ends of equation (*) from $k = n_0$ to $k = N$, we obtain by repeated use of Theorem 6.5

$$0 \le \sum_{k=n_0}^{N} F_{k-n_0} = \sum_{k=n_0}^{N} f(k) - \int_{n_0}^{N+1} f(t)dt < f(n_0) - f(N+1) < f(n_0). \qquad (**)$$

In (**) the next-to-the-last inequality is a strict inequality since in (*) $f(k)$ would equal $f(k+1)$ for all $k \ge n_0$ only if $f(t)$ were a constant function for all $t \ge n_0$. We exclude this uninteresting possibility. Figure 6.11 gives a geometric interpretation of (**).

Since the F_{k-n_0}'s are generally positive, their partial sums are positive, increasing, and bounded from above by the constant $f(n_0)$. Hence, by Theorem 2.2, $\sum_{k=n_0}^{\infty} F_{k-n_0}$ automatically converges; that is, the limit in (***) exists (in $\mathbf{R}^1$).

$$\lim_{N\to\infty} \left[\sum_{k=n_0}^{N} f(k) - \int_{n_0}^{N+1} f(t)dt \right] \qquad (***)$$

Suppose that the improper integral $I(f)$ converges to $L \in \mathbf{R}^1$. It follows from (***) that $\sum_{k=n_0}^{\infty} f(k)$ must also converge, and from (**) its sum will be strictly less than $L + f(n_0)$. On the other hand, if $I(f)$ diverges, then so must $\sum_{k=n_0}^{\infty} f(k)$, in order for the limit in (***) to be real. ■

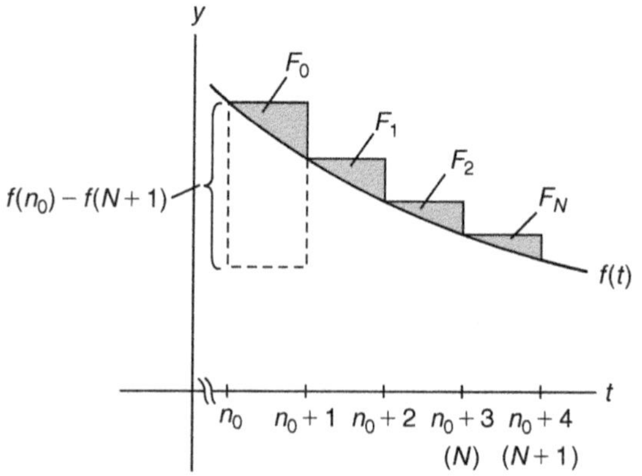

FIGURE 6.11

Geometric interpretation of MacLaurin's Integral Test.

■ Example 6.29

The series $\sum_{k=1}^{\infty} \frac{1}{k^{3/2}}$ converges, for $\lim_{x\to\infty} \int_1^x \frac{dt}{t^{3/2}} = \lim_{x\to\infty} \left[\frac{-2}{\sqrt{x}} + 2\right] = 2$. The sum of the series, according to Theorem 6.23, is less than $2 + \frac{1}{1^{3/2}} = 3$. It is known that the sum is, in fact, approximately 2.6124 (Liang and Todd, 1972).

Similarly, it is straightforward to show that for $r > 0$ we have $\sum_{k=0}^{\infty} \frac{1}{(r+k)^2} < \frac{r+1}{r^2}$ (verify!). ■

6.8 RIEMANN INTEGRALS IN $\mathbf{R}^n$

This concluding section will be just an introduction to the topic herein. Only a few ideas will be presented, and there will be no discussion of methods for the evaluation of multiple integrals. The development of the Riemann integral in $\mathbf{R}^n$, $n > 1$, proceeds along lines roughly analogous to those for the integral in $\mathbf{R}^1$. We shall illustrate this explicitly with integrals in $\mathbf{R}^2$ (**double integrals**).

Let a function $f(\mathbf{p})$, $\mathbf{p} = (x, y)$, be defined and bounded on a closed, bounded set $\mathfrak{D}$ in the plane and let $\mathbf{T}$ be a closed, bounded rectangular region such that $\mathbf{T} \supset \mathfrak{D}$ (Figure 6.12). On $\mathbf{T}$ we erect a rectangular partition P by drawing $r + 1$ lines parallel to the y-axis that pass through the x-axis at points $a = a_0 < a_1 < a_2 < \cdots < a_r = b$ and $s + 1$ lines parallel to the x-axis that pass through the y-axis at points $c = c_0 < c_1 < c_2 < \cdots < c_s = d$. Rectangle $\mathbf{T}$ is thus partitioned into $n = rs$ smaller, closed subrectangles $\{\mathbf{R}_i\}_{i=1}^n$, with areas $\Delta A_1, \Delta A_2, \ldots, \Delta A_n$.

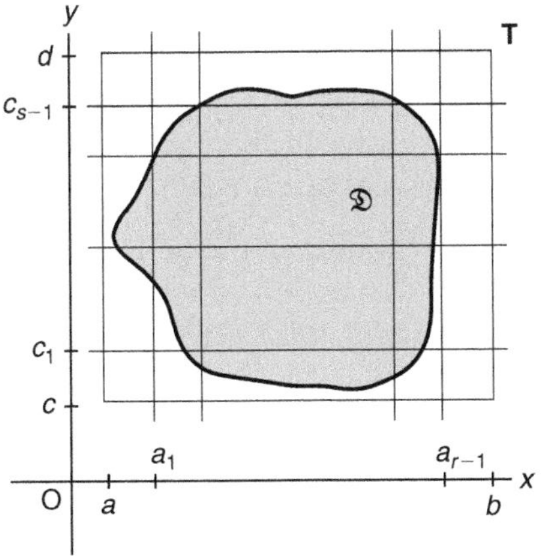

FIGURE 6.12
Construction of a partition of a closed, bounded rectangle in $\mathbf{R}^2$.

In each subrectangle, $\mathbf{R}_i$, a randomly chosen point $\mathbf{p}_i$, $i = 1, 2, \ldots, n$, serves as a tag for that subrectangle, thereby providing us with a tagged partition $\overline{P}$ of rectangle $\mathbf{T}$. Additionally, we let d_i denote the length of a diagonal of the ith subrectangle. The norm $\mathfrak{N}$ of $\overline{P}$ is $\mathfrak{N} = \max_i \{d_i\}$. Since some rectangles of $\mathbf{T}$ may consist entirely of points exterior to $\mathfrak{D}$, we adopt the following artifice. A new function F on $\mathbf{T}$ is defined by

$$F(\mathbf{p}) = \begin{cases} f(\mathbf{p}) & \mathbf{p} \in \mathfrak{D} \\ 0 & \mathbf{p} \notin \mathfrak{D}. \end{cases}$$

Then the Riemann sum S for f, given $\mathfrak{D}$, $\overline{P}$, and f, is

$$S(\mathfrak{D}, \overline{P}, f) = \sum_{i=1}^{n} F(\mathbf{p}_i)\Delta A_i.$$

Definition. *Let $f(p)$ be defined and bounded on $\mathfrak{D} \subset \mathbf{T} \subset \mathbf{R}^2$, and let F be defined as earlier. Let $\varepsilon > 0$ be given. Then f is* **Riemann-integrable** *on $\mathfrak{D}$ iff there is a constant I and a number $\delta > 0$ (in general, dependent on ε) such that for all rectangular partitions P of norm $\mathfrak{N} < \delta$ and for all sets of tags $\{\mathbf{p}_i\}_{i=1}^{n}$, we have*

$$|S(\mathfrak{D}, \overline{P}, f) - I| < \varepsilon.$$

The number I is called the **Riemann double integral** *of f over $\mathfrak{D}$, and we write*

$$I = \iint_{\mathfrak{D}} f,$$

or, equivalently, as $\iint_{\mathfrak{D}} f(x, y)dxdy$.

Many of the theorems on double integrals are obvious analogs of those for integrals in $\mathbf{R}^1$, and the proofs of the former are generally quite similar to those for the latter. The numbering of the next few theorems, which are given without proof, reflects the analogies. As in the one-dimensional case, so with integrals in $\mathbf{R}^2$, we have that if $\lim_{\mathfrak{N}\to 0} S(\mathfrak{D}, \overline{P}, f) = \lim_{\mathfrak{N}\to 0} \sum_{i=1}^{n} F(p_i)\Delta A_i$ exists, then it is unique.

Theorem 6.1′. *If $f(x, y)$ is Riemann-integrable over $\mathfrak{D}$, then $f(x, y)$ is bounded on $\mathfrak{D}$.*

Theorem 6.2′. *If $f(x, y)$, $g(x, y)$ are Riemann-integrable over $\mathfrak{D}$ and $f(x, y) \leq g(x, y)$ there, then $\iint_{\mathfrak{D}} f \leq \iint_{\mathfrak{D}} g$.*

Theorem 6.4′. *If $f(x, y)$, $g(x, y)$ are Riemann-integrable over $\mathfrak{D}$ and $c \in \mathbf{R}^1$, then*

(i) $\iint_{\mathfrak{D}} cf = c\iint_{\mathfrak{D}} f$;

(ii) $\iint_{\mathfrak{D}}(f + g) = \iint_{\mathfrak{D}} f + \iint_{\mathfrak{D}} g$.

Theorem 6.6′. *If $f(x, y)$ is Riemann-integrable over $\mathfrak{D}$ to I and the value of f at one point $\mathbf{p} = (x, y)$ in $\mathfrak{D}$ is changed, then f is still Riemann-integrable over $\mathfrak{D}$ and the value of the integral is still I.*

As with the integral in $\mathbf{R}^1$, the double integral can be understood equivalently via the approach of Darboux. Let P be any untagged partition of a rectangle $\mathbf{T}$ that contains the closed, bounded region $\mathfrak{D}$ on which f is defined and bounded. We define F on $\mathbf{T}$ as we did previously, and we denote

$$M = \sup_{\mathbf{p}} F(\mathbf{p}), \quad m = \inf_{\mathbf{p}} F(\mathbf{p}),$$

where $\mathbf{p} \in \mathbf{T}$. Similarly, for each subrectangle $\mathbf{R}_i$ we define

$$M_i = \sup_{\mathbf{p}_i} F(\mathbf{p}_i), \quad m_i = \inf_{\mathbf{p}_i} F(\mathbf{p}_i), \quad \mathbf{p}_i \in \mathbf{R}_i.$$

The **upper Darboux sum** and the **lower Darboux sum** that correspond to f and to the partition P are, respectively,

$$U(P, f) = \sum_{i=1}^{n} M_i \Delta A_i, \quad L(P, f) = \sum_{i=1}^{n} m_i \Delta A_i.$$

The following two real numbers then exist (why?):

$$\overline{\iint_{\mathfrak{D}}} f = \inf_{P} U(P, f), \quad \underline{\iint_{\mathfrak{D}}} f = \sup_{P} L(P, f),$$

where inf, sup mean with respect to all possible partitions P of $\mathbf{T}$.

Definition. *A function f is **Darboux-integrable** on a closed, bounded region $\mathfrak{D}$ of $\mathbf{R}^2$ iff $\underline{\iint_{\mathfrak{D}}} f = \overline{\iint_{\mathfrak{D}}} f$, and their common value is the **Darboux integral** of f over $\mathfrak{D}$, which we write simply as (D) $\iint_{\mathfrak{D}} f$.*

As we did with one-dimensional Riemann integrals, so we can consider refinements of partitions P of sets $\mathbf{T}$ in $\mathbf{R}^2$. This leads (by reasoning analogous to that given in Section 6.4) ultimately to:

Theorem 6.12′. *If either (R) $\iint_{\mathfrak{D}} f$ or (D) $\iint_{\mathfrak{D}} f$ exists, then so does the other, and the two integrals are equal.*

The analog of the very useful Theorem 6.5 requires some care in its statement, and we need to digress briefly for a brief discussion of area (Courant and John, 1974; Knopp, 1969). The **characteristic function** for a set $\mathfrak{D}$, denoted by $I_{\mathfrak{D}}(\mathbf{p})$, is defined by

$$I_{\mathfrak{D}}(\mathbf{p}) = \begin{cases} 1 & \mathbf{p} \in \mathfrak{D} \\ 0 & \mathbf{p} \notin \mathfrak{D}. \end{cases}$$

We define the area of $\mathfrak{D} \subset \mathbf{R}^2$ in terms of the characteristic function for $\mathfrak{D}$.

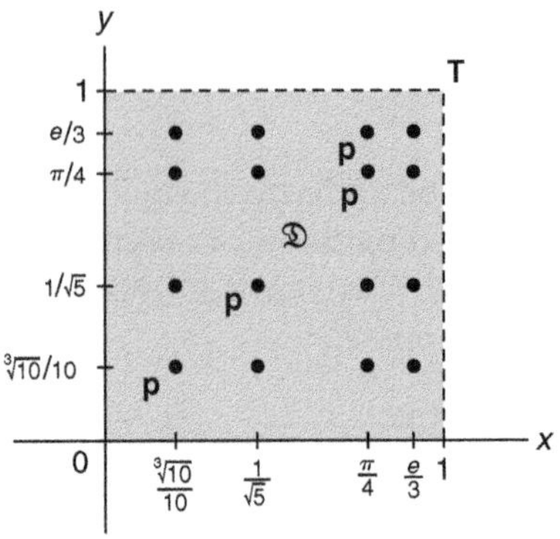

FIGURE 6.13
A set without area.

Definition. *A bounded set* $\mathfrak{D}$ *in* $\mathbf{R}^2$ *has* **area** *iff its characteristic function is integrable over* $\mathfrak{D}$. *The area,* $A(\mathfrak{D})$, *is the value of the double integral* $\iint_{\mathfrak{D}} I_{\mathfrak{D}}$. *In the trivial case where* $\mathfrak{D}$ *is the empty set,* $\emptyset \subset \mathbf{R}^2$, *we define (or set)* $A(\emptyset) = 0$.

■ Example 6.30

Let $\mathfrak{D}$ be the set of those points $\mathbf{p} = (x, y)$, both of whose coordinates are irrational numbers, that are contained in the closed, unit square $\mathbf{T}: 0 \leq x \leq 1, 0 \leq y \leq 1$ (Figure 6.13). Let P be an arbitrary partition of $\mathbf{T}$. Then in every subrectangle $\mathbf{R}_i$, no matter how fine may be P, there are points $\mathbf{p}_i$ with both coordinates irrational (so $\mathbf{p}_i \in \mathfrak{D}$) and there are points $\mathbf{p}_i$ in which at least one coordinate is rational (so $\mathbf{p}_i \notin \mathfrak{D}$). Hence, for any P the upper Darboux integral of $I_{\mathfrak{D}}$ has the value 1 (corresponding to each $\mathbf{p}_i \in \mathfrak{D}$)

$$\iint_{\mathfrak{D}} I_{\mathfrak{D}} = \inf_{P} U(P, I_{\mathfrak{D}}) = 1,$$

and the lower Darboux integral of $I_{\mathfrak{D}}$ has the value 0 (corresponding to each $\mathbf{p}_i \notin \mathfrak{D}$). It follows that $I_{\mathfrak{D}}$ is not integrable over $\mathfrak{D}$, so $\mathfrak{D}$ does not have area. ■

The concept just defined as *area*, when extended in the obvious way to $\mathbf{R}^3$, is known as **volume**. When extended to arbitrary Euclidean spaces, the concept is known as **Jordan content** (after the French mathematician **Camille Jordan** (1838–1922)). A set $\mathfrak{D} \subset \mathbf{R}^n$ with Jordan content is said to be **Jordan-measurable.**[10]

---

[10] Jordan content, when generalized to even more arbitrary spaces, is known as **measure** (Royden, 1988; Borden, 1998), of which Lebesgue measure is a special case.

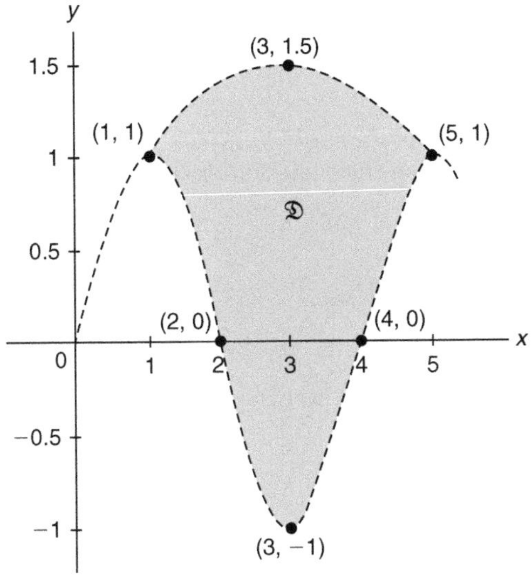

FIGURE 6.14
A region $\mathfrak{D}$ in $\mathbf{R}^2$ with Jordan content.

The **frontier** (or **boundary**) of a set $\mathfrak{D} \subset \mathbf{R}^n$ consists of all points $\mathbf{p}$ such that any deleted ball around $\mathbf{p}$, no matter how small, contains points of $\mathfrak{D}$ *and* points not belonging to $\mathfrak{D}$. An important theorem is the result that a set $\mathfrak{D} \subset \mathbf{R}^n$ has Jordan content iff its frontier has Jordan content 0 (Olmsted, 1961).

■ Example 6.31

A popular trick question: What is the area of a circle of radius 1? ■

■ Example 6.32

The shaded region $\mathfrak{D}$ in Figure 6.14 consists of all points $\mathbf{p} = (x, y)$ for which $1 < x < 5$ and $\sin(\pi x/2) < y < 1 + \left[\sqrt{4 - (x-3)^2}\right] / 4$. The frontier of $\mathfrak{D}$ is the union of the two simple arcs. It has Jordan content (area) 0 (can you see why?); hence, $\mathfrak{D}$ has Jordan content. ■

■ Example 6.33

The frontier of the set $\mathfrak{D}$ in Example 6.30 is a set with easily calculated positive Jordan content. We may conclude in this case that the set $\mathfrak{D}$ itself *does not* have Jordan content (area) (Exercise 6.64). ■

The generalization of Example 6.32 is that any bounded set in $\mathbf{R}^2$ whose frontier consists of a finite number of rectifiable, simple arcs has Jordan content (area). This, of course, is a common situation.

Continuing the sequence of theorem analogs that began with Theorem 6.1′, we suppose that f is defined on the bounded, Jordan-measurable set $\mathfrak{D} \subset \mathbf{R}^2$, and that $\mathfrak{D}$ can be represented as a union $\mathfrak{D}_1 \cup \mathfrak{D}_2$.

Theorem 6.5 (Finite Additivity)′. *Let f be Riemann-integrable on each of the Jordan-measurable sets $\mathfrak{D}_1$ and $\mathfrak{D}_2$, and also on their intersection $\mathfrak{D}_1 \cap \mathfrak{D}_2$, which has Jordan content (area) 0. Then f is Riemann-integrable on $\mathfrak{D} = \mathfrak{D}_1 \cup \mathfrak{D}_2$ and*

$$\iint_{\mathfrak{D}} f = \iint_{\mathfrak{D}_1} f + \iint_{\mathfrak{D}_2} f.$$

Proof. Let $\mathbf{S} = \mathfrak{D}_1 \cap \mathfrak{D}_2$; then $\mathfrak{D} = (\mathfrak{D}_1 \backslash \mathbf{S}) \cup \mathfrak{D}_2$ and $(\mathfrak{D}_1 \backslash \mathbf{S}) \cap \mathfrak{D}_2 = \emptyset$.

Next, we define F by

$$F(\mathbf{p}) = \begin{cases} f(\mathbf{p}) & \mathbf{p} \in \mathfrak{D}_1 \backslash \mathbf{S} \\ 0 & \mathbf{p} \notin \mathfrak{D}_1 \backslash \mathbf{S}. \end{cases}$$

Let $\mathfrak{D}$ be contained in a bounded rectangle $\mathbf{T}$. We then have, from Theorem 6.4′(ii) and the definition of F,

$$\begin{aligned} \iint_{\mathfrak{D}} f &= \iint_{\mathfrak{D}} [F + (f - F)] \\ &= \iint_{\mathfrak{D}} F + \iint_{\mathfrak{D}} (f - F) \\ &= \iint_{\mathfrak{D}_1 \backslash \mathbf{S}} F + \iint_{\mathfrak{D}_2} (f - F) \\ &= \iint_{\mathfrak{D}_1 \backslash \mathbf{S}} f + \iint_{\mathfrak{D}_2} f. \end{aligned} \qquad (*)$$

Similarly, since $\mathfrak{D}_1 = \mathfrak{D}_1 \backslash \mathbf{S} \cup \mathbf{S}$ and $(\mathfrak{D}_1 \backslash \mathbf{S}) \cap \mathbf{S} = \emptyset$, then

$$\iint_{\mathfrak{D}_1} f = \iint_{\mathfrak{D}_1 \backslash \mathbf{S}} f + \iint_{\mathbf{S}} f.$$

Substitution of this into equation (*) yields

$$\iint_{\mathfrak{D}} f = \iint_{\mathfrak{D}_1} f + \iint_{\mathfrak{D}_2} f - \iint_{\mathbf{S}} f. \qquad (**)$$

Equation (**) shows that f is Riemann-integrable over $\mathfrak{D}$. By Theorem 6.1′, f is bounded on $\mathfrak{D}$: $|f(\mathbf{p})| < C$, $\mathbf{p} \in \mathfrak{D}$. Since $\mathbf{S} = \mathfrak{D}_1 \cap \mathfrak{D}_2$ has zero area, then a partition P on $\mathbf{T}$ of small enough norm $\mathfrak{N}$ can be erected so that, if any $\varepsilon > 0$ is given, there is a corresponding $\delta > 0$ such that if $\mathfrak{N} < \delta$, then the sum of the areas $\sum_{i=1}^{n} \Delta A_i$ of the rectangles that cover $\mathbf{S}$ is less than ε/C. A corresponding Riemann sum for f on $\mathbf{S}$ then satisfies

$$\left|\sum_{i=1}^{n} f(\mathbf{p})\Delta A_i\right| \leq \sum_{i=1}^{n} C\Delta A_i < C\left(\frac{\varepsilon}{C}\right) = \varepsilon.$$

As $\varepsilon > 0$ is arbitrarily small, then in the limit as $\mathfrak{N} \to 0$ we have $\iint_{\mathbf{S}} f = 0$. Equation (**) then reduces to the desired result. ■

Theorem 6.13′. *If f is bounded and continuous on a bounded set $\mathfrak{D} \subset \mathbf{R}^2$ that has Jordan content, then $\iint_{\mathfrak{D}} f$ exists.*[11]

■ Example 6.34

Let $f : \mathbf{R}^2 \to \mathbf{R}^1$ be defined by $f(\mathbf{p}) = \left(\frac{\cos x_1}{x_2}\right) e^{x_1+x_2}$, $\mathbf{p} = (x_1, x_2)$, and let $\mathfrak{D}$ be the disk $\mathfrak{D} = \left\{(x_1, x_2) : (2 - x_1)^2 + (2 - x_2)^2 < 1\right\}$. Then f is bounded on $\mathfrak{D}$: $|f(\mathbf{p})| \leq \frac{1}{1} \cdot e^{3+3} < 404$. It is also continuous on $\mathfrak{D}$ because it is the product of functions continuous on $\mathfrak{D}$. Set $\mathfrak{D}$ itself is bounded and has Jordan content (area) π. Hence, by Theorem 6.13′, $\iint_{\mathfrak{D}} f$ exists. ■

EXERCISES

Section 6.1

6.1. This Exercise and the next one deal with the topic of antidifferentiation. Suppose that F is continuous on $[a, b]$, and is an antiderivative for the constant function $K(x) = 0$, $x \in (a, b)$. Show that these facts imply that $F(x) = C$ for all $x \in [a, b]$ and for some $C \in \mathbf{R}^1$.

6.2. Suppose that G, H are two functions continuous on $[a, b]$ and that $G'(x) = H'(x)$ for all $x \in (a, b)$. From these hypotheses, prove that any two antiderivatives of a function $f : [a, b] \to \mathbf{R}^1$ must differ by a constant.

6.3. The three main methods of antidifferentiation are (i) integration by substitution, (ii) integration by parts, and (iii) use of partial fractions. For each f, find $D^{-1}f$, antiderivatives of the following functions.

(a) $f(t) = 4^{\sqrt{2t+1}}$;

---

[11]The conclusion of Theorem 6.13′ also holds if the set of discontinuities of f has Jordan content 0. A generalization of this that involves, as expected, some measure theory, is known as (the important) **Lebesgue's Criterion** for existence of a multiple Riemann integral (Brown, 1936). This can be regarded as a generalization of Theorem 6.14.

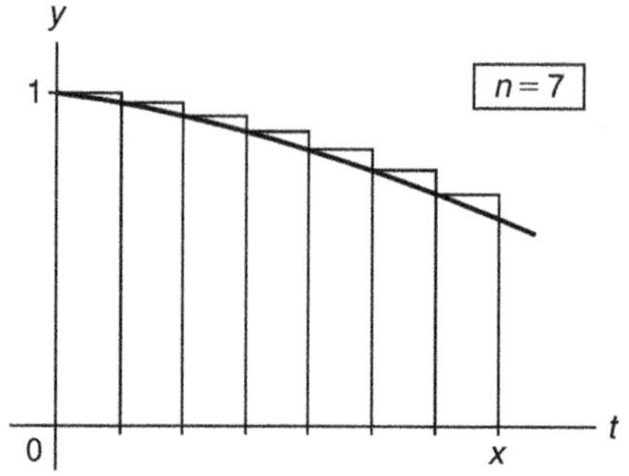

FIGURE 6.15
A plausible Leibniz type of integration of $\cos t$.

(b) $f(t) = \mathrm{Tan}^{-1}(2t)$;
(c) $f(t) = (1 + t^3)^{-1}$;
(d) $f(t) = \sqrt{t^2 + a^2}$;
(e) $f(t) = \sinh^3 t$.

6.4. This argument could have been used by Leibniz in order to integrate $\cos t$ (Mathews and Shultz, 1989). Figure 6.15 shows a graph of $\cos t$ from $t = 0$ to $t = x$, $0 < x < \pi/2$. The x-axis is partitioned there into n equal subintervals. Above each subinterval is erected the circumscribing rectangle of base x/n.

(a) First prove **Lagrange's Identity**

$$\sum_{k=0}^{n-1} \cos(k\theta) = \frac{1}{2} + \frac{\sin((2n-1)\theta/2)}{2\sin(\theta/2)}.$$

(b) Use part (a) in order to write an expression for the sum of the areas of n circumscribing rectangles in a diagram such as that shown.

(c) Leibniz would see that the desired result could be obtained by requiring the rectangles to be as "thin as possible." Interpret this in modern terms and carry it out, thereby showing that $\int_0^x \cos t \, dt = \sin x$.

6.5. Explain why a function that is not bounded on an interval $[a, b]$ is not Riemann-integrable.

6.6. For each of the following functions, the interval of integration, the partition $\{a_k\}_{k=0}^n$, and the tags $\{x_j\}_{j=1}^n$ are given. Compute the Riemann sums S and compare these with the corresponding values of the Riemann integral.

(a) $f(x) = \cos x$; $[0, \pi/2]$ $\quad a_k = k\pi/16$

$x_{j+1} = (1 + 3j)\pi/48$

$k(j) = 0, 1, 2, \ldots, 8(7)$

(b) $f(x) = \sqrt{1 + x}$; $[3, 8]$ $\quad a_k = (k + 6)/2$

$x_{j+1} = (13 + 2j)/4$

$k(j) = 0, 1, 2, \ldots, 10(9)$

(c) $f(x) = x^{-2}$; $[1, 243/32]$ $\quad$ $a_k = (3/2)^k$

$x_{j+1} = (11/9)(3/2)^j$

$k(j) = 0, 1, 2, \ldots, 5(4)$

6.7. The American mathematician **N. Wiener** (1894–1964) defined the integral as follows (Sklar, 1960):

$$\int_a^b f = \lim_{n\to\infty} \sum_{k=1}^{n} f\left(a + \frac{k(b-a)}{n}\right)\left(\frac{b-a}{n}\right),$$

provided that the limit exists.

(a) Now let f be the Dirichlet function: $f(x) = \begin{cases} 1 & x \in \mathbf{Q} \\ 0 & x \notin \mathbf{Q}. \end{cases}$ Show that by Wiener's definition, one has $\int_0^{\sqrt{2}} f = 0$.

(b) Indicate why the integral $\int_0^{\sqrt{2}} f$ does not exist in the Riemann sense.

Section 6.2

6.8. Write out the proof of Theorem 6.2. Use the Theorem to prove that the Riemann integral $\int_1^2 \frac{\sin t}{t} dt$, which does exist, is less than $2/3$. In fact, the integral has the approximate value of 0.65933, but this requires more work.

6.9. There is no known antiderivative in closed form for $f(t) = \ln(\ln t)$.

(a) Establish that on the interval $[e, 4]$ we have

$$-\frac{1}{2}(\ln t)^2 + 2\ln t - \frac{3}{2} \le \ln(\ln t) \le \frac{1}{3}(\ln t)^3 - \frac{3}{2}(\ln t)^2 + 3\ln t - \frac{11}{6}.$$

(b) Apply Theorem 6.2 and show that

$$-2(\ln 4)^2 + 12\ln 4 - 18 + 2e < \int_e^4 \ln(\ln t)dt < \frac{4}{3}(\ln 4)^3 - 10(\ln 4)^2$$

$$+ 32\ln 4 - \frac{118}{3} + 4e,$$

that is, $0.2285 < \int_e^4 \ln(\ln t)dt < 0.2354$.

A numerical integration yields 0.2341 as an approximate value of the integral.

6.10. In Example 6.3, show that $F'(x) \le 0$ on $[0, 1]$. Would this example have been useful to Newton, who in 1680 might have been interested in the antidifferentiation of e^{-x^2}?

6.11. **(a)** How do you know that it is legitimate to apply the Mean-Value Theorem to each subinterval in the proof of Theorem 6.3?

(b) Prove part (ii) of Theorem 6.4.

6.12. Make use of Theorem 6.5 and evaluate $\int_{1/2}^{2} \frac{\ln t}{1+t^2} dt$.

6.13. Refer to Exercise 6.7 and to Wiener's value for $\int_0^{\sqrt{2}} f$. Now look at Wiener's values for $\int_0^1 f$ and $\int_1^{\sqrt{2}} f$, and tell why we are unhappy with Wiener's definition of the integral.

6.14. Fill in the last few details in the proof of Theorem 6.6.

6.15. Let T be a cubic or quartic polynomial in t with no repeated factors, and let $R(t, \sqrt{T})$ be a rational function of t, $\sqrt{T}$. Then a Riemann integral of the form $\int_0^x R(t, \sqrt{T})dt$ is an **elliptic integral** if it cannot be expressed in terms of elementary functions.[12]

An elliptic integral is a new kind of transcendental function.[13] There are three basic kinds of elliptic integrals,[14] expressed here in terms of the function $T(t) = (1 - t^2)(1 - mt^2)$, $0 < m < 1$:

First Kind:	$F(x, m) = \int_0^x dt/\sqrt{T(t)}$
Second Kind:	$E(x, m) = \int_0^x (1 - mt^2)dt/\sqrt{T(t)}$
Third Kind:	$\Pi(x, n, m) = \int_0^x (1 - nt^2)^{-1}dt/\sqrt{T(t)}, \quad n \neq 0.$

(a) Transform these three kinds of elliptic integrals into corresponding equivalent expressions by means of the substitutions $t = \sin\theta$, $x = \sin\phi$.

(b) Now consider the ellipse $(x^2/a^2) + (y^2/b^2) = 1$, where $a = 1$, $b = 1/2$. We rewrite this parametrically by substituting $x = a\sin\theta$ and $y = b\cos\theta$. The perimeter C of the ellipse is then given by the parametric arclength formula as

$$C = 4\int_0^{\pi/2}\left[\left(\frac{dx}{d\theta}\right)^2 + \left(\frac{dy}{d\theta}\right)^2\right]^{1/2} d\theta.$$

Hence, show that C is given by an elliptic integral of the second kind.

(c) An infinite series expansion of $E(1, m)$ is (Abramowitz and Stegun, 1965):

$$E(1, m) = \frac{\pi}{2}\left[1 - \left(\frac{1}{2}\right)^2\frac{m}{1} - \left(\frac{1\cdot 3}{2\cdot 4}\right)^2\frac{m^2}{3} - \left(\frac{1\cdot 3\cdot 5}{2\cdot 4\cdot 6}\right)^2\frac{m^3}{5} - \left(\frac{1\cdot 3\cdot 5\cdot 7}{2\cdot 4\cdot 6\cdot 8}\right)^2\frac{m^4}{7} - \cdots\right].$$

Use the first seven terms of this series to estimate the perimeter C of the ellipse in part (b).

6.16. It was not realized by Legendre when he proposed his tripartite classification of elliptic integrals (Exercise 6.15) that the three classes are not actually independent of each other (Niven, 1943). As before, we define $T(t) = (1 - t^2)(1 - mt^2)$.

---

[12]These are rational functions, trigonometric and hyperbolic functions and their inverses, and exponential and logarithmic functions. Our definition of an elliptic integral is from the classic work (Whittaker and Watson, 1963).

[13]Proved in 1834 by the French mathematician **Joseph Liouville** (1809–1882).

[14]The tripartite classification is due to **Adrien-Marie Legendre** (1752–1833), and appeared in his book *Exercices de calcul intégral* (1811). Notation for elliptic integrals is not uniform.

(a) Show that $\frac{d}{dt}\left[t/\sqrt{T(t)}\right]$ is given by

$$\frac{1}{\sqrt{T(t)}}\left[-1+\frac{1}{1-t^2}+\frac{1}{1-mt^2}\right].$$

(b) Integrate the result in (a) from $t = 0$ to $t = x$ to obtain

$$\frac{x}{\sqrt{T(x)}} = -F(x,m) + \Pi(x,1,m) + \Pi(x,m,m).$$

(c) Show that $\frac{d}{dt}\left[mt\sqrt{(1-t^2)/(1-mt^2)}\right]$ is given by

$$\frac{1}{\sqrt{T(t)}}\left[1-mt^2-\frac{1-m}{1-mt^2}\right].$$

(d) Integrate the result in (c) as you did in (b).

(e) Combine the results of parts (b) and (d) to show that elliptic integrals of the first and second kinds can be expressed as functions of those of the third kind.

(f) Given that $\Pi\left(\frac{\sqrt{3}}{2},\frac{1}{2},\frac{1}{2}\right) = 1.3822$ and $\Pi\left(\frac{\sqrt{3}}{2},1,\frac{1}{2}\right) = 1.9511$, compute $F\left(\frac{\sqrt{3}}{2},\frac{1}{2}\right)$ and $E\left(\frac{\sqrt{3}}{2},\frac{1}{2}\right)$.

Section 6.3

6.17. In Example 6.6, establish that $\int_a^b [\sigma(x)-\rho(x)]dx = \frac{2}{n}$.

6.18. **(a)** Complete the proof of Theorem 6.9.

(b) Tell how Example 6.6 could have been done by using Theorem 6.9 instead of Theorem 6.8.

6.19. **(a)** Write out the proof of the Corollary 6.9.1.

(b) Explain how you know that each integral in Example 6.7 exists.

6.20. If on the interval $[0, 1]$ the function f is not finitely piecewise monotonic, does this mean that f is then not Riemann-integrable there? Discuss.

Section 6.4

6.21. Verify the entries in Example 6.8 for $k = 3, 8$.

6.22. Fill in the missing details in the proof of Lemma 6.4.1. To illustrate this Lemma, add two more division points to the partition in Example 6.8 and then compute $L(P^*, f), U(P^*, f)$.

6.23. Let f be a function bounded on the finite interval $[a, b]$, and suppose that $\{L_n\}_{n=1}^{\infty}$ is a sequence of lower Darboux sums $L(P, f)$ and $\{U_n\}_{n=1}^{\infty}$ is a corresponding sequence of upper Darboux sums. If the sequences are arbitrary but $\lim_{n\to\infty}[U_n - L_n] = 0$ holds, then prove that f is integrable on $[a, b]$.

6.24. A student asserts that if f is defined on and is strictly bounded on $[a, b]$, $|f(x)| < K$, and because for any x we have $f(x) = |f(x)|$, and if $|f(x)|$ is Riemann-integrable on $[a, b]$, then f is itself Riemann-integrable on $[a, b]$. If this assertion is true, prove it by use of theorems in the present section. If it is not true, discuss a counterexample with ideas in the present section.

6.25. Regarding some parts of Example 6.10:

Part (b). If the integrand were $\sin x$, the value of the integral would be 0. Determine the smallest positive number k such that $\int_0^{k\pi} \sin(\sqrt{x})dx$ is 0.

Part (c). Account for the existence (in $\mathbf{R}^1$) of this integral.

Part (d). Show that this integral is an elliptic integral of the first kind (Exercise 6.15).

6.26. Complete the proof of the Mean-Value Theorem for Integrals.

6.27. The following proof of the Mean-Value Theorem for Integrals is similar in spirit to our alternative proof of Theorem 5.7. Let f be continuous and, hence, integrable on $[a, b]$. Define the determinantal function $D(x)$ by (Putney, 1953)

$$D(x) = \begin{vmatrix} G(x) & H(x) & 1 \\ G(a) & H(a) & 1 \\ G(b) & H(b) & 1 \end{vmatrix},$$

where $G(x) = \int_a^x f(t)dt$ and $H(x) = x - a$.

(a) Why is $D(x)$ continuous on $[a, b]$?

(b) Why are $D(a), D(b)$ both zero?

(c) Why does it follow that there is a number $c, a < c < b$, such that $D'(c) = 0$? Expand $D'(c)$ to see what this means.

6.28. Let f be continuous on $[a, x], x \geq a$, and be differentiable at $a, f'(a) \neq 0$. Regard a as fixed but x as variable, and for each x let c_x designate a number in (a, x) such that the Mean-Value Theorem for Integrals holds. Certainly as $x \to a$, then $c_x \to a$ also. Now consider

$$H_a(x) = \frac{\int_a^x f(t)dt - xf(a) + af(a)}{(x-a)^2}.$$

Prove that $\lim\limits_{x \to a} \frac{c_x - a}{x - a} = \frac{1}{2}$ (Jacobson, 1982).

6.29. **(Generalized Mean-Value Theorem for Integrals)** Suppose that f, g are continuous on $[a, b]$ and $g(x) \geq 0$ there. Prove that there is a $c \in [a, b]$ such that

$$\int_a^b f(x)g(x)dx = f(c)\int_a^b g(x)dx.$$

6.30. Refer to Exercise 6.28; let $f(t) = \cos t$ and $a = 0$, and let $I(x)$ denote $\int_a^x f(t)dt$. Complete the following table. Do we have a violation of Jacobson's theorem? Read Bao-Lin (1997) and make a report on its contents.

x	$I(x)$	$f(c_x)$	c_x	$\frac{c_x - a}{x - a}$
1				
1/4				
1/16				
1/64				
1/128				

6.31. (Bunyakovski's Inequality)[15] Refer to Exercise 1.35 for guidance. If f, g are continuous on $[a, b]$, $b > a$, prove that

$$\left| \int_a^b f(x)g(x)dx \right| \leq \left[\int_a^b [f(x)]^2 dx \int_a^b [g(x)]^2 dx \right]^{1/2}.$$

In some spaces of functions, Bunyakovski's Inequality takes the place of the classical Cauchy-Schwarz Inequality (see Section 1.5).

Section 6.5

6.32. Refer back to material that followed Theorem 6.15.

(a) Show that the derivative of $f(x) = \begin{cases} x^2 \sin(x^{-2}) & 0 < x \leq 1 \\ 0 & x = 0 \end{cases}$ is

$$f'(x) = \begin{cases} 2x^2 \sin(x^{-2}) - 2x^{-1} \cos(x^{-2}) & 0 < x \leq 1 \\ 0 & x = 0. \end{cases}$$

(b) Verify that the integrand in Example 6.13 is the derivative of

$$f(x) = \sqrt{x^2 - a^2} - a \operatorname{Sec}^{-1}(x/a), \quad x \geq a > 0.$$

(c) Obtain the $f(x)$ in part (b) on your own.

6.33. Determine $\lim_{n \to \infty} \left[\frac{n+1}{n^2+1} + \frac{n+2}{n^2+4} + \frac{n+3}{n^2+9} + \cdots + \frac{n+n}{n^2+n^2} \right]$. Where did you use Theorem 6.15?

6.34. Determine $\lim_{x \to 0^+} x^5 \left[\int_0^x t\sqrt{\sin t}\, dt \right]^{-2}$.

6.35. (Riemann Integration by Substitution)[16] Let $t(x)$ be differentiable on (a, b) and let both t, t' be continuous on $[a, b]$. Then the range of $t(x)$ is an interval $\mathbf{I}$, and if $f(t)$ is continuous on $\mathbf{I}$, then

$$\int_a^b f[t(x)]t'(x)dx = \int_{t(a)}^{t(b)} f(t)dt.$$

(a) Explain how the existence of $\mathbf{I}$ is a consequence of Theorems 6.12 and 6.13. Also explain how you know that $f[t(x)]$ is continuous on $[a, b]$.

(b) Choose any $t_0 \in \mathbf{I}$ and define $F(t) = \int_{t_0}^t f(y)dy$. Why is this permitted? Why does $F'(t) = f(t)$ for all $t \in \mathbf{I}$?

(c) Next, define $h(x) = F[t(x)]$. Why does $h'(x) = F'[t(x)]t'(x) = f[t(x)]t'(x)$? How do we then know that $f[t(x)]t'(x)$ is continuous on $[a, b]$?

---

[15] Hardy says that this inequality was first given (in 1859) by the versatile Russian mathematician V.Ya. Bunyakovski (1804–1889) (Grigorian, 1978; Hardy, Littlewood, and Pólya, 1967).

[16] Many writers prefer to call this "Change of Variable."

(d) Hence, what is the justification for writing

$$\int_a^b f[t(x)]t'(x)dx = \int_a^b h'(x)dx = h(b) - h(a)?$$

(e) Fill in the last tiny steps.

6.36. Prove that for any $c \geq 0$ we have $\int_0^{c/\sqrt{1+c^2}} \frac{dx}{\sqrt{1-x^2}} = \int_0^c \frac{dt}{1+t^2}$.

6.37. **(Second Mean-Value Theorem for Integrals)** If f, g are continuous on $[a, b]$ and $g'(t)$ is nonnegative and continuous on $[a, b]$, then there is a $c \in [a, b]$ such that

$$\int_a^b f(t)g(t)dt = g(a)\int_a^c f(t)dt + g(b)\int_c^b f(t)dt.$$

To prove this, begin with the left-hand side, define $F(t) = \int_a^t f(x)dx$, and make use of Exercise 6.29 and the Fundamental Theorem of the Calculus—B.

6.38. Figure 6.16 shows a sector of the unit circle centered at the origin. We *define* the measure of $\angle QOP$ to be twice the area of sector OPQ. Denote by $m > 0$ the slope of OP, and by $A(m)$ the area of OPQ. Let $\overline{OP'} = x_0$ and be variable. We have $A(m) =$ area $\Delta OPP' +$ area segment $PP'Q$.

(a) Show, with the aid of the Fundamental Theorem, that $A'(m) = \frac{1}{2}(1 + m^2)^{-1}$ and, hence, the measure of $\angle QOP$ is $y = \int_0^m \frac{dt}{1+t^2}$.

(b) Guided by the elementary geometry of the picture, we know (after the fact) that $\tan \angle QOP = \tan y = m$, so we make the reasonable *definition* that $\mathrm{Tan}^{-1} m = \int_0^m \frac{dt}{1+t^2}$. Thinking of m now as any real number x, not necessarily connected with geometry,

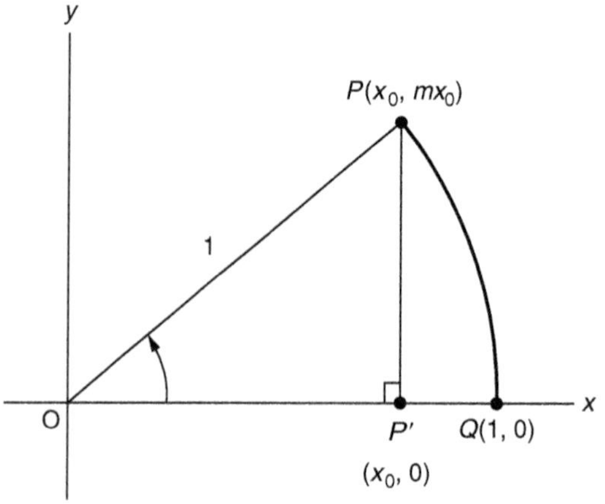

FIGURE 6.16

Introduction of the arctangent function.

we see that the integral defines a unique number for each real x. Why? How does the Fundamental Theorem tell us, additionally, that $\mathrm{Tan}^{-1}x = \int_0^x \frac{dt}{1+t^2}$ is strictly increasing?

(c) How do we now know that $\mathrm{Tan}^{-1}x$ has an inverse function that is continuous and strictly increasing? We denote this inverse function by $\tan x$, the **tangent**.

(d) Again, from the figure, we write $\cos y = 1/\sqrt{1+m^2}$ and $\sin y = m/\sqrt{1+m^2}$ as reasonable *definitions* for two new functions, called **cosine** and **sine**. What is $\sin y/\cos y$, and what is $d(\tan y)/dy$?

This development continues in Exercises 6.50 and 6.51.

6.39. Let f and its first $n+1$ derivatives be continuous on $[a, b]$. Choose any x in the interval $a < x \le b$ and define

$$F_n(t) = -f(t) + f(x) - (x-t)f'(t) - \frac{(x-t)^2}{2!}f''(t) - \cdots - \frac{(x-t)^n}{n!}f^{(n)}(t).$$

(a) Work out $F_n'(t)$ and then apply Theorem 6.15 to obtain

$$f(x) = f(a) + (x-a)f'(a) + \frac{(x-a)^2}{2!}f''(a) + \cdots + \frac{(x-a)^n}{n!}f^{(n)}(a)$$
$$+ \int_a^x \frac{(x-t)^{(n)}}{n!}f^{(n+1)}(t)dt.$$

Thus, we have the **Integral Form of the Remainder** in Taylor's Theorem.

(b) Now make use of Exercise 6.29 and deduce Lagrange's Form of the Remainder (see Theorem 5.11).

(c) Finally, show how to obtain Cauchy's Form of the Remainder,

$$R_n(x) = \frac{(x-a)(x-c)^n f^{(n+1)}(c)}{n!},$$

for some $c \in (a, x)$. For some problems, Cauchy's form is more useful than Lagrange's form (see discussion after Theorem 5.14).

6.40. The papers (Botsko, 1991; van de Lune, 1975) are very pertinent to this section. Read and digest them, and write up a report on your reading.

Section 6.6*

6.41. For use in the proof of Theorem 6.18, derive the result

$$\frac{d}{dh}\int_{a_k-h}^{a_k+h} f(x)dx = f(a_k+h) + f(a_k-h).$$

6.42. Verify in the proof of Theorem 6.18 that

$$g_k^{(4)}(h) = \frac{-h}{3}\left[f^{(4)}(a_k+h) + f^{(4)}(a_k-h)\right] - \frac{1}{3}\left[f^{(3)}(a_k+h) - f^{(3)}(a_k-h)\right].$$

6.43. Refer to Example 6.3. Show that application of Simpson's Rule with $n = 8$ gives the estimated value of $\int_0^1 e^{-t^2} dt \approx 0.74683$. The tabulated value is 0.746824 (Abramowitz and Stegun, 1965).

6.44. The **Debye functions**, $\{D(n, x)\}_{n=1}^{\infty}$, are important in theories of the specific heat of metals:

$$D(n, x) = \int_0^x \frac{t^n}{e^t - 1} dt, \quad |x| < 2\pi.$$

(a) Establish that on the interval $[0, 1]$ we have

$$\frac{t}{1 + \frac{1}{2}t} \geq \frac{t^2}{e^t - 1} \geq \frac{t}{e - 1},$$

and, hence, $0.3781 > D(2, 1) > 0.2910$. The tabulated value of the function $D(2, x)$ at $x = 1$ is 0.353939 (Abramowitz and Stegun, 1965).

(b) Although this integral is improper at the lower limit, the impropriety is removable by the appropriate assignment of value to the integrand at $t = 0$. Do this and then apply Simpson's Rule ($n = 8$) to estimate $D(2, 1)$.

6.45. The **error function**, erf x, is important in probability and statistics:

$$\operatorname{erf} x = \frac{2}{\sqrt{\pi}} \int_0^x e^{-t^2} dt, \quad x \geq 0.$$

(a) Prove that $\lim_{x \to \infty} \operatorname{erf} x$ exists (in $\mathbf{R}^1$).

(b) The tabulated value of erf 1.76 (to 6 places) is 0.987190 (Abramowitz and Stegun, 1965). Apply Simpson's Rule ($n = 8$) to estimate erf 1.76.

6.46. In Exercise 6.15 it is better to leave the expression for C in the form with θ as the variable of integration, since this form of the integral is not improper. We then have $E(1, \frac{3}{4}) = C/4$. Apply Simpson's Rule ($n = 10$) to estimate $E(1, \frac{3}{4})$. Compute the percent error; the tabulated value of $E(1, \frac{3}{4})$ is 1.211056 (Abramowitz and Stegun, 1965).

Section 6.7

6.47. Prove part (b) of Theorem 6.19.

6.48. **(a)** In Example 6.19, prove that (b) diverges.

(b) In Example 6.20, prove that (c), (d) converge.

6.49. Review Example 6.23. Then prove that the integral in Example 6.19 (d) has a value less than $5\sqrt{2}/2$.

6.50. We have from Exercise 6.38 that if $y = \operatorname{Tan}^{-1} x = \int_0^x \frac{dt}{1+t^2}$, then $x = \operatorname{Tan} y$, $1/\sqrt{1 + x^2} = \cos y$, and $x/\sqrt{1 + x^2} = \sin y$.

(a) Show that if $t = u^{-1}$ and if M is large, then

$$y = \operatorname{Tan}^{-1} M = 2\int_0^1 \frac{dt}{1+t^2} - \int_0^{1/M} \frac{du}{1+u^2}.$$

How does this immediately show that the improper integral $\int_0^\infty \frac{dt}{1+t^2}$ converges?

(b) We now define the number π by the relation

$$\frac{1}{2}\pi = \int_0^\infty \frac{dt}{1+t^2}.$$

Use part (a) plus an application of Simpson's Rule ($n = 12$) to estimate π.

(c) Show that $\lim_{x\to-\infty} \int_0^x \frac{dt}{1+t^2} = \frac{-1}{2}\pi$. This and the definition in part (b), incidentally, clear up two facts that we took for granted in Example 6.21. Hence, in summary, we have that if $-\frac{1}{2}\pi < y < \frac{1}{2}\pi$, then $-\infty < \tan y < \infty$.

(d) We have immediately that $\cos 0 = 1$ and $\sin 0 = 0$. The values of $\cos y, \sin y$, and $y = \pm\frac{1}{2}\pi$ are *defined* to be the limits of these functions as $x \to \pm\infty$. What are these limits?

6.51. Refer to the previous exercise.

(a) Since $x = \tan y$, show that $\frac{dx}{dy} = (\tan y)' = 1 + x^2$, and that if we define $\sec y = 1/\cos y$, then $(\tan y)' = \sec^2 y$.

(b) If we write $(\cos y)' = \frac{d(\cos y)}{dx}\frac{dx}{dy}$, then show that $(\cos y)' = -\sin y$.

(c) Similarly, show that $(\sin y)' = \cos y$. How does this show that $\sin y$ is monotonic on $(-\pi/2, \pi/2)$? Is $\cos y$ also monotonic on this interval?

(d) Outside of $(-\pi/2, \pi/2)$ for $\tan y$ and outside of $[-\pi/2, \pi/2]$ for $\cos y$ and $\sin y$, we *define* $\tan(y+\pi) = \tan y$, $\cos(y+\pi) = -\cos y$, $\sin(y+\pi) = -\sin y$. Prove that $\sin y$ is continuous at $y = \pi/2$.

6.52. **(Laplace Transforms)** If $f(t)$ is a given function, then a function $F(s)$ obtained as the following improper integral of the first kind,

$$F(s) = L[f(t)] = \int_a^\infty K(s,t) f(t)\,dt,$$

is known as an **integral transform** of f. When the **kernel** of the transform, $K(s,t)$, is e^{-st} and $a = 0$, the transform is called a **Laplace transform**.[17] Some conditions on f and on s are naturally needed for convergence.

Let $F_n(s)$ be the Laplace transform of $\sin^n(kt)$, where $k \neq 0$, n is a nonnegative integer, and $s > 0$. Show that for any $n \geq 2$ we have

$$F_n(s) = \frac{k^2 n(n-1)}{s^2 + k^2 n^2} F_{n-2}(s).$$

Now work out $F_0(s), F_1(s), F_2(s), F_3(s), F_4(s)$. Extra credit if you can obtain an explicit expression for $F_n(s)$, $n \geq 1$.

---

[17] After the French mathematical physicist **Pierre-Simon Laplace** (1749–1827).

Integral transforms are of great utility in mathematics, and the literature is voluminous. If you would like to read a little of it, please consult any of the following: Borden (1998), Dence (2007), Dettman (1969), Sneddon (1995), Watson (1981), and Widder (1941).

6.53. For each improper integral determine if it converges, diverges but has a Cauchy principal value, or diverges and does not have a Cauchy principal value.

(a) $\int_0^1 \frac{dt}{\sqrt{t+2t^3}}$;

(b) $\int_0^\infty \frac{dt}{t^{3/2}}$;

(c) $\int_{-\infty}^\infty \frac{t}{\sqrt{t^2+1}}\,dt$;

(d) $\int_0^1 \frac{t^n dt}{\sqrt{1-t^2}}, \quad n \in \mathbf{N}$;

(e) $\int_1^\infty \sin(\sqrt{t})dt$;

(f) $\int_0^{\pi/2} t\sqrt{\sec t}\,dt$;

(g) $\int_0^{\pi/2} \frac{dt}{\sqrt{1-\sin t}}$;

(h) $\int_2^\infty \frac{\sin(t^{-1})}{\sqrt{t}}\,dt$;

(i) $\int_1^\infty t\sin(t^{-3})\,dt$;

(j) $\int_{-1}^1 \frac{2^{t^4}}{t^3}\,dt$.

6.54. Verify the values of the following improper integrals:

(a) $\int_0^1 \sqrt{\frac{1+t}{1-t}}dt = \frac{\pi}{2} + 1$;

(b) $\int_0^1 t\ln(t)dt = -1/4$;

(c) $\int_0^\infty \frac{dt}{t^4+1} = \frac{\pi\sqrt{2}}{4}$;

(d) $\int_0^1 \frac{t^2 dt}{\sqrt{1-t}} = \frac{16}{15}$;

(e) $(P)\int_{-\infty}^\infty \frac{1+t}{1+t^2}\,dt = \pi$;

(f) $\int_{\sqrt{7}-3}^1 \frac{dt}{\sqrt{t^2+6t+2}} = \frac{1}{2}\ln 7$.

6.55. The **gamma function** can be defined by

$$\Gamma(\alpha) = \int_0^\infty t^{\alpha-1}e^{-t}dt.$$

(a) Determine for which $\alpha \in \mathbf{R}$ the integral converges.

(b) Show that for any α in part (a), we have $\Gamma(\alpha+1) = \alpha\Gamma(\alpha)$.

(c) What values of $\Gamma(\alpha)$ are obtained when α is a natural number?

6.56. Refer to the previous exercise.

(a) Although we have not yet had the pertinent theory (see, later, Exercise 7.56), present a plausibility argument for the assertion

$$\Gamma'(\alpha) = \int_0^\infty t^{\alpha-1}e^{-t}\ln t\,dt.$$

What do you then conjecture is $\Gamma''(\alpha)$?

(b) How does the functional form of $\Gamma(\alpha)$, $\Gamma''(\alpha)$ tell you that $\Gamma(\alpha)$ has a relative minimum at some $\alpha = \alpha_0$? How do you deduce that $1 < \alpha_0 < 2$?

(c) From parts (a) and (b) and from Exercise 6.55, sketch semiquantitatively the graph of $\Gamma(\alpha)$ versus α for $0 < \alpha \leq 4$.

(d) Show that the expression for $\Gamma'(\alpha)$ in part (a) can be rewritten as

$$\Gamma'(\alpha) = \int_0^1 \left[t^{\alpha-1}e^{-t} - t^{-\alpha-1}e^{-1/t}\right] \ln t \, dt.$$

The integrand approaches 0 at both endpoints.

(e) It is known that the α_0 in part (b) is less than 1.47. Write a computer program for a series of Simpson's Rule calculations ($n = 20$) over a range of values of α. See if you can find an interval in which $\Gamma'(\alpha)$ must be 0 at some point.

6.57. Use the Integral Test to decide which of the following series converge:

(a) $\sum_{k=1}^{\infty} \frac{k}{k^3+1}$;

(b) $\sum_{k=2}^{\infty} \frac{1}{k\sqrt{k-1}}$;

(c) $\sum_{k=1}^{\infty} \frac{\sec^{-1}(2k)}{k^2}$;

(d) $\sum_{k=0}^{\infty} \frac{k^3}{2^k}$.

6.58. This exercise shows the use of an improper integral for the evaluation of a finite sum of binomial coefficients (Dence, 2010).

(a) For any $n \in N$, let $f_n(s) \cong \frac{s^n}{2\sqrt{1-s}}$. Then let $w = 1 - s$ and show that

$$\int_0^1 f_n(s)ds = \sum_{k=0}^{n} \binom{n}{k} \frac{(-1)^k}{2k+1}.$$

(b) On the other hand, in $f_n(s)$ make the substitution $s = \sin^2 t$. Now show that

$$\int_0^1 f_n(s)ds = \frac{4^n (n!)^2}{(2n+1)!},$$

and, thus, we have obtained

$$S_{2n+1} = \sum_{k=0}^{n} \binom{n}{k} \frac{(-1)^k}{2k+1} = \frac{4^n (n!)^2}{(2n+1)!}.$$

(c) Determine $\lim_{n\to\infty} S_{2n+1}$.

Section 6.8

6.59. Write out the proof of Theorem 6.1′.

6.60. Write out the proof of Theorem 6.2′.

6.61. Write out the proof of Theorem 6.4′ (ii).

6.62. Let $\mathbf{T}$ be the rectangular region $\mathbf{T} = \{(x_1, x_2) : 1 \leq x_1 \leq 3, 1 \leq x_2 \leq 3\}$, and on this erect a partition $\overline{P}$ in which there are 11 equally spaced lines parallel to the x_2-axis and 11 equally spaced lines parallel to the x_1-axis. Let $\mathfrak{D}$ be the closed unit disk centered at $(2, 2)$, and tag each subrectangle $\mathbf{R}_i \subset \mathbf{T}$ by choosing the geometric center of $\mathbf{R}_i$. Define $f : \mathfrak{D} \to \mathbf{R}^1$ to be $f(\mathbf{x}) = f((x_1, x_2)) = x_1 + x_2 - 1$. Compute $S(\mathfrak{D}, \overline{P}, f)$, which will be an estimate of $\iint_{\mathfrak{D}} (x_1 + x_2 - 1)$. It can be shown that the double integral has the value of $3\pi \approx 9.42$.

6.63. Refer to Example 6.31.

(a) Give a definition in terms of a Riemann sum for a set $\mathbf{S}$ in $\mathbf{R}^2$ to have Jordan content (area) 0.

(b) Tell how, corresponding to any given $\varepsilon > 0$, you could explicitly construct a partition P, that you could use in your definition in part (a) in order to show that the circle of radius 1 has Jordan content 0.

6.64. Explain how you know, as stated in Example 6.33, that consideration of $Bd(\mathfrak{D})$ in Example 6.30 shows that $\mathfrak{D}$ itself is not Jordan-measurable.

6.65. The analog of the definition given in the opening paragraph of Section 6.3 is this:

Definition. *A **step-function** $\sigma(x, y)$ is a function that is defined on some closed, bounded rectangle $\mathbf{T}$ and which has constant values σ_i on the interiors of the subrectangles $\{\mathbf{R}_i\}_{i=1}^n$ relative to some partition P of $\mathbf{T}$.*

Prove that any step-function defined on $\mathbf{T}$ is Riemann-integrable and that the double integral has the value $\iint_{\mathbf{T}} \sigma(x, y) = \sum_{i=1}^n \sigma_i \Delta A_i$, where ΔA_i is the area of the ith subrectangle. (Do not assume that Theorem 6.13' has been established yet.)

6.66. (**Mean-Value Theorem for Double Integrals**) Let $\mathfrak{D} \subset \mathbf{R}^2$ be compact, connected, and have Jordan content, and suppose that $f(x, y)$ is continuous on $\mathfrak{D}$. Prove that there is a point $\mathbf{p}$ in the interior of $\mathfrak{D}$ such that

$$\iint_{\mathfrak{D}} f(x, y) = f(\mathbf{p}) A(\mathfrak{D}),$$

where $A(\mathfrak{D})$ is the area of $\mathfrak{D}$. This generalizes Corollary 6.13.1.

SUPPLEMENTARY PROBLEMS

Additional problems on integration for your enrichment and for challenge can be found in Appendix C.

Good luck!

REFERENCES

Cited Literature

Abramowitz, M. and Stegun, I.A. (eds.), *Handbook of Mathematical Functions*, Dover Publications, NY, 1965. Very useful reference work on tabulated data; used in our Exercises 6.15, 6.43–6.46, and in our Example 6.16.

Bao-Lin, Z., "A Note on the Mean Value Theorem for Integrals," *Amer. Math. Monthly*, **104**, 561–562 (1997). Proof of an interesting theorem on the constant c in the MVT for Integrals; the source of our Exercise 6.30.

Borden, R.S., *A Course in Advanced Calculus*, Dover Publications, NY, 1998, pp. 199–208, 344–358. The citations are, respectively, to discussions of measure and discussions of transforms (Fourier and Laplace); cited in our Exercise 6.52. An easy introduction to Lebesgue measure can be found in Boas (pp. 195–200).

Botsko, M.W., "A Fundamental Theorem of Calculus that Applies to All Riemann Integrable Functions," *Math. Mag.*, **64**, 347–348 (1991). The author's result: If f is Riemann-integrable on $[a, b]$ and g is a function that satisfies a Lipschitz condition and $g'(x) = f(x)$ almost everywhere, then $\int_a^b f = g(b) - g(a)$. Cited in our Exercise 6.40.

Boyer, C.B., *The History of the Calculus and Its Conceptual Development*, Dover Publications, NY, 1959, pp. 189–202. Read these pages for a thumbnail history of Newton's creation of the calculus.

Brown, A.B., "A Proof of the Lebesgue Condition for Riemann Integrability," *Amer. Math. Monthly*, **43**, 396–398 (1936). This paper is accessible and would make a nice reading assignment for a motivated student.

Burk, F., Goel, S.K. and Rodríguez, D.M., "Using Riemann Sums in Evaluating a Familiar Limit," *Coll. Math. J.*, **17**, 170–171 (1986). Nice illustration of the fact that those "dreaded" Riemann sums are good for more than just integration theory.

Chatterji, S.D., "A Frequent Oversight Concerning the Integrability of Derivatives," *Amer. Math. Monthly*, **95**, 758–761 (1988). It is not always true that $\int_a^b g' = g(b) - g(a)$, even if g' is defined and bounded on $[a, b]$.

Courant, R. and John, F., *Introduction to Calculus and Analysis*, Vol. 2, John Wiley, NY, 1974, pp. 515–523. These authors' two-volume work is a high-quality exposition that is worth owning. The indicated pages are an introduction to Jordan measure, and would make a nice outside reading assignment.

Cunningham, Jr., F., "The Two Fundamental Theorems of Calculus," *Amer. Math. Monthly*, **72**, 406–407 (1965). A pedagogically important paper; the inspiration for our Section 6.5.

Dence, T.P., "Some Half-Row Sums from Pascal's Triangle via Laplace Transforms," *Coll. Math. J.*, **38**, 205–209 (2007). Unexpected combinatoric results obtained from a novel application of Laplace transforms; cited in our Exercise 6.52.

Dence, T.P., "On an Identity Involving Binomial Coefficients via arctan (x)," *Math. Gaz.*, **94** (2010) (to appear).

Dettman, J.W., *Mathematical Methods in Physics and Engineering*, 2nd ed., McGraw-Hill, NY, 1969, pp. 368–418. The indicated pages are Chapter 8, "Integral Transform Methods," which emphasizes the use of Fourier and Laplace transforms to solve ordinary differential, partial differential, and integral equations. There is also a brief look at finite Fourier sine and cosine transforms, Hankel and Mellin transforms. Cited in our Exercise 6.52.

Freudenthal, H., "Georg Friedrich Bernhard Riemann," in Gillispie, C.C. (ed.), *Dictionary of Scientific Biography*, Vol. XI, Charles Scribner's Sons, NY, 1975, pp. 447–456. Very nice summary of Riemann's all-too-brief life's work in mathematics and mathematical physics.

Grabiner, J.V., *The Origins of Cauchy's Rigorous Calculus*, MIT Press, Cambridge, 1981. See the author's Chapter 6 (pp. 140–163) on the origins of Cauchy's theory of the definite integral.

Grabiner, J.V., "Was Newton's Calculus a Dead End? The Continental Influence of MacLaurin's Treatise of Fluxions," *Amer. Math. Monthly*, **104**, 393–410 (1997). A fine, scholarly article on one of Scotland's most brilliant sons, Colin MacLaurin.

Grigorian, A.T., "Viktor Yakovlevich Bunyakovsky," in Gillispie, C.C. (ed.), *Dictionary of Scientific Biography*, Vol. XV, Supplement I, Charles Scribner's Sons, NY, 1978, pp. 66–67. This mathematician deserves to be better known; cited in our Exercise 6.31.

Hardy, G.H., *A Course of Pure Mathematics*, 10th ed., Cambridge University Press, Cambridge, 1967, pp. 351–353. Our presentation of MacLaurin's Integral Test (Theorem 6.23) is drawn from these pages.

Hardy, G.H., Littlewood, J.E. and Pólya, G., *Inequalities*, Cambridge University Press, Cambridge, 1967, pp. 16, 132–134. The indicated pages refer to Bunyakovsky's Inequality; pp. 126–132 are a useful prelude.

Herstein, I.N., *Topics in Algebra*, 2nd ed., Xerox College Publishing, Lexington, MA, 1975, pp. 207–214. Read these few pages on extension fields; the pertinent theorem for us is Herstein's Theorem 5.14 on algebraic elements.

Hofmann, J.E., "Gottfried Wilhelm Leibniz," in Gillispie, C.C. (ed.), *Dictionary of Scientific Biography*, Vol. VIII, Charles Scribner's Sons, NY, 1973, pp. 149–168. Leibniz's universality was comparable to that of Aristotle's.

Jacobson, B., "On the Mean Value Theorem for Integrals," *Amer. Math. Monthly*, **89**, 300–301 (1982). The source of our Exercise 6.28.

Kline, M., *Mathematical Thought from Ancient to Modern Times*, Oxford University Press, NY, 1972, pp. 344–364. These pages give a brief summary of seventeenth-century work on calculus, including that of Isaac Barrow.

Knopp, M.I., *Theory of Area*, Markham Publishing Co., Chicago, 1969. A rigorous and very accessible study of the restriction to $\mathbf{R}^2$ of the theory of Jordan content. Worth owning, if you can find a copy.

Kristensen, E., Poulsen, E.T. and Reich, E., "A Characterization of Riemann-Integrability," *Amer. Math. Monthly*, **69**, 498–505 (1962). The authors' principal result: If f is bounded on $[a, b]$ and $\lim_{\mathfrak{N}\to 0} \sum_{k=1}^{n} f(a_{k-1})\Delta a_k$ exists and is finite, then f is Riemann-integrable on $[a, b]$. See, also, Gillespie, D.C., "The Cauchy Definition of a Definite Integral," *Ann. Math.*, **17**, 61–63 (1915).

Leonard, I.E. and Duemmel, J., "More—and Moore—Power Series Without Taylor's Theorem," *Amer. Math. Monthly*, **92**, 588–589 (1985). A very simple method for obtaining power series expansions for $\sin x$, $\cos x$, e^x, e^{-x} that uses no sophisticated concepts. The source of our Example 6.2.

Liang, J.J.Y. and Todd, J., "The Stieltjes Constants," *J. Res. Nat. Bur. Stand.*, **76B**, 161–178 (1972). The authors computed the first 19 coefficients in the Laurent expansion of the Riemann zeta function by use of the Euler-MacLaurin Summation Formula. We've approximated $\zeta(3/2)$ by the finite series $\frac{1}{(3/2)-1} + \sum_{k=0}^{6} A_k(\frac{3}{2} - 1)^k$; rounded off to six decimals, $\zeta(3/2)$ is 2.612461. Cited in our Example 6.29.

Mathews, J.H. and Shultz, H.S., "Riemann Integral of cos x," *Coll. Math. J.*, **20**, 237 (1989). A Leibniz-like evaluation of $\int \cos x\, dx$; the source of our Exercise 6.4.

Niven, I., "On Elliptic Integrals," *Amer. Math. Monthly*, **50**, 41–42 (1943). It is interesting that the elementary observation noted here had gone undetected during the more than 150 years of existence of elliptic integrals. The source of our Exercise 6.16.

Olds, C.D., "The Fresnel Integrals," *Amer. Math. Monthly*, **75**, 285–286 (1968). A clever evaluation of $C(\infty)$ and $S(\infty)$, carried out in the complex plane.

Olmsted, J.M.H., *Advanced Calculus*, Prentice-Hall, Englewood Cliffs, NJ, 1961, pp. 335–347. The citation is to a discussion of Jordan content (in $\mathbf{R}^2$).

Putney, T., "Proof of the First Mean Value Theorem of the Integral Calculus," *Amer. Math. Monthly*, **60**, 113–114 (1953). The source of our Exercise 6.27.

Richert, A., "A Non-Simpsonian Use of Parabolas in Numerical Integration," *Amer. Math. Monthly*, **92**, 425–426 (1985). The procedure here leads to an absolute error in $\int_a^b f(x)dx$ that is about $\frac{1}{11}$ of that in Theorem 6.18.

Royden, H.L., *Real Analysis*, 3rd ed., Macmillan, NY, 1988, pp. 54–74. If you are already tempted to step beyond the Riemann integral, then these pages will give you a nice introduction to Lebesgue measure, a generalization of Jordan content.

Sklar, A., "On the Definition of the Riemann Integral," *Amer. Math. Monthly*, **67**, 897–900 (1960). Author shows a way that the Riemann integral can be defined, rigorously and simply, as an ordinary limit of an ordinary sequence of numbers, and it is easy to then show from this definition that Finite Additivity holds, which is not true of *N*. Wiener's definition. The source of our Exercise 6.7.

Sneddon, I.N., *Fourier Transforms*, Dover Publications, NY, 1995. This monograph begins with three chapters of basic theory, and then follows this with several chapters of applications to various physical problems, including vibrations, conduction of heat in solids, hydrodynamics, atomic and nuclear physics, and mechanical stress. Cited in our Exercise 6.52.

van de Lune, J., "A Note on the Fundamental Theorem for Riemann-Integrals," *Amer. Math. Monthly*, **82**, 918–919 (1975). If $f: [a,b] \to \mathbf{R}^1$ is differentiable on $[a,b]$, then f' is Riemann-integrable over $[a,b]$ iff there is a Riemann-integrable function ϕ: $[a,b] \to \mathbf{R}^1$ such that $f(x) = f(a) + \int_a^x \phi(t)dt, a \le x \le b$. Cited in Exercise 6.40.

Velleman, D.J., "Simpson Symmetrized and Surpassed," *Math. Mag.*, **77**, 31–45 (2004). Very stimulating paper; this could spawn undergraduate research projects in various directions, or it could be a suitable paper for senior seminar.

Watson, E.J., *Laplace Transforms and Applications*, Van Nostrand Reinhold, NY, 1981. A modest-sized (190 pp.), accessible monograph on these important transforms; cited in our Exercise 6.52.

Weber, H. (ed.), *Bernhard Riemann's gesammelte mathematische Werke und wissenschaftlicher Nachlass*, 2nd ed., Dover Publications, NY, 1953, pp. 227–271. These pages constitute Riemann's 1854 article "*Ueber die Darstellbarkeit einer Funktion durch eine trigonometrische Reihe*"; the definition of the integral appears on p. 239.

Whittaker, E.T. and Watson, G.N., *A Course of Modern Analysis*, 4th ed., Cambridge University Press, Cambridge, 1963, pp. 512–535. The indicated pages are a rigorous introduction to elliptic integrals. This book is one of the real classics in mathematics, and you should become acquainted with it. Cited in our Exercise 6.15.

Widder, D.V., *The Laplace Transform*, Princeton University Press, Princeton, 1941. Another true classic, deeper and more comprehensive than the smaller work by E.J. Watson. Widder begins with an informative chapter on Stieltjes integrals. Cited in our Exercise 6.52.

Additional Literature

Bartle, R.G., "Return to the Riemann Integral," *Amer. Math. Monthly*, **103**, 625–632 (1996).

Boas, Jr., R.P., *A Primer of Real Functions*, 4th ed., Mathematical Association of America, Washington, D.C., 1996.

Burk, F., *A Garden of Integrals*, Mathematical Association of America, Washington, D.C., 2007.

Cavalcante, R. and Todor, T.D., "A Lost Theorem: Definite Integrals in an Asymptotic Setting," *Amer. Math. Monthly*, **115**, 45–56 (2008).

Dunham, W., "Touring the Calculus Gallery," *Amer. Math. Monthly*, **112**, 1–19 (2005).

Ferguson, R.P., "An Application of Stieltjes Integration to the Power Series Coefficients of the Riemann Zeta Function," *Amer. Math. Monthly*, **70**, 60–61 (1963).

González-Velasco, E.A., "James Gregory's Calculus in the Geometriae Pars Universalis," *Amer. Math. Monthly*, **114**, 565–576 (2007).

Havil, J., *Gamma*, Princeton University Press, Princeton, 2003, pp. 53–60 (the gamma function).

Huber, G., "Gamma Function Derivation of n-Sphere Volumes," *Amer. Math. Monthly*, **89**, 301–302 (1982).

Kilmer, S.J., "Integration by Parts and Infinite Series," *Math. Mag.*, **81**, 51–55 (2008).

LaVita, J.A., "A Necessary and Sufficient Condition for Riemann Integration," *Amer. Math. Monthly*, **71**, 193–196 (1964).

McShane, E.J., "A Unified Theory of Integration," *Amer. Math. Monthly*, **80**, 349–359 (1973).

Nanjundiah, T.S., "Note on the Beta and Gamma Functions," *Amer. Math. Monthly*, **76**, 411–413 (1969).

Pfeffer, W.F., *The Riemann Approach to Integration*, Cambridge University Press, Cambridge, 1993.

http://de. wikipedia.org/wiki/Bernhard_Riemann.

Staib, J.H., "The Integration of Inverse Functions," *Math. Mag.*, **39**, 223–224 (1966).

Talman, L.A., "Simpson's Rule Is Exact for Quintics," *Amer. Math. Monthly*, **113**, 144–155 (2006).

Thomson, B.S., "Rethinking the Elementary Real Analysis Course," *Amer. Math. Monthly*, **114**, 469–490 (2007).

Uetrecht, J.A., "A Clamped Simpson's Rule," *Coll. Math. J.*, **19**, 43–52 (1988).

Wendel, J.G., "Note on the Gamma Function," *Amer. Math. Monthly*, **55**, 563–564 (1948) (the classical asymptotic relation).

Widder, D.V., *Advanced Calculus*, Prentice-Hall, NY, 1947, pp. 303–323 (the gamma function).

2

PART

Advanced Topics

CHAPTER 7

Commutation of Limit Operations

"Weierstrass's discovery [of uniform convergence] was the earliest, and he alone fully realized its far-reaching importance as one of the fundamental ideas of analysis."

Godfrey H. Hardy

CONTENTS

Reviewed in this chapter	Pointwise convergence of a sequence of functions; Cauchy sequence.
New in this chapter	Uniform convergence of a sequence of functions; continuity of limit functions; integration and differentiation of series; Leibniz's Theorem.

7.1 UNIFORM CONVERGENCE

Let $\mathfrak{C}(\mathbf{D})$ denote the set of real-valued functions continuous on $\mathbf{D} \subseteq \mathbf{R}^1$. By analogy to the definition given in the second paragraph of Section 3.6, we say that a sequence $\{f_k\}_{k=0}^{\infty}, f_k \in \mathfrak{C}(\mathbf{D})$ **converges pointwise** on $\mathbf{D}$ to some function f iff for each $x \in \mathbf{D}$ we have $\lim_{k\to\infty} f_k(x) = f(x)$, that is, given $\varepsilon > 0$ and given $x_0 \in \mathbf{D}$, there is an $N \in \mathbf{N}$ such that $k > N$ implies $f_k(x_0) \in \mathbf{B}(f(x_0); \varepsilon)$. It is plausible that f itself should be a member of $\mathfrak{C}(\mathbf{D})$. A simple example shows this assertion to be false.

■ Example 7.1

The sequence $\{f_k(x)\}_{k=0}^{\infty}$ is defined for each k by $f_k(x) = [\sin(\pi x/2)]^k, \mathbf{D} = [0, 1]$. Each $f_k(x)$ clearly belongs to $\mathfrak{C}(\mathbf{D})$. For this sequence we have

$$\lim_{k\to\infty} f_k(x) = f(x) = \begin{cases} 0 & 0 \le x < 1 \\ 1 & x = 1. \end{cases}$$

This is plainly discontinuous at $x = 1$, so f is not an element of $\mathfrak{C}(\mathbf{D})$. ■

The preceding result is reinforced by consideration of the graphs of some of the f_k's. The limit function f and members f_1, f_2, f_6, f_{12} are shown qualitatively in Figure 7.1.

The definitions of pointwise convergence of a sequence of functions and of continuity of a function f at a cluster point $x_0 \in \mathbf{D}$ indicate that the continuity of the limit function f of the sequence would be guaranteed if, at any cluster

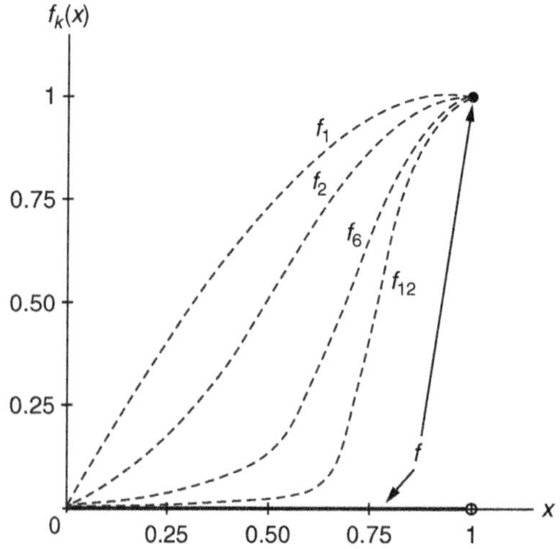

FIGURE 7.1

Some functions f (- - -) and the limit function f_k (—).

point x_0,

$$f(x_0) = \lim_{x \to x_0} f(x) = \lim_{x \to x_0} \left[\lim_{k \to \infty} f_k(x) \right]$$

and, *simultaneously*,

$$f(x_0) = \lim_{k \to \infty} f_k(x_0) = \lim_{k \to \infty} \left[\lim_{x \to x_0} f_k(x) \right].$$

That is, commutation of two limit operations is needed.

■ Example 7.2

The two limit operations for the preceding sequence do not commute everywhere on $\mathbf{D} = [0, 1]$, for $\lim_{k \to \infty} \left[\lim_{x \to 1^-} \{\sin(\pi x/2)\}^k \right] = \lim_{k \to \infty} 1^k = 1$ but, in contrast, we have $\lim_{x \to 1^-} \left[\lim_{k \to \infty} \{\sin(\pi x/2)\}^k \right] = \lim_{x \to 1^-} f(x) = 0$. ■

The reason that the two limit operations in Example 7.1 do not commute everywhere on $\mathbf{D}$ is that for some $\varepsilon > 0$ there fails to be an $N \in \mathbf{N}$ such that for all $k > N$ and *all* $x \in \mathbf{D}$, $|f_k(x) - f(x)| < \varepsilon$ holds. This is implicit in Figure 7.1.

■ Example 7.3

Suppose that there were such an N for the sequence in Example 7.1; let $\varepsilon > 0$ be given. The number $x_0 = (2/\pi)\sin^{-1}\left(\varepsilon^{1/(N+1)}\right)$ lies in $(0, 1)$. Then for any $x \in (x_0, 1)$ we have

$$\begin{aligned} |f_{N+1}(x) - f(x)| &\geq \left| \left[\sin\left(\frac{\pi x_0}{2} \right) \right]^{N+1} - 0 \right| \\ &= \left[\sin\left\{ \frac{\pi}{2} \frac{2}{\pi} \sin^{-1}\left(\varepsilon^{1/(N+1)}\right) \right\} \right]^{N+1} \\ &= \varepsilon, \end{aligned}$$

which is a contradiction. ■

We are led to make the following definition, which has become standard (Hardy, 1919).

Definition. *A sequence* $\{f_k\}_{k=0}^{\infty}$ *of functions* ***converges uniformly*** *(or is* ***uniformly convergent****) on* $\mathbf{D} \subseteq \mathbf{R}^1$ *to a function* $f\colon \mathbf{D} \to \mathbf{R}^1$ *iff for each* $\varepsilon > 0$ *there is an* $N \in \mathbf{N}$ *such that for all* $x \in \mathbf{D}$ *and all* $k > N$ *we have* $|f_k(x) - f(x)| < \varepsilon$, *that is,* $f_k(x) \in \mathbf{B}(f(x); \varepsilon)$.

The crucial part of the definition is that N depends *only* upon ε and *not* upon x. It is also clear that uniform convergence of a sequence $\{f_k(x)\}_{k=0}^{\infty}$ on the set $\mathbf{D} \subseteq \mathbf{R}^1$ implies pointwise convergence of the sequence on $\mathbf{D}$. The converse is generally false; however, a result known as **Dini's Theorem** provides additional hypotheses under which pointwise convergence will imply uniform convergence (Sprecher, 1987). The definition of uniform convergence is interpretable geometrically by the generic curves in Figure 7.2.

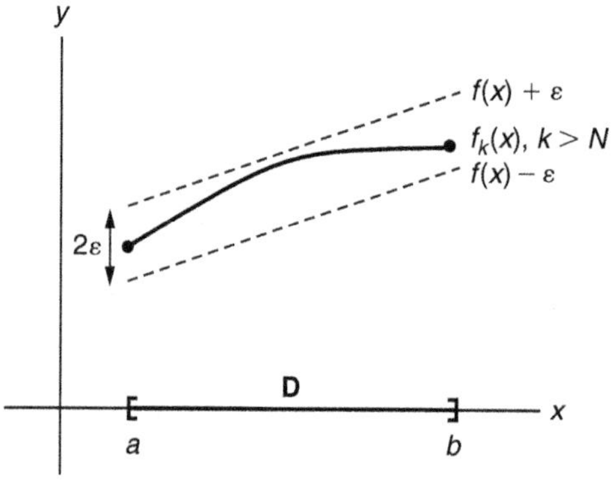

FIGURE 7.2
Uniform convergence of functions $f_k : \mathbf{D} \to \mathbf{R}^1, \mathbf{D} = [a, b]$.

It is plausible that uniform convergence might be important in other interchanges of pairs of limit operations. For example, consider these two cases:

CASE (A): Let $\{f_k\}_{k=1}^{\infty}$ be defined by $f_k(x) = kxe^{-kx^2}, \mathbf{D} = [0, 1]$.

(a) Each f_k is continuous on $\mathbf{D}$ and $\lim_{k\to\infty} f_k(x) = 0$ for all $x \in \mathbf{D}$.

(b) We observe that $\int_0^1 \left[\lim_{k\to\infty} f_k(x)\right] dx = 0$, whereas $\lim_{k\to\infty} \int_0^1 f_k(x)dx = \lim_{k\to\infty} \left[\frac{-1}{2} \int_0^1 \left(-2kxe^{-kx^2}\right) dx\right] = \lim_{k\to\infty} \frac{-1}{2}\left(e^{-k} - e^0\right) = \frac{1}{2}$.

CASE (B): Let $\{f_k\}_{k=1}^{\infty}$ be defined by $f_k(x) = k^{-1}\cos(kx), \mathbf{D} = [0, 20]$.

(a) Each f_k is continuous on $\mathbf{D}$ and $\lim_{k\to\infty} f_k(x) = 0$ for all $x \in \mathbf{D}$.

(b) We observe that $\frac{d}{dx}\left[\lim_{k\to\infty} f_k(x)\right] = 0$, whereas $\lim_{k\to\infty} f_k'(x) = \lim_{k\to\infty} [-\sin(kx)] = 0$ for only finitely many $x \in \mathbf{D}$ and fails to exist everywhere else in $\mathbf{D}$.

■ Example 7.4

Suppose the sequence in Case (A) had been conjectured to be uniformly convergent on $\mathbf{D} = [0, 1]$. Then if we choose $\varepsilon = 1/4$ arbitrarily, there should be an $N \in \mathbf{N}$ such that for all $k > N$ and for all $x \in \mathbf{D}$ we would have

$$\left| f_k(x) - f(x) \right| = \left| kxe^{-kx^2} - 0 \right| < 1/4,$$

and, in particular, $2Nxe^{-2Nx^2} < 1/4$. But the number $x = \frac{1}{2\sqrt{N}}$ lies in $[0, 1]$ and we have the contradiction:

$$2N\frac{1}{2\sqrt{N}}e^{-2N\left(\frac{1}{4N}\right)} = \sqrt{\frac{N}{e}} > 0.60.$$

Hence, the sequence in (A) cannot be uniformly convergent on $\mathbf{D}$. See Theorem 7.3 and a follow-up comment immediately after Example 7.8. ■

■ Example 7.5

The sequence $\{f_k\}_{k=1}^{\infty}$ in Case (B) converges uniformly on $[0, 20]$ to $f(x) = 0$, since if $\varepsilon > 0$ is given, then $|f_k(x) - f(x)| = |k^{-1}\cos(kx)| < k^{-1} < \varepsilon$ for all $x \in [0, 20]$ and for all $k > N = \lceil \varepsilon^{-1} \rceil$. However, we note that $\{f_k'\}_{k=1}^{\infty}$ is not uniformly convergent on $[0, 20]$ because $\{-\sin(kx)\}_{k=1}^{\infty}$ is not even pointwise convergent everywhere on $[0, 20]$. See Section 7.5 for a follow-up comment. ■

Recall from Section 2.7 the definition of a Cauchy sequence of real numbers. As expected, there is an analog for sequences of functions.

Definition. *A sequence of functions* $\{f_k\}_{k=0}^{\infty}$ *is termed* ***uniformly Cauchy*** *on* $\mathbf{D} \subseteq \mathbf{R}^1$ *iff, given any* $\varepsilon > 0$*, there is an* $N \in \mathbf{N}$ *such that for all* $x \in \mathbf{D}$ *and for all* $n > m > N$ *we have*

$$\left| f_n(x) - f_m(x) \right| < \varepsilon.$$

Theorem 7.1 (Cauchy Criterion for Uniform Convergence). *A sequence of functions* $\{f_k\}_{k=0}^{\infty}$ *is uniformly convergent on* $\mathbf{D} \subseteq \mathbf{R}^1$ *iff it is uniformly Cauchy on* $\mathbf{D}$.

Proof. ($\rightarrow$) Assume that $\{f_k\}_{k=0}^{\infty}$ is uniformly convergent on $\mathbf{D} \subseteq \mathbf{R}^1$. This part of the proof is left to you.

($\leftarrow$) Assume that $\{f_k\}_{k=0}^{\infty}$ is uniformly Cauchy on $\mathbf{D} \subseteq \mathbf{R}^1$. Choose any $a \in \mathbf{D}$. Then if $\varepsilon > 0$ is given, there is an $N \in \mathbf{N}$ such that for any $n > m > N$ we have $|f_n(a) - f_m(a)| < \varepsilon$. Hence, $\{f_k(a)\}_{k=0}^{\infty}$ is Cauchy in $\mathbf{R}^1$, and by Theorem 2.16 it converges. But a was arbitrary, so $\{f_k\}_{k=0}^{\infty}$ is pointwise convergent on $\mathbf{D}$ to some function f.

Again, if $\varepsilon > 0$ is given, there is an $N \in \mathbf{N}$ such that for all $n > m > N$ and for all $x \in \mathbf{D}$ we have

$$|f_n(x) - f_m(x)| < \varepsilon/2,$$

that is, $f_m(x) - \frac{\varepsilon}{2} < f_n(x) < f_m(x) + \frac{\varepsilon}{2}$. Keep m fixed but let $n \to \infty$; we then obtain

$$f_m(x) - \frac{\varepsilon}{2} \leq f(x) \leq f_m(x) + \frac{\varepsilon}{2}.$$

Hence, for all $x \in \mathbf{D}$ and all $m > N$ we have

$$|f_m(x) - f(x)| \leq \frac{\varepsilon}{2} < \varepsilon.$$

This says that $\{f_k\}_{k=1}^{\infty}$ is uniformly convergent on $\mathbf{D}$. ■

7.2 LIMIT INTERCHANGE FOR CONTINUITY

We now resume more directly the thread of discussion initiated at the start of the previous section. We shall prove that continuity is preserved in the space $\mathfrak{C}(\mathbf{D})$ under the stipulation of uniform convergence, and that this implies the commutation of two limit operations.

Theorem 7.2. *If* $\{f_k\}_{k=0}^{\infty}$, $f_k \in \mathfrak{C}(\mathbf{D})$ *converges uniformly to* f *on* $\mathbf{D} \subseteq \mathbf{R}^1$, *then* $f \in \mathfrak{C}(\mathbf{D})$.

Proof. Choose any $x_0 \in \mathbf{D}$ and let $\varepsilon > 0$ be given. Then by hypothesis there is an $N \in \mathbf{N}$ such that for all $x \in \mathbf{D}$

$$|f_N(x) - f(x)| < \varepsilon/3. \qquad (*)$$

Since each f_k is continuous on $\mathbf{D}$, then there is a $\delta > 0$ such that $|x - x_0| < \delta$ implies

$$|f_N(x) - f_N(x_0)| < \varepsilon/3. \qquad (**)$$

We can also rewrite $(*)$ for the special case $x = x_0$:

$$|f_N(x_0) - f(x_0)| < \varepsilon/3. \qquad (***)$$

Finally, the Triangle Inequality yields, when (*), (**), and (***) are combined, that $|x - x_0| < \delta$ implies

$$\begin{aligned}\left|f(x) - f(x_0)\right| &= \left|\{f(x) - f_N(x)\} + \{f_N(x) - f_N(x_0)\} + \{f_N(x_0) - f(x_0)\}\right| \\ &\leq \left|f_N(x) - f(x)\right| + \left|f_N(x) - f_N(x_0)\right| + \left|f_N(x_0) - f(x_0)\right| \\ &= \varepsilon.\end{aligned}$$

Thus, f is continuous at x_0, and since $x_0 \in \mathbf{D}$ was arbitrary, then f is continuous on $\mathbf{D}$. ■

■ Example 7.6

The contrapositive of Theorem 7.2 yields immediately the result that the sequence discussed in Example 7.1 cannot be uniformly convergent. ■

Theorem 7.2 may be rephrased as giving a *sufficient* condition for the commutation relation

$$\lim_{x\to x_0}\left[\lim_{k\to\infty} f_k(x)\right] = \lim_{k\to\infty}\left[\lim_{x\to x_0} f_k(x)\right]. \qquad (****)$$

It is not a necessary condition, for we can construct a sequence $\{f_k\}_{k=0}^{\infty}$ of functions continuous on $\mathbf{D} = [0, 1]$ that is not uniformly convergent and, yet, the equality above holds (Exercise 7.11).

Under the conditions of the Theorem, however, $\lim_{k\to\infty} f_k(x) = f(x)$ on the left-hand side of equation (****) because the f_k's converge pointwise on $\mathbf{D}$. Then $\lim_{x\to x_0} f(x) = f(x_0)$ because we now know from the Theorem that $f \in \mathfrak{C}(\mathbf{D})$. On the other hand, on the right-hand side of equation (****), $\lim_{x\to x_0} f_k(x) = f_k(x_0)$ because each $f_k \in \mathfrak{C}(\mathbf{D})$. Then $\lim_{k\to\infty} f_k(x_0) = f(x_0)$ because the f_k's converge pointwise on $\mathbf{D}$. Finally, since both double limit operations yield $f(x_0)$, then they are equal.

The Theorem is beautiful. We might ask, however, if it is at all practical. Why not just show directly that a limit function f is continuous and dispense with consideration of uniform convergence? But how would you show this if it is not mechanically possible to obtain a simple, closed form for f? For example, f might emerge as a nontrivial series. See Theorems 7.1′, 7.2′, shortly, as well as Example 7.14.

Ideas for extension of the material so far: (a) In what sense can it be said that $\mathfrak{C}(\mathbf{D})$ is a complete metric space? (b) Can $\mathfrak{C}(\mathbf{D})$, as defined, be profitably generalized to spaces of continuous mappings from one metric space into another metric space? (c) Is there a valid analog in $\mathfrak{C}(\mathbf{D})$ of the famous Bolzano-Weierstrass Theorem for $\mathbf{R}^n$? We leave these questions, and others of this ilk, for you to anticipate in a future course.[1]

7.3 INTEGRATION OF SEQUENCES AND SERIES

In connection with the type of interchange of limit operations alluded to in Example 7.4, we shall now restrict our discussion to sequences of functions defined and Riemann-integrable on a compact interval $[a, b]$. The following Lemma, which properly belonged in Chapter 6, will be needed in Theorem 7.3 and again in Theorem 7.9.

Lemma 7.3.1. *If f is defined and integrable on the finite interval $[a, b]$, $b > a$, then $|f|$ is also integrable on $[a, b]$, and we then have*

$$\left| \int_a^b f(x)dx \right| \leq \int_a^b |f(x)|\, dx.$$

Proof. Let $P = \{a_k\}_{k=0}^{\infty}$ be a partition of $[a, b]$, and choose any subinterval $\mathbf{I}_k = [a_{k-1}, a_k]$ and any pair of points $x_1, x_2 \in \mathbf{I}_k$. Since f is integrable, then by Theorem 6.1 it is bounded on $[a, b]$, and so is $|f|$.

Next, for each $\mathbf{I}_k$ we denote the following four numbers:

$$\begin{cases} M_k = \sup_{x\in\mathbf{I}_k} f(x) & M_k^* = \sup_{x\in\mathbf{I}_k} |f(x)| \\ m_k = \inf_{x\in\mathbf{I}_k} f(x) & m_k^* = \inf_{x\in\mathbf{I}_k} |f(x)|. \end{cases}$$

There are four cases to consider (see Figure 7.3):

CASE 1: $0 \leq m_k \leq M_k$. Then $M_k^* = M_k$ and $m_k^* = m_k$, and so $M_k^* - m_k^* = M_k - m_k$.

CASE 2: $m_k \leq 0 \leq -m_k \leq M_k$. Then $M_k^* = M_k$ and $m_k^* \geq 0 \geq m_k$, and so $M_k^* - m_k^* \leq M_k - m_k$.

CASE 3: $m_k \leq 0 \leq M_k \leq -m_k$. Then, as M_k is not negative and $-m_k^*$ cannot be positive, we have $M_k \geq -m_k^*$ and then $-m_k - M_k \leq m_k^* - m_k$, which implies $M_k^* - M_k \leq m_k^* - m_k$ because $M_k^* = -m_k$. Rearrangement then gives $M_k^* - m_k^* \leq M_k - m_k$.

CASE 4: $m_k \leq M_k \leq 0$. Then $M_k^* = -m_k \geq m_k^* = -M_k$, so $m_k^* - M_k^* = -M_k - (-m_k)$, that is, $M_k^* - m_k^* = M_k - m_k$.

---

[1] Or, you can take a peek at Phillips (1984) and Sprecher (1987).

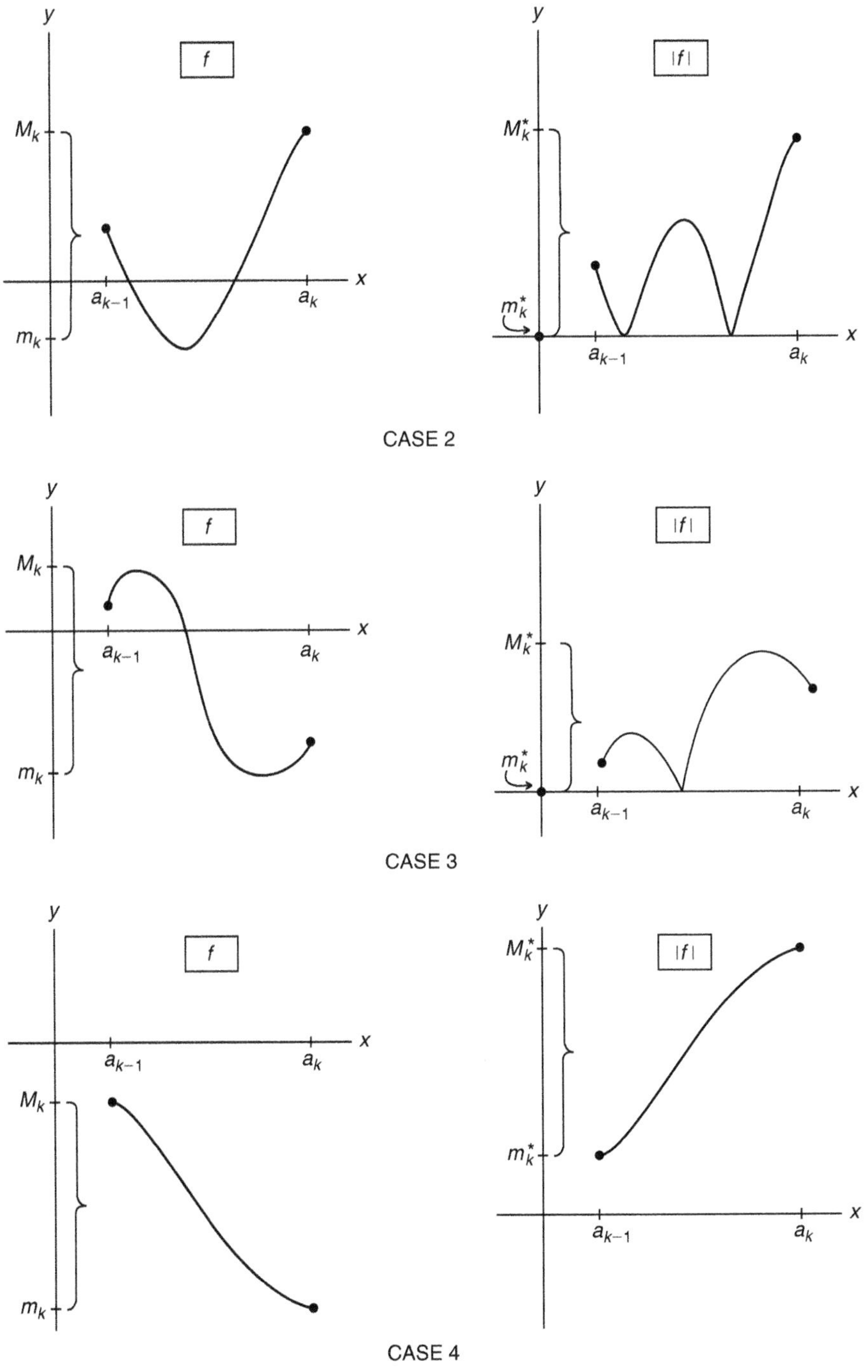

FIGURE 7.3
Suprema and infima for f and $|f|$.

The four cases collectively imply, since each $\Delta a_k = a_k - a_{k-1} > 0$, that

$$\left(M_k^* - m_k^*\right) \Delta a_k \le \left(M_k - m_k\right) \Delta a_k.$$

Summing over all k, we obtain, in the symbolism of Section 6.4,

$$U(P, |f|) - L(P, |f|) \le U(P, f) - L(P, f). \qquad (*)$$

This applies to all partitions of $[a, b]$ because the choice of P was arbitrary. By Theorem 6.11 there is, corresponding to each $\varepsilon > 0$, a $\delta > 0$ such that for all P of norm $\mathfrak{N} < \delta$, we have $U(P, f) - L(P, f) < \varepsilon$. By Theorem 6.10 this says that $|f|$ is integrable on $[a, b]$.

Finally, since $-|f| \le f \le |f|$ for all $x \in [a, b]$, then from Theorem 6.2 we have

$$-\int_a^b |f(x)|\, dx \le \int_a^b f(x)dx \le \int_a^b |f(x)|\, dx,$$

and this is equivalent to $\left|\int_a^b f(x)dx\right| \le \int_a^b |f(x)|\, dx$, since the integrand in the integral on the right-hand side is never negative. ■

■ Example 7.7

Let $f(x) = x\cos x$ and $[a, b] = [-\pi/2, \pi/4]$. We obtain

$$\begin{aligned}
\int_{-\pi/2}^{\pi/4} |f(x)|\, dx &= \int_{-\pi/2}^{0} -x\cos x\, dx + \int_{0}^{\pi/4} x\cos x\, dx \\
&= [-x\sin x - \cos x]\Big|_{-\pi/2}^{0} + [x\sin x + \cos x]\Big|_{0}^{\pi/4} \\
&= \left(1 + \frac{\pi}{4}\right)\frac{\sqrt{2}}{2} + \frac{\pi}{2} - 2 \\
&\approx 0.833,
\end{aligned}$$

and

$$\begin{aligned}
\left|\int_{-\pi/2}^{\pi/4} f(x)dx\right| &= \left|\int_{-\pi/2}^{\pi/4} x\cos x\, dx\right| \\
&= \left|[x\sin x + \cos x]_{-\pi/2}^{\pi/4}\right| \\
&= \left|\left(1 + \frac{\pi}{4}\right)\frac{\sqrt{2}}{2} - \frac{\pi}{2}\right| \\
&\approx 0.308 \\
&< 0.833.
\end{aligned}$$

■

Theorem 7.3. *If $\{f_k\}_{k=0}^{\infty}$ is a uniformly convergent sequence of functions integrable on the finite interval $[a,b]$, $b > a$, then $f = \lim_{k\to\infty} f_k$ is also integrable on $[a,b]$ and we have*

$$\int_a^b f(x)dx = \lim_{k\to\infty} \int_a^b f_k(x)dx.$$

Proof. We first prove that f is integrable on $[a,b]$. For each $k \in \mathbf{N}$ let $\varepsilon_k = \sup_{x\in[a,b]} |f_k(x) - f(x)|$. It follows, since $\lim_{k\to\infty} fk(x) = f(x)$, that

$$f_k(x) - \varepsilon_k \le f(x) \le f_k(x) + \varepsilon_k. \qquad (*)$$

The upper and lower Darboux integrals then satisfy the following inequalities (refer to Lemma 6.4.3):

$$\int_a^b [f_k(x) - \varepsilon_k]\,dx \le \underline{\int_a^b} f(x)dx \le \overline{\int_a^b} f(x)dx \le \int_a^b [f_k(x) + \varepsilon_k]\,dx.$$

The difference of the two middle integrals cannot exceed the difference of the two terminal integrals; we then obtain

$$0 \le \overline{\int_a^b} f(x)dx - \underline{\int_a^b} f(x)dx \le 2\varepsilon_k(b-a).$$

As $\lim_{k\to\infty} f_k(x) = f(x)$, then $\lim_{k\to\infty} \varepsilon_k = 0$ from the definition of ε_k, and the upper and lower Darboux integrals are then equal. We conclude (by definition) that f is then integrable on $[a,b]$.

We next prove the equality in the statement of the theorem. Since f and each f_k are integrable, then so is $f_k - f$, and from Lemma 7.3.1 it follows that $|f_k - f|$ is also integrable on $[a,b]$. Let $\varepsilon > 0$ be given. There is an $N \in \mathbf{N}$ such that for all $k > N$ and for all $x \in [a,b]$ we have

$$|f_k(x) - f(x)| < \frac{\varepsilon}{2(b-a)}.$$

Invoking Theorem 6.2, we obtain for all f_k, $k > N$,

$$\int_a^b |f_k(x) - f(x)|\,dx \le \int_a^b \frac{\varepsilon\,dx}{2(b-a)} = \frac{\varepsilon}{2}.$$

Finally, from Lemma 7.3.1 again, we have

$$\left| \int_a^b f_k(x)dx - \int_a^b f(x)dx \right| = \left| \int_a^b [f_k(x) - f(x)]dx \right| \leq \int_a^b \left| f_k(x) - f(x) \right| dx \leq \frac{\varepsilon}{2},$$

or equivalently, for every $\varepsilon > 0$,

$$\int_a^b f(x)dx - \frac{\varepsilon}{2} \leq \int_a^b f_k(x)dx \leq \int_a^b f(x)dx + \frac{\varepsilon}{2}.$$

Hence, $\int_a^b f(x)dx = \lim_{k\to\infty} \int_a^b f_k(x)dx$. ■

■ Example 7.8

Consider the sequence $\{f_k\}_{k=1}^{\infty}$, $f_k(x) = k\left[k + \left(1 + k^{-1}\right)\ln x\right]^{-1}$, $x \in [2, 7]$. We have

$$\lim_{k\to\infty} f_k(x) = \lim_{k\to\infty} \left[1 + \left(k^{-1} + k^{-2}\right)\ln x\right]^{-1} = 1,$$

so the sequence converges pointwise to $f(x) = 1$. Then

$$\begin{aligned} \left| f_k(x) - f(x) \right| &= \frac{\left(1 + k^{-1}\right)\ln x}{k + \left(1 + k^{-1}\right)\ln x} \\ &< \frac{\left(1 + k^{-1}\right)(2)}{k + \left(1 + k^{-1}\right)(1/2)} \quad \text{on } [2, 7] \\ &= \frac{2k + 2}{k^2 + k/2 + 1/2} \\ &< \frac{4k + 2}{k(k + 1/2)} \\ &< 4k^{-1}, \end{aligned}$$

and $\left| f_k(x) - f(x) \right| < \varepsilon$ if $4k^{-1} < \varepsilon$ iff $k > 4\varepsilon^{-1}$. Hence, the sequence converges uniformly on [2, 7], and from Theorem 7.3 we obtain

$$\lim_{k\to\infty} \int_2^7 f_k(x) = \int_2^7 1 \cdot dx = 5.$$

■

Like Theorem 7.2, Theorem 7.3 is a *sufficient* condition for the commutation of the limit operations.[2] So although the absence of uniform convergence of $\{f_k\}_{k=0}^{\infty}$ on a compact interval $[a, b]$ does not necessarily mean that commutation of two limit operations is invalid (Exercise 7.14), we should not be surprised when, as in Example 7.4, failure of commutation does occur.

Since sequences and series are closely related, we should expect the material so far in this section to have natural analogs for series of functions.

Definition. *A series* $\sum_{k=0}^{\infty} f_k(x)$ *is* ***uniformly convergent*** *on a set* $\mathbf{D}$ *iff the sequence of partial sums* $\{S_n(x)\}_{n=0}^{\infty}$, $S_n = \sum_{k=0}^{n} f_k(x)$ *is uniformly convergent to a function* $f : \mathbf{D} \to \mathbf{R}^1$.

To emphasize the analogies, we renumber the previous theorems with primes for statements about series. Their proofs follow immediately from the corresponding statements for sequences.

Theorem 7.1′ (Cauchy Criterion for Uniform Convergence). *A series* $\sum_{k=0}^{\infty} f_k(x)$ *is uniformly convergent on* $\mathbf{D} \subseteq \mathbf{R}^1$ *iff, given any* $\varepsilon > 0$, *there is an* $N \in \mathbf{N}$ *such that for all* $n > m > N$ *and all* $x \in \mathbf{D}$ *we have* $\left|\sum_{k=m+1}^{n} f_k(x)\right| < \varepsilon$.

Theorem 7.2′. *If the series* $\sum_{k=0}^{\infty} f_k(x), f_k \in \mathfrak{C}(\mathbf{D})$, *converges uniformly to* f *on* $\mathbf{D} \subseteq \mathbf{R}^1$, *then* $f \in \mathfrak{C}(\mathbf{D})$.

Theorem 7.3′. *If the series* $\sum_{k=0}^{\infty} f_k(x)$ *is uniformly convergent on the finite interval* $[a, b], b > a$, *and each partial sum* $S_n(x) = \sum_{k=0}^{n} f_k(x)$ *is integrable on* $[a, b]$, *then* $\sum_{k=0}^{\infty} f_k(x) = \lim_{n\to\infty} S_n(x)$ *is also integrable on* $[a, b]$, *and we have* $\int_a^b \left[\sum_{k=0}^{\infty} f_k(x)\right] dx = \sum_{k=0}^{\infty} \int_a^b f_k(x)dx$.

■ Example 7.9

(Weierstrass's M-test)[3] For $\sum_{k=0}^{\infty} f_k(x)$, $x \in \mathbf{D}$, let $F_k = \sup_{x\in\mathbf{D}} |f_k(x)|$, and let $\sum_{k=0}^{\infty} M_k$ be any convergent series of nonnegative numbers such that each $F_k \leq M_k$. If $\varepsilon > 0$ is given, then by Exercise 3.6 and the Triangle Inequality there is an $N \in \mathbf{N}$ such that for any $n > m > N$ we have uniformly on $\mathbf{D}$

$$\left|\sum_{k=m+1}^{n} f_k(x)\right| \leq \sum_{k=m+1}^{n} F_k \leq \sum_{k=m+1}^{n} M_k < \varepsilon.$$

By Theorem 7.1′, we conclude that $\sum_{k=0}^{\infty} f_k(x)$ converges uniformly on $\mathbf{D}$ to some function $f : \mathbf{D} \to \mathbf{R}^1$.

---

[2]The sufficiency of uniform convergence for commutation of the two limit operations had been overlooked by Cauchy and not treated by him in his 1823 book, *Cours d'analyse*. Cauchy's work (and errors) inspired Abel.

[3]Weierstrass's discovery of uniform convergence and his use of the *M*-test (in lectures) date from the 1840s.

There are other tests more delicate than the M-test (Exercise 7.22).

It need hardly be stressed that Weierstrass's M-test (or any similar test) is a *sufficient* condition for uniform convergence. ■

■ Example 7.10

The series $\sum_{k=1}^{\infty} f_k(x)$, $f_k(x) = x^{k-1}/\left(k^{3/2} \cdot 3^k\right)$ has a radius of convergence $R = 3$; by the p-Series Test (Exercise 3.23(b)) and by Theorem 3.9, it converges at $x = 3, -3$, respectively. Weierstrass's M-test applies if we take $M_k = k^{-3/2}$, so we conclude that $\sum_{k=1}^{\infty} f_k(x)$ converges uniformly on $[-3, 3]$ to some function $f(x)$. ■

■ Example 7.11

The function $f(x)$ in the previous example is continuous on $[-3, 3]$, according to Theorem 7.2′. Further, each partial sum $S_n(x) = \sum_{k=1}^{n} \left[x^{k-1}/\left(k^{3/2} \cdot 3^k\right)\right]$ is integrable on $[-3, 3]$ (Theorem 6.13), so by Theorem 7.3′ we have (verify!)

$$\int_{-3}^{3} \left[\sum_{k=1}^{\infty} \frac{x^{k-1}}{k^{3/2} \cdot 3^k}\right] dx = \lim_{n\to\infty} \int_{-3}^{3} \left[\sum_{k=1}^{n} \frac{x^{k-1}}{k^{3/2} \cdot 3^k}\right] dx$$

$$= \lim_{n\to\infty} \sum_{k=1}^{n} \int_{-3}^{3} \frac{x^{k-1}}{k^{3/2} \cdot 3^k}\, dx$$

$$= \lim_{n\to\infty} \sum_{k=1}^{n} \left[1 - (-1)^k\right] \frac{1}{k^{5/2}}$$

$$= 2\left(1 - \frac{\sqrt{2}}{8}\right) \sum_{k=1}^{\infty} k^{-5/2}.$$ [4]

■

7.4 INTEGRATION OF POWER SERIES

Power series, such as the series in Example 7.10, have several nice features. For example, any power series in x, $x \in \mathbf{R}^1$, has a radius of convergence, R (Theorem 5.12). The following basic property of power series is not shared with other kinds of series of functions.

---

[4] The last series shown is a particular case of the **Riemann zeta function**: $\zeta(5/2)$ (see Exercise 3.44). Its value may be estimated as 1.34149, from information in Apostol (1985). The definite integral then has an estimated value of 2.20869. Despite what Exercise 3.23(b) might suggest, we can show without the use of any heavy machinery (Osler, 2008) that a broader definition of $\zeta(s)$ leads to simple, rational values for $\zeta(0)$ and $\zeta(-n)$, $n \in \mathbf{N}$.

Theorem 7.4. *Let $R > 0$ be the finite radius of convergence of $\sum_{n=0}^{\infty} c_n(x-a)^n$. Then for any $r \in [0, R)$ the series automatically converges uniformly to a continuous function $f(x)$ for all x that satisfy $-r \leq x - a \leq r$.*

Proof. If $|x-a| \leq r < R$, then for each n we have $|c_n(x-a)^n| \leq |c_n|\, r^n$. By hypothesis, $\sum_{n=0}^{\infty} |c_n|\, r^n$ converges, so by the Weierstrass M-test (Example 7.9) the series $\sum_{n=0}^{\infty} c_n(x-a)^n$ converges uniformly for $|x-a| \leq r$ to some function $f(x)$. By Theorem 7.2′ $f(x)$ is continuous for $|x-a| \leq r$. ■

■ Example 7.12

The series $\sum_{k=0}^{\infty}(-1)^k \frac{x^{k+1}}{k+1}$ has $R = 1$. By Theorem 7.4 it converges uniformly to a continuous function on any compact interval $\mathbf{I} \subset (-1, 1)$. We see, from Example 5.18, that this continuous function is $f(x) = \ln(1+x)$. ■

The series in Example 7.12 diverges at $x = -1$, and Weierstrass's M-test is of no help in deciding if the uniform convergence can be extended to $x = 1$. However, by Exercise 3.6 there is for each $\varepsilon > 0$ a corresponding $N \in \mathbf{N}$ such that $n > m > N$ implies $\left|\sum_{k=m+1}^{n}(-1)^k \frac{1}{k+1}\right| < \varepsilon$ (why?). So it is conceivable, but not yet proved, that for each $\varepsilon > 0$ an N can be found such that for all $x \in [r, 1]$, $-1 < r \leq 1$, $n > m > N$ implies $\left|\sum_{k=m+1}^{n}(-1)^k \frac{x^k}{k+1}\right| < \varepsilon$. If so, then by Theorems 7.1′ and 7.2′ the original series in Example 7.12 would be uniformly convergent and continuous on $[r, 1]$, $-1 < r \leq 1$. See Theorem 7.6, shortly.

Since the partial sums of a power series are everywhere continuous, they are integrable. Combination of Theorems 7.3′ and 7.4 then yields the following result. Note the restriction to a finite interval.

Theorem 7.5. *If $R > 0$ is the finite radius of convergence of the power series*

$$\sum_{k=0}^{\infty} c_k(x-a)^k \text{ and if } [r_1, r_2] \subset (-R, R), \text{ then}$$

$$\int_{r_1+a}^{r_2+a}\left[\sum_{k=0}^{\infty} c_k(x-a)^k\right] dx = \sum_{k=0}^{\infty}\int_{r_1+a}^{r_2+a} c_k(x-a)^k\, dx.$$

Proof. The proof is left to you. ■

In short, we can integrate a power series term-by-term on any compact interval inside its interval of convergence. The discussion after Example 7.12 raises the possibility that in some cases the range of integration can be extended to one or both of the endpoints of the interval of convergence. In general, the behavior of a power series there is more delicate than at interior points.

By way of illustration, let us consider again the arctangent function from Exercise 6.38:

$$\mathrm{Tan}^{-1}x = \int_0^x \frac{dt}{1+t^2}.$$

For $|t| < 1$, the Taylor series representation of the integrand is

$$\frac{1}{1+t^2} = 1 - t^2 + t^4 - t^6 + \cdots = \sum_{k=0}^{\infty} (-1)^k t^{2k}.$$

For any $x \in [0, R), R = 1$, the t-series is uniformly convergent on $[0, x]$, and term-by-term integration on $[0, x]$ is valid, by Theorem 7.5:

$$\begin{aligned}\mathrm{Tan}^{-1}x &= \int_0^x \left[\sum_{k=0}^{\infty} (-1)^k t^{2k}\right] dt = \sum_{k=0}^{\infty} \int_0^x (-1)^k t^{2k}\, dt \\ &= \sum_{k=0}^{\infty} (-1)^k \frac{x^{2k+1}}{2k+1}. \qquad (*)\end{aligned}$$

But what can we say when $x = 1$? The original series (in t) diverges at $t = 1$, so the uniform convergence of that series cannot extend to $t = 1$. This is a worse situation than the $x = 1$ case in Example 7.12!

To answer the question just posed, we look at the integrated series (*). Let us call this series $F(x)$:

$$F(x) = \begin{cases} \mathrm{Tan}^{-1}x & 0 \le x < 1 \\ \sum_{k=0}^{\infty} (-1)^k (2k+1)^{-1} & x = 1. \end{cases}$$

So $\mathrm{Tan}^{-1}x$ and $F(x)$ agree on the interval $[0, 1)$. From Corollary 6.16.1 we know that $\mathrm{Tan}^{-1}x$ is continuous at $x = 1$. In order to say that $\lim_{x\to 1^-} [F(x) - \mathrm{Tan}^{-1}x] = 0$, it is necessary that F have one-sided continuity at $x = 1$ (strict continuity of F at $x = 1$ is not possible because $\sum_{k=0}^{\infty} (-1)^k \frac{x^{2k+1}}{2k+1}$ diverges at any $x > 1$). From Theorem 7.2′ it follows that F will be continuous on $[0, 1]$ if it is uniformly convergent on $[0, 1]$.

The Norwegian mathematician **Abel**[5] proved a theorem that is just what we need here and also clears up what was alluded to after Example 7.12. The

[5] See footnote 2, Chapter 3; see also Ore (1970, 1974).

elementary proof (next) uses only the definition of a uniformly convergent series and a couple of ideas from Chapter 3 (Buck, 1978).

Theorem 7.6 (Abel's Limit Theorem). *Suppose that* $\sum_{k=0}^{\infty} c_k(x-a)^k$ *has radius of convergence* $R, 0 < R < \infty$, *and that the series* $\sum_{k=0}^{\infty} c_k R^k$ *converges to* L. *Then* $\lim_{u \to R^-} \sum_{k=0}^{\infty} c_k u^k = L$.

Proof. For each $n \in \mathbf{N}$ let $C_n = \sum_{k=n}^{\infty} c_k R^k$, from which we have $C_n - C_{n+1} = c_n R^n$. The convergence of $\sum_{k=0}^{\infty} c_k R^k$ to some $L \in \mathbf{R}^1$ implies that if $\varepsilon > 0$ is given, then for all sufficiently large n, $\left|\sum_{k=n}^{\infty} c_k R^k\right| = \left|L - \sum_{k=0}^{n-1} c_k R^k\right| < \varepsilon$, that is, $\lim_{n\to\infty} C_n = 0$. Now let $x - a \in [0, R)$; we obtain from the definition of the C_n's

$$\left|\sum_{k=n+1}^{\infty} c_k(x-a)^k\right| = \left|(C_{n+1} - C_{n+2})\left(\frac{x-a}{R}\right)^{n+1} + (C_{n+2} - C_{n+3})\left(\frac{x-a}{R}\right)^{n+2}\right.$$
$$\left. + (C_{n+3} - C_{n+4})\left(\frac{x-a}{R}\right)^{n+3} + \cdots\right|$$
$$= \left|C_{n+1}\left(\frac{x-a}{R}\right)^{n+1} + \left\{C_{n+2}\left(\frac{x-a}{R}\right)^{n+1}\left[\frac{x-a}{R} - 1\right]\right.\right.$$
$$\left.\left. + C_{n+3}\left(\frac{x-a}{R}\right)^{n+2}\left[\frac{x-a}{R} - 1\right] + \cdots\right\}\right|.$$

The last line follows from the fact that for any $x - a \in [0, R)$, the series $\sum_{k=n+1}^{\infty} c_k(x-a)^k$ and its equivalent are absolutely convergent (Ratio Test), and from Riemann's Rearrangement Theorem (Theorem 3.11) any rearrangement of terms in an absolutely convergent series leaves the sum unchanged. Hence, by repeated use of the Triangle Inequality

$$\left|\sum_{k=n+1}^{\infty} c_k(x-a)^k\right| = \left|C_{n+1}\left(\frac{x-a}{R}\right)^{n+1} + \left[\frac{x-a}{R} - 1\right]\sum_{k=n+1}^{\infty} C_{k+1}\left(\frac{x-a}{R}\right)^k\right|$$
$$\le |C_{n+1}|\left(\frac{x-a}{R}\right)^{n+1} + \left|\left[\frac{x-a}{R} - 1\right]\sum_{k=n+1}^{\infty} C_{k+1}\left(\frac{x-a}{R}\right)^k\right|$$
$$\le |C_{n+1}|\left(\frac{x-a}{R}\right)^{n+1} + \left[1 - \frac{x-a}{R}\right]\left(\frac{x-a}{R}\right)^{n+1}\sum_{k=n+1}^{\infty} |C_{k+1}|\left(\frac{x-a}{R}\right)^{k-n-1}. \quad (*)$$

Let $\varepsilon > 0$ be given. Then there is an $N \in \mathbf{N}$ such that $n > N$ implies $|C_n| < \varepsilon/2$. Thus, for any $x - a \in [0,\ R)$ it follows from (*) that for $n > N$

$$\begin{aligned}\left|\sum_{k=n+1}^{\infty} c_k\,(x-a)^k\right| &\le \frac{\varepsilon}{2}\left(\frac{x-a}{R}\right)^{n+1} + \left[1-\frac{x-a}{R}\right]\left(\frac{x-a}{R}\right)^{n+1}\frac{\varepsilon}{2}\sum_{k=0}^{\infty}\left(\frac{x-a}{R}\right)^k \\ &= \frac{\varepsilon}{2}\left(\frac{x-a}{R}\right)^{n+1}\left[1+\left(1-\frac{x-a}{R}\right)\frac{1}{1-\left(\frac{x-a}{R}\right)}\right] \\ &= \varepsilon\left(\frac{x-a}{R}\right)^{n+1} \\ &< \varepsilon.\end{aligned}$$

This last inequality is also true when $x - a = R$, for then $\left|\sum_{k=n+1}^{\infty} c_k\,(x-a)^k\right|$ is $|C_{n+1}|$, which is less than $\varepsilon/2$ for $n > N$.

By definition, we conclude that $\sum_{k=0}^{\infty} c_k\,(x-a)^k$ is uniformly convergent on $[0, R]$. The statement

$$\lim_{u\to R-}\sum_{k=0}^{\infty} c_k u^k = \sum_{k=0}^{\infty} c_k R^k = L$$

then follows from Theorem 7.2′. ■

The following remarks are pertinent to the proof of Theorem 7.6:

1. The result that $\sum_{k=0}^{\infty} c_k(x-a)^k$ is uniformly convergent on $[0, R]$ is not a restriction, for Theorem 7.4 already has told us that $\sum_{k=0}^{\infty} c_k(x-a)^k$ is uniformly convergent also on $[r, 0]$, $-1 < r < 0$. Hence, the uniform convergence extends to the entire interval $[r, R]$.

2. It is clear from the mechanics of the proof of Theorem 7.6 that the uniform convergence could have been proved for $[r, R]$, $-1 < r < 0$, but at the slight expense of additional clutter by more absolute value signs.

3. The validity of the theorem is in no way dependent upon the origin of the series $\sum_{k=0}^{\infty} c_k(x-a)^k$. Of present interest is the case where this series has arisen by integration of a prior series.

4. If $\sum_{k=0}^{\infty} c_k(x-a)^k$ has radius of convergence $R, 0 < R < \infty$, then so does the integrated series $\sum_{k=0}^{\infty} \frac{c_k}{k+1}(x-a)^{k+1}$. This follows if

$R = \left\{ \lim_{k\to\infty} \left| \frac{c_{k+1}}{c_k} \right| \right\}^{-1}$ holds, since then by the Ratio Test

$$\begin{aligned}
\lim_{k\to\infty} \left| \frac{c_{k+1}}{c_k} \frac{k+1}{k+2} \frac{(x-a)^{k+2}}{(x-a)^{k+1}} \right| &= |x-a| \left[\lim_{k\to\infty} \frac{k+1}{k+2} \right] \left[\lim_{k\to\infty} \left| \frac{c_{k+1}}{c_k} \right| \right] \\
&= |x-a| \left[\lim_{k\to\infty} \left| \frac{c_{k+1}}{c_k} \right| \right] \\
&= |x-a| \cdot R^{-1} \\
&< 1 \text{ iff } |x-a| < R.
\end{aligned}$$

If the ratio of the $|c_k|$'s does not have a limit, then we must fall back on the use of the lim sup and employ Theorem 5.13. In order to implement this for the integrated series $\sum_{k=0}^{\infty} \frac{c_k}{k+1} (x-a)^{k+1}$, we need the subsidiary result that if a sequence $\{a_k\}_{k=0}^{\infty}$ converges to some $A > 0$ and $\{u_k\}_{k=0}^{\infty}$ is any real-valued sequence, then $\limsup_{k\to\infty} (a_k u_k) = A \limsup_{k\to\infty} u_k$ (Exercise 7.28). In this more complicated situation, we still obtain for the integrated series $\sum_{k=0}^{\infty} \frac{c_k}{k+1} (x-a)^{k+1}$ a radius of convergence of R, if the initial series $\sum_{k=0}^{\infty} c_k (x-a)^k$ has a radius of convergence of R.

In view of Remarks (3) and (4), the following useful corollary of Theorem 7.6 now results:

Corollary 7.6.1. *If* $f(x) = \sum_{k=0}^{\infty} c_k (x-a)^k$ *has radius of convergence* R, $0 < R < \infty$, *and* $\sum_{k=0}^{\infty} c_k R^k$ *diverges to* $\pm\infty$ *but* $\sum_{k=0}^{\infty} \frac{c_k R^{k+1}}{k+1}$ *converges, then*

$$\int_a^{a+R} f(x)dx = \sum_{k=0}^{\infty} \frac{c_k R^{k+1}}{k+1}.$$

Proof. The proof is left to you. ■

■ Example 7.13

Returning to Example 7.12, we can apply Theorem 7.6 and conclude that the series is uniformly convergent on $[r, 1]$, $-1 < r \leq 1$, and, hence, continuous there. It follows that $\ln 2 = \sum_{k=0}^{\infty} (-1)^k (k+1)^{-1}$. This series is poor for the estimation of ln 2; 20 terms give only 0.66877. But if the series in Example 7.12 is integrated twice and 0 is used as the lower limit in each integration, then we can obtain (Exercise 7.30 outlines a slightly different procedure)

$$\ln 2 = \frac{5}{8} + \frac{1}{2}\sum_{k=0}^{\infty}(-1)^k \frac{1}{(k+1)(k+2)(k+3)}.$$

Twenty terms of this series now give $\ln 2 \approx 0.69312$. ■

■ Example 7.14

Returning to our discussion of $\mathrm{Tan}^{-1}x$, we conclude from Corollary 7.6.1 that

$$F(1) = \int_0^1 \frac{dt}{1+t^2} = \sum_{k=0}^{\infty}(-1)^k \frac{1}{2k+1} = \mathrm{Tan}^{-1}1.$$

Additionally, letting $t = u^{-1}$, we obtain

$$\int_0^1 \frac{dt}{1+t^2} = \int_\infty^1 \frac{-u^2 du}{1+u^{-2}} = \int_1^\infty \frac{du}{1+u^2}$$

$$= \int_0^\infty \frac{du}{1+u^2} - \int_0^1 \frac{du}{1+u^2}$$

$$= \frac{\pi}{2} - \int_0^1 \frac{du}{1+u^2} \qquad \text{(by definition of } \pi\text{)}.$$

Hence, we arrive finally at the interesting series

$$\mathrm{Tan}^{-1}1 = \frac{\pi}{4} = 1 - \frac{1}{3} + \frac{1}{5} - \frac{1}{7} + \cdots = \sum_{k=0}^{\infty}(-1)^k \frac{1}{2k+1}.$$

This historically important series was arrived at independently by the German mathematician **Leibniz** (1646–1716), the Scottish mathematician **James Gregory** (1638–1675), and the South Indian mathematician **Kerala Gargya Nilakantha** (ca. 1450–1550) (Roy, 1990). It is possible that in the future the series may also be discovered among the ancient writings of some Japanese or Chinese mathematician. ■

■ Example 7.15

An application of Abel's Limit Theorem was used in Vernescu (2008) in order to sum an interesting series. Newton's Binomial Theorem gives, for $x \in (-1, 1)$,

$$\left(1-x^2\right)^{-1/2} - 1 = \sum_{n=1}^{\infty} \frac{1 \cdot 3 \cdot 5 \cdots (2n-1)}{2^n \cdot n!} x^{2n}. \qquad (*)$$

For any nonzero $x \in (-1,\ 1)$, we obtain from $(*)$

$$-\frac{d}{dx} \ln\left[1+\sqrt{1-x^2}\right] = \frac{1}{x\sqrt{1-x^2}} - \frac{1}{x}$$
$$= \sum_{n=1}^{\infty} \frac{1 \cdot 3 \cdot 5 \cdots (2n-1)}{2^n \cdot n!} x^{2n-1}. \qquad (**)$$

The series on the right is uniformly convergent on any compact interval $[a,\ b] \subset (0,\ 1)$, so by Theorem 7.5 we obtain

$$-\ln\left[1+\sqrt{1-b^2}\right] + \ln\left[1+\sqrt{1-a^2}\right]$$
$$= \sum_{n=1}^{\infty} \frac{1 \cdot 3 \cdot 5 \cdots (2n-1)}{2n\,(2^n)\,n!} b^{2n} - \sum_{n=1}^{\infty} \frac{1 \cdot 3 \cdot 5 \cdots (2n-1)}{2n\,(2^n)\,n!} a^{2n}. \qquad (***)$$

The function $F(x) = \sum_{n=1}^{\infty} \frac{1 \cdot 3 \cdot 5 \cdots (2n-1)}{2n(2^n)n!} x^{2n}$ is continuous on $[0, 1)$, and equals 0 at $x = 0$. At $x = 1$, the series for $F(x)$ converges to some $L \in \mathbf{R}^1$ by the Ratio Test. Hence, by Theorem 7.6, equation $(***)$ becomes, for $[0, 1]$,

$$-\ln 1 + \ln 2 = \lim_{x \to 1^-} F(x) - 0 = L,$$

that is,

$$2L = 2F(1) = \lim_{x \to 1^-} \sum_{n=1}^{\infty} \frac{1 \cdot 3 \cdot 5 \cdots (2n-1)}{(n)(2^n)(n!)} x^{2n}$$
$$= \sum_{n=1}^{\infty} \frac{1 \cdot 3 \cdot 5 \cdots (2n-1)}{(n)\,(2^n)\ n!}$$
$$= \ln 4.$$

■

7.5 DIFFERENTIATION

We saw in Example 7.5 that there is not a close analog of Theorem 7.3 for the differentiation of a sequence of functions. The example implied a connection with the failure of the sequence of *derivatives* to converge uniformly. The following theorem, a *sufficient* condition for commutation of two limit operations, is both plausible and true.

Theorem 7.7. *If $\{f_k\}_{k=0}^{\infty}$ is a sequence of functions continuously differentiable*[6] *on $[a, b]$ and pointwise convergent there to f, and if $\{f_k'\}_{k=0}^{\infty}$ converges uniformly on $[a, b]$, then for any $x \in [a, b]$ we have*

$$\frac{d}{dx}\left[\lim_{k\to\infty} f_k(x)\right] = f'(x) = \lim_{k\to\infty} f_k'(x).$$

Proof. Let $r \in [a, b]$ be arbitrary but fixed, and let $x \in [a, b]$. Suppose that $\lim_{k\to\infty} f_k'(x) = h(x)$ on $[a, b]$. Since each f_k' is continuous on $[a, b]$ and $\{f_k'\}_{k=0}^{\infty}$ is uniformly convergent on $[a, b]$, then by Theorem 7.3 we have

$$\lim_{k\to\infty} \int_r^x f_k'(t) = \int_r^x h(t)dt.$$

But by the Fundamental Theorem of the Calculus—A (Theorem 6.15), the left-hand side can be rewritten as $\lim_{k\to\infty} [f_k(x) - f_k(r)]$. Hence, we have by hypothesis

$$f(x) - f(r) = \int_r^x h(t)dt.$$

Finally, using the Fundamental Theorem of the Calculus—B (Theorem 6.16), we obtain

$$f'(x) = \frac{d}{dx}\int_r^x h(t)dt = h(x) = \lim_{k\to\infty} f_k'(x).$$

■

Although the theorem is a theorem about differentiation, its proof drew on ideas from integration. This was facilitated by the f_k's being continuous on $[a, b]$. The hypothesis of the theorem can be weakened so that integration is no longer useful; the proof is somewhat more involved (Sprecher, 1987).

The principal use of Theorem 7.7 is its adaptation for the differentiation of series of differentiable functions. The specific case of power series is given in Theorem 7.8.

---

[6]Denotes that each f_k is differentiable on $[a, b]$ and each corresponding f_k' is continuous on $[a, b]$. Differentiability at a, b means, of course, sided differentiability.

Theorem 7.7′. *Let $\{f_k\}_{k=0}^{\infty}$ be a sequence of functions continuously differentiable on $[a, b]$, and suppose that $\sum_{k=0}^{\infty} f_k$ converges pointwise to f on $[a, b]$ and that $\sum_{k=0}^{\infty} f_k'$ converges uniformly there. Then for each $x \in [a,\ b]$*

$$f'(x) = \frac{d}{dx}\sum_{k=0}^{\infty} f_k(x) = \sum_{k=0}^{\infty} f_k'(x).$$

Proof. For each n, let $g_n = \sum_{k=0}^{n} f_k$ and $g_n' = \sum_{k=0}^{n} f_k'$, and go on to apply Theorem 7.7. The completion of the proof is left to you. ■

■ Example 7.16

Let $\{f_k(x)\}_{k=1}^{\infty}$ be defined by $f_k(x) = \frac{\cos(kx)}{k^{5/2}}$, $-1 \le x \le 1$. Each $f_k(x)$ is continuously differentiable on $[-1, 1]$. The Weierstrass M-test shows that $\sum_{k=1}^{\infty} f_k(x)$ converges uniformly (and, thus, pointwise) on $[-1, 1]$. A second application of the M-test shows that $\sum_{k=1}^{\infty} f_k'(x) = \sum_{k=1}^{\infty} \frac{-\sin(kx)}{k^{3/2}}$ converges uniformly on $[-1, 1]$. We conclude from Theorem 7.7′ that

$$\frac{d}{dx}\sum_{k=1}^{\infty}\frac{\cos(kx)}{k^{5/2}} = -\sum_{k=1}^{\infty}\frac{\sin(kx)}{k^{3/2}}.$$ ■

■ Example 7.17

Let $\{f_k(\theta)\}_{k=1}^{\infty}$ be defined by $f_k(\theta) = \frac{\cos(k\theta)}{2^{2k-1}}$. It can be established that $1 + \sum_{k=1}^{\infty} f_k(\theta)$ converges pointwise on $[-\pi, \pi]$ to $\frac{15}{17-8\cos(\theta)}$ (Exercise 7.41). Additionally, we have that the series

$$\sum_{k=1}^{\infty}\frac{-k\sin(k\theta)}{2^{2k-1}}$$

is uniformly convergent on $[-\pi, \pi]$ (Exercise 7.42). We conclude from Theorem 7.7′ that

$$\sum_{k=1}^{\infty}\frac{-k\sin(k\theta)}{2^{2k-1}} = \frac{d}{d\theta}\left(\frac{15}{17-8\cos\theta}\right) = \frac{-120\sin\theta}{(17-8\cos\theta)^2}.$$

For example, at $\theta = \sin^{-1}(4/5)$, the series sums to $\frac{-2400}{3721}$. ■

We now apply Theorem 7.7′ to power series. For this purpose we need to know that if $\sum_{k=0}^{\infty} c_k(x-a)^k$ has radius of convergence R, $0 < R < \infty$, then so does the derived series $\sum_{k=1}^{\infty} kc_k(x-a)^{k-1}$. This can be established by use of the lim sup in a manner similar to that for the integrated series (see Point

(4) in Section 7.4, as well as Exercise 7.28). You are requested to work this out in Exercise 7.43. An alternative proof that is worth reading because of its elementary nature (it does not use the lim sup concept) is found in Apostol (1952).

■ Example 7.18

The series $\ln(1+\frac{u}{a}) = \sum_{k=1}^{\infty} \frac{(-1)^{k+1}}{ka^k} u^k$, $a > 0$, has $R = a$ (verify!). Hence, so does the series $\sum_{k=0}^{\infty} \frac{(-1)^k}{a^{k+1}} u^k$. ■

■ Example 7.19

If a series $\sum_{k=0}^{\infty} c_k(x-a)^k$ has radius of convergence R and converges at one of the endpoints, then the derived series may or may not converge at that endpoint. Compare $\sum_{k=1}^{\infty} (-1)^{k-1} \frac{x^k}{k}$, which has $R = 1$ and converges at $x = 1$, with the series $\sum_{k=0}^{\infty} (-1)^k x^k$, which also has $R = 1$ but does not converge at $x = 1$. ■

Theorem 7.8. *If $f(x) = \sum_{k=0}^{\infty} c_k(x-a)^k$ has radius of convergence R, $0 < R < \infty$, then for any $r \in (0, R)$ we have*

$$f'(x) = \sum_{k=1}^{\infty} kc_k (x-a)^{k-1}, \quad x - a \in [-r, r].$$

Proof. Each $g_k(x) = c_k(x-a)^k$ is continuously differentiable on $|x-a| \le r$. By hypothesis, $\sum_{k=0}^{\infty} g_k(x)$ converges pointwise to $f(x)$ on $|x-a| \le r$, so from remarks earlier, $\sum_{k=1}^{\infty} g_k'(x)$ also converges pointwise on $|x-a| \le r$. From Theorem 7.4 we may then conclude that $\sum_{k=1}^{\infty} g_k'(x)$ converges uniformly on $|x-a| \le r$. Hence, by Theorem 7.7′ the function $f(x)$ is differentiable on $|x-a| \le r$, and we have

$$f'(x) = \frac{d}{dx} \sum_{k=0}^{\infty} g_k(x) = \sum_{k=1}^{\infty} g_k'(x) = \sum_{k=1}^{\infty} kc_k (x-a)^{k-1}.$$

■

Corollary 7.8.1. *If $f(x) = \sum_{k=0}^{\infty} c_k(x-a)^k$ is as in Theorem 7.8, then f is infinitely differentiable on $|x-a| \le r$.*

■ Example 7.20

Suppose that for all $|x-a| < R$ we have

$$\sum_{k=0}^{\infty} b_k(x-a)^k = f(x) = \sum_{k=0}^{\infty} c_k(x-a)^k.$$

Letting $x = a$, we obtain $b_0 = c_0$. Now differentiate across the equation; from Theorem 7.8 we deduce

$$\sum_{k=1}^{\infty} kb_k(x-a)^{k-1} = f'(x) = \sum_{k=1}^{\infty} kc_k(x-a)^{k-1},$$

for all $|x - a| < R$. Letting $x = a$ here, we now obtain $b_1 = c_1$. By mathematical induction, together with Corollary 7.8.1, we conclude that for each $k \in \mathbf{N} \cup \{0\}$ we have $b_k = c_k$, that is, the power series representation of f (about a) is unique. ■

■ Example 7.21

For any $n \in \mathbf{N}$, Corollary 7.8.1 implies that $f^{(n)}(x) = \sum_{k=0}^{\infty} c_k \frac{d^n}{dx^n}(x-a)^k$. Fix n; then we have

$$\frac{d^n}{dx^n}(x-a)^k = \begin{cases} 0, \text{identically} & k < n \\ 0 \quad \text{at} \quad x = a & k > n \\ n! & k = n. \end{cases}$$

Thus, $f^{(k)}(a) = c_k k!$, that is, $c_k = \frac{f^{(k)}(a)}{k!}$ and the unique power series representation of f about a, indicated in Example 7.20, is just the Taylor series for f. ■

■ Example 7.22

Let $f(x) = \left(1 + x^2\right)^{1/2}$; by Newton's Binomial Theorem we have

$$f(x) = 1 + \sum_{k=1}^{\infty} \binom{1/2}{k} x^{2k},$$

and in Exercise 5.61 it was established that the right-hand side represents f if $x^2 < 1$. In view of Example 7.21 (where we take $a = 0$), we deduce that

$$\binom{1/2}{k} = \frac{(1/2)(1/2-1)(1/2-2)\cdots(1/2-k+1)}{k!}$$
$$= \frac{f^{(k)}(0)}{k!}.$$

Thus, it is unnecessary to do tedious differentiations, for Newton has given us the pattern. The first few terms in the expansion of $f(x)$, $0 \le x < 1$, are

$$\left(1 + x^2\right)^{1/2} \approx 1 + \frac{1}{2}x^2 - \frac{1}{8}x^4 + \frac{1}{16}x^6 - \frac{5}{128}x^8.$$ ■

■ Example 7.23

Let $f(x) = \left(1 + x^2\right)^{1/2}$ and let r satisfy $0 < r < 1$. By Theorem 7.8 and Example 7.22, we have for all $x \in [-r, r]$

$$f'(x) = x(1 + x^2)^{-1/2} = \sum_{k=1}^{\infty} 2k \binom{1/2}{k} x^{2k-1}.$$

At $x \neq 0$,

$$(1 + x^2)^{-1/2} = \sum_{k=1}^{\infty} 2k \binom{1/2}{k} x^{2k-2}$$

$$= \sum_{k=0}^{\infty} 2(k+1) \binom{1/2}{k+1} x^{2k}$$

$$= 1 - \frac{1}{2}x^2 + \frac{3}{8}x^4 - \frac{5}{16}x^6 + \cdots,$$

which is valid also at $x = 0$. What happens if you do a Cauchy multiplication of this series and the one in Example 7.22? ■

7.6 WEIERSTRASS'S FUNCTION; LEIBNIZ'S THEOREM

We shall close out this chapter and this text by presenting two differentiation topics of a somewhat specialized nature: (A) an example of a function (as promised in Section 5.1) that is continuous everywhere in $\mathbf{R}^1$ but is differentiable nowhere there, and (B) a theorem that, fittingly, connects the two fundamental operations of calculus—differentiation and integration—and that dates back to the time of the German father of calculus, Leibniz.

(A)

Consider the following function, a particular case of a family of functions examined by Weierstrass (Titchmarsh, 1939):

$$f(x) = \sum_{k=0}^{\infty} \left(\frac{2}{3}\right)^k \cos\left(9^k \pi x\right), \quad x \in \mathbf{R}^1.$$

No term of this series exceeds in magnitude $(2/3)^k$ and since $\sum_{k=0}^{\infty} (2/3)^k = 3$, then by the M-test this series converges uniformly on $\mathbf{R}^1$, and from Theorem 7.2′ the function $f(x)$ is continuous on $\mathbf{R}^1$.

The Cauchy quotient for f is now written

$$\begin{aligned}\frac{f(x+h)-f(x)}{h} &= \sum_{k=0}^{m-1}\left(\frac{2}{3}\right)^k \frac{\cos[9^k\pi(x+h)]-\cos(9^k\pi x)}{h} \\ &\quad+\sum_{k=m}^{\infty}\left(\frac{2}{3}\right)^k \frac{\cos[9^k\pi(x+h)]-\cos(9^k\pi x)}{h} \\ &= R_m+S_m,\end{aligned}$$

where $m \in \mathbf{N}$ is arbitrary. From the Mean-Value Theorem, each numerator satisfies

$$\left|\cos\left[9^k\pi(x+h)\right]-\cos(9^k\pi x)\right| = 9^k\pi\left|h\ \sin(9^k\pi c_k)\right| \leq 9^k\pi\,|h|,$$

for some $c_k \in (x, x+h)$. Hence, by summation we obtain

$$|R_m| \leq \sum_{k=0}^{m-1}\left(\frac{2}{3}\right)^k 9^k\pi = \pi\left(\frac{6^m-1}{6-1}\right) < \pi\cdot\frac{6^m}{5}.$$

Next, we obtain a lower bound for $|S_m|$. We can write uniquely

$$9^m x = a_m + \gamma_m,$$

where $\frac{-1}{2} \leq \gamma_m < \frac{1}{2}$ and $a_m \in \mathbf{Z}$; then define the sequence $\{h_m\}_{m=1}^{\infty}$ by $h_m = \left(1-\gamma_m\right)/9^m$. We then have

$$0 < h_m \leq \frac{1-(-1/2)}{9^m} = \frac{3}{2\cdot 9^m},$$

and as $k \geq m$ for the terms in the series for S_m, then

$$9^k\pi\left(x+h_m\right) = 9^{k-m}\pi\left[9^m\left(x+h_m\right)\right] = 9^{k-m}\pi(a_m+1).$$

Since 9 is odd, it is clear that

$$\begin{aligned}\cos\left[9^k\pi(x+h_m)\right] &= \cos\left[9^{k-m}\pi(a_m+1)\right] \\ &= (-1)^{a_m+1},\end{aligned}$$

and also

$$
\begin{aligned}
\cos\left(9^k \pi x\right) &= \cos\left[9^{k-m}\pi\left(a_m + y_m\right)\right] \\
&= \cos\left(9^{k-m}\pi a_m\right)\cos\left(9^{k-m}\pi y_m\right) \\
&\quad - \sin\left(9^{k-m}\pi a_m\right)\sin\left(9^{k-m}\pi y_m\right) \\
&= \cos\left(9^{k-m}\pi a_m\right)\cos\left(9^{k-m}\pi y_m\right) \\
&= (-1)^{a_m}\cos\left(9^{k-m}\pi y_m\right).
\end{aligned}
$$

Substitution into the definition of S_m now yields

$$
\begin{aligned}
S_m &= \sum_{k=m}^{\infty}\left(\frac{2}{3}\right)^k \frac{(-1)^{a_m+1} - (-1)^{a_m}\cos(9^{k-m}\pi y_m)}{h_m} \\
&= \frac{(-1)^{a_m+1}}{h_m}\sum_{k=m}^{\infty}\left(\frac{2}{3}\right)^k\left[1 + \cos\left(9^{k-m}\pi y_m\right)\right].
\end{aligned}
$$

All terms of this series are nonnegative, so its sum exceeds the first term and we have

$$
\begin{aligned}
|S_m| &> \frac{1}{|h_m|}\left(\frac{2}{3}\right)^m [1+0] \\
&\geq \left[\left(2\cdot 9^m\right)/3\right]\left(\frac{2}{3}\right)^m \\
&= \frac{2}{3}\cdot 6^m.
\end{aligned}
$$

Finally, using the Triangle Inequality in the form $|x-y| \geq ||x|-|y||$ (Exercise 1.33(a)), we have

$$
\begin{aligned}
\left|\frac{f(x+h_m)-f(x)}{h_m}\right| &= |S_m + R_m| \\
&= |S_m - (-R_m)| \\
&\geq ||S_m| - |-R_m|| \\
&\geq \left|\frac{2}{3}\cdot 6^m - \frac{\pi}{5}\cdot 6^m\right| \\
&= \left(\frac{2}{3} - \frac{\pi}{5}\right)\cdot 6^m.
\end{aligned}
$$

As $m \to \infty$, then on the left-hand side $h_m \to 0$, as desired. But the right-hand side diverges to ∞; hence, $f'(x)$ does not exist in $\mathbf{R}^1$ for any $x \in \mathbf{R}^1$.

Weierstrass's unusual function was presented in lecture form before the Berlin Academy of Sciences in 1872, although Weierstrass may have known it, or at least suspected its existence, much earlier. It electrified the mathematical community and made many members even more leery of overly trusting intuition and nonanalytical reasoning.[7]

(B)

Leibniz's Theorem for the differentiation of definite integrals is of great utility. Here we shall restrict our considerations to proper Riemann integrals (but see Exercise 7.55). Several proofs of the theorem are available; our proof makes contact with compactness (Section 4.7) and uniform continuity (Section 4.9). Other proofs appear later (Exercises 7.51, 7.52). Recall from Section 5.10 the notation for partial derivatives.

Theorem 7.9 (Leibniz's Theorem). *If $f, D_2 f$ are defined and continuous on the closed, bounded rectangle $\mathbf{D} = \{(x, y) : a_1 \le x \le a_2, b_1 \le y \le b_2\}$, then $\frac{d}{dy}\int_{a_1}^{a_2} f(x, y)dx = \int_{a_1}^{a_2} (D_2 f)\, dx$.*

Proof. By Theorem 6.13, the integral $g(y) = \int_{a_1}^{a_2} f(x, y)dx$ exists. Then if $y_0, y_0 + k \in [b_1, b_2]$, the Cauchy quotient can be written as

$$\frac{g(y_0+k) - g(y_0)}{k} = \int_{a_1}^{a_2} \frac{f(x, y_0+k) - f(x, y_0)}{k} dx$$

$$= \int_{a_1}^{a_2} \left[(D_2 f)(x, c_0)\right] dx, \qquad (*)$$

by the Mean-Value Theorem, where $c_0 \in (y_0, y_0 + k)$. The number c_0 depends in some complicated way upon x. In spite of this, $(D_2 f)(x, c_0)$ is a continuous function of x because $D_2 f$ is continuous on all of $\mathbf{D}$. Accordingly, the integral in equation (*) exists.

From the Heine-Borel Theorem (Theorem 4.8), we know that $\mathbf{D}$ is compact. Since $D_2 f$ is continuous on all of $\mathbf{D}$, then by Theorem 4.16 it is uniformly continuous on $\mathbf{D}$. Thus, if $\varepsilon > 0$ is given, then there is a $\delta > 0$ such that for all $x \in [a_1, a_2]$ and whenever $|y - y_0| < \delta$, we have

$$\left|(D_2 f)(x, y) - (D_2 f)(x, y_0)\right| < \frac{\varepsilon}{1 + (a_2 - a_1)}. \qquad (**)$$

[7] Weierstrass communicated the general case of his family of functions to German colleague **Paul du Bois-Reymond** (1831–1889), who subsequently published it in 1875. Many more examples soon followed, and the literature on such functions is now large. Some remarks, with numerous original references, can be found in Hawkins (1975).

If we restrict k in equation (*) so that $0 < |k| < \delta$, then $c_0 \in (y_0, y_0 + k)$ implies $c_0 \in (y_0 - \delta, y_0 + \delta)$, and an application of Lemma 7.3.1 to the integration of equation (**) yields

$$\left| \int_{a_1}^{a_2} \left[(D_2 f)(x, c_0) - (D_2 f)(x, y_0) \right] dx \right|$$

$$\leq \int_{a_1}^{a_2} \left| \left[(D_2 f)(x, c_0) - (D_2 f)(x, y_0) \right] \right| dx$$

$$< \int_{a_1}^{a_2} \frac{\varepsilon}{1 + (a_2 - a_1)} dx < \varepsilon. \qquad (***)$$

Clearly, as $\varepsilon \to 0$, then $\delta \to 0$ and also $k \to 0$. Finally, as $k \to 0$, then $c_0 \to y_0$ in equation (***), and equation (*) becomes

$$\lim_{k \to 0} \frac{g(y_0 + k) - g(y_0)}{k} = \lim_{c_0 \to y_0} \int_{a_1}^{a_2} \left[(D_2 f)(x, c_0) \right] dx$$

or

$$g'(y_0) = \int_{a_1}^{a_2} \left[(D_2 f)(x, y_0) \right] dx.$$

Since $y_0 \in [b_1, b_2]$ was arbitrary, then

$$\frac{d}{dy} \int_{a_1}^{a_2} f(x, y) dx = \int_{a_1}^{a_2} (D_2 f)\, dx.$$

■

■ Example 7.24

Let $F(y) = \int_0^{\pi/2} f(x, y) dx$, where

$$f(x, y) = \begin{cases} \dfrac{\mathrm{Tan}^{-1}(y \tan x)}{\tan x} & 0 < x < \frac{\pi}{2}, \quad 0 < y \leq 5 \\ y & x = 0, \quad 0 < y \leq 5 \\ 0 & x = \frac{\pi}{2}, \quad 0 < y \leq 5 \\ 0 & y = 0. \end{cases}$$

We can verify that f is continuous on $\mathbf{D} = \left\{(x, y) : 0 \le x \le \frac{\pi}{2},\ 0 \le y \le 5\right\}$, and that $D_2 f$ is also continuous on $\mathbf{D}$. From Leibniz's Theorem we have (verify!)

$$F'(y) = \int_0^{\pi/2} [(D_2 f)(x, y)]dx = \frac{\pi}{2}\frac{1}{1+y}.$$

Then integration gives $F(y) = \int_0^y \frac{\pi}{2}\frac{dt}{1+t} = \frac{\pi}{2}\ln(1+y)$. Setting $y = 1$, we obtain

$$F(1) = \int_0^{\pi/2} x \cot x \, dx = \frac{\pi}{2}\ln 2,$$

which is not at all easy to obtain by direct integration (Wiener, 2001) (Exercise 7.49). ■

The differentiation of definite integrals ought to be useful in the solution of differential equations. Suppose that $y(t)$ is unknown but that $y'(t)$ is presumed continuously differentiable on some domain $\mathbf{D}$. The following expression might appear in a differential equation,

$$\frac{d}{dt}\frac{dy}{dt} + y(t),$$

from which it is apparent that attempted integration of this will not take us very far. We hunt for an integrating factor that will allow us to make progress. A simple choice is $\sin(t)$, and we find that

$$\begin{aligned}\int \left[\frac{d}{dt}\frac{dy}{dt} + y(t)\right]\sin t\,dt &= \int \frac{d}{dt}\frac{dy}{dt}\sin t\,dt + \int y(t)\sin t\,dt \\ &= \left[\sin t\frac{dy}{dt} - \int \frac{dy}{dt}\cos t\,dt\right] \\ &\quad + \left[-y(t)\cos t + \int \frac{dy}{dt}\cos t\,dt\right] \\ &= \sin t\frac{dy}{dt} - y(t)\cos t + C. \qquad (*)\end{aligned}$$

The object is to isolate $y(t)$; that is, to obtain values of $y(t)$ for arbitrary choices $t = x$ in $\mathbf{D}$. In order that the right-hand side of equation (*) shall lead to this objective for us, we do the following: (a) replace $\sin(t)$ by $\sin(x - t)$, (b) impose initial conditions $y(0) = 0$ and $y'(0) = 0$, and (c) carry out the integration in equation (*) from $t = 0$ to $t = x$. Equation (*) is then modified slightly and reduces to (verify!)

$$\int_0^x \left[\frac{d}{dt}\frac{dy}{dt} + y(t)\right] \sin(x-t)\, dt = y(x). \qquad (**)$$

■ Example 7.25

Find a particular solution of

$$\frac{d^2y}{dx^2} + 2y(x) = x^2, \quad -2 < x < 2, \quad y(0) = y'(0) = 0.$$

Reasoning from equation (**) and allowing for the coefficient of 2, we write for the integral representation of a particular solution

$$y(x) = \frac{1}{\sqrt{2}} \int_0^x \sin\left[\sqrt{2}\,(x-t)\right] t^2 dt.$$

The variable x appears in the integrand and in the upper limit. Let us, therefore, define $u(x) \cong x$, $v(x) \cong x - t$, and write

$$y(x) = \Phi(u, v) = \frac{1}{\sqrt{2}} \int_0^u \sin\left[\sqrt{2}\, v\right] t^2 dt,$$

To verify that this satisfies the differential equation, we proceed to differentiate Φ:

$$\frac{dy(x)}{dx} = \frac{d\Phi(u, v)}{dx} = \frac{\partial \Phi}{\partial u}\frac{du}{dx} + \frac{\partial \Phi}{\partial v}\frac{dv}{dx},$$

from a Chain Rule. Since $u(x) = x$, then the first term is handled by the Fundamental Theorem of the Calculus—B:

$$\frac{\partial \Phi}{\partial u} = \frac{1}{\sqrt{2}} \sin\left[\sqrt{2}(x-x)\right] x^2 = 0.$$

The second term is handled by Leibniz's Theorem:

$$\frac{\partial \Phi}{\partial v} = \frac{1}{\sqrt{2}} \int_0^u \sqrt{2} \cos\left[\sqrt{2}\, v\right] t^2 dt,$$

and so, $\frac{dy(x)}{dx} = \int_0^u \cos\left[\sqrt{2}\, v\right] t^2 dt.$

A second differentiation now gives

$$\frac{d^2y(x)}{dx^2} = \cos\left[\sqrt{2}(x-x)\right]x^2 + \int_0^x \left[-\sqrt{2}\sin\left[\sqrt{2}\, v\right]t^2\right]dt$$
$$= x^2 - \sqrt{2}\int_0^x \sin\left[\sqrt{2}\, v\right] t^2 dt.$$

Finally, making the appropriate substitutions into the differential equation, we obtain as desired.

$$\frac{d^2y(x)}{dx^2} + 2y(x) = \left[x^2 - \sqrt{2}\int_0^x \sin\left[\sqrt{2}\, v\right] t^2 dt\right]$$
$$+ 2\cdot\frac{1}{\sqrt{2}}\int_0^x \sin\left[\sqrt{2}\, v\right] t^2 dt$$
$$= x^2.$$

The function $\frac{1}{\sqrt{2}}\sin\left[\sqrt{2}(x-t)\right]$ is known as the **Green's function** for the operator $D^2 + 2$; the Green's function technique is very important in applications. ■

EXERCISES

Section 7.1

7.1. Define the sequence $\{f_k(x)\}_{k=1}^{\infty}$ by $f_k(x) = \frac{kx}{1+kx^2}$, $x \in [0, 1]$.
- **(a)** In the spirit of Figure 7.1, sketch f_1, f_2, f_4, f_9 and the limit function f on a common pair of axes.[8]
- **(b)** Discuss the possible uniform convergence of $\{f_k(x)\}_{k=1}^{\infty}$ on $[0, 1]$.

7.2. Show that the sequence $\{f_k(x)\}_{k=1}^{\infty}, f_k(x) = \frac{2k}{3+2kx}$, converges uniformly on $[1, \infty)$ but does not do so on $(0, \infty)$.

7.3. Suppose that $\{f_k(x)\}_{k=1}^{\infty}$ and $\{g_k(x)\}_{k=1}^{\infty}$ are uniformly convergent on a common domain $\mathbf{D} \subseteq \mathbf{R}^1$. Prove that $\{h_k(x)\}_{k=1}^{\infty}$, where $h_k(x) = f_k(x) + g_k(x)$, is also uniformly convergent on $\mathbf{D}$.

---

[8] Diagrams like Figure 7.1 were made popular in the 1890s by the Harvard-based American mathematician **William Fogg Osgood** (1864–1943).

7.4. In the spirit of Case (A), construct a sequence of continuous functions $\{f_k\}_{k=1}^{\infty}, f_k : [0,1] \to \mathbf{R}^1$, such that $\lim_{k\to\infty} f_k(x)$ is the zero function but $\lim_{k\to\infty} \int_0^1 f_k(x) = \infty$.

7.5. Regarding Case (B),

(a) Explain how you know that $\lim_{k\to\infty} f_k'(x) = 0$ for only finitely many $x \in [0, 20]$ and that the limit fails to exist everywhere else in $[0, 20]$;

(b) If $\{f_k\}_{k=1}^{\infty}$ in (B) had been defined by $f_k(x) = k^{-2}\cos(kx)$, would $\lim_{k\to\infty} f_k'(x)$ then have equaled $\frac{d}{dx}\left[\lim_{k\to\infty} f_k(x)\right]$ for all $x \in \mathbf{D}$? Is $\{f_k(x)\}_{k=1}^{\infty}$ now uniformly convergent on $\mathbf{D}$?

7.6. Suppose that the sequence $S = \{f_k\}_{k=1}^{\infty}$ converges uniformly on $\mathbf{D} \subseteq \mathbf{R}^1$. Prove that any subsequence of S also converges uniformly on $\mathbf{D}$.

7.7. Review in Section 4.9 the definition of a function being uniformly continuous on a set $\mathbf{S}$. Now suppose that F is uniformly continuous on $\mathbf{S} = \mathbf{R}^1$. Define the sequence $\{f_k\}_{k=1}^{\infty}$ by $f_k(x) = F(x + 2k^{-3/2})$. Use Theorem 7.1 to prove that $\{f_k\}_{k=1}^{\infty}$ is uniformly convergent on $\mathbf{R}^1$.

7.8. Complete the proof of Theorem 7.1.

Section 7.2

7.9. Explain how Theorem 7.2 provides the explicit answer to the implicit question in Exercise 7.1(b).

7.10. Prove the following slightly more general version of Theorem 7.2: If $\mathbf{M}, \mathbf{M}'$ are metric spaces and $\{f_k\}_{k=0}^{\infty}$ is a sequence of functions from $\mathbf{M}$ to $\mathbf{M}'$ that converges uniformly on some open ball in $\mathbf{M}$ centered at $\mathbf{p}_0 \in \mathbf{M}$, and if each f_k is continuous at $\mathbf{p}_0$, then $\lim_{k\to\infty} f_k$ is also continuous at $\mathbf{p}_0$.

7.11. Define the sequence $\{f_k\}_{k=1}^{\infty}$ as follows:

$$f_k(x) = \begin{cases} 2kx & 0 \le x \le \frac{1}{2k} \\ 2 - 2kx & \frac{1}{2k} < x \le \frac{1}{k} \\ 0 & \frac{1}{k} < x \le 1. \end{cases}$$

(a) On a common pair of axes sketch f_1, f_2, f_4.

(b) Does the sequence have a limit function, and if so, is it continuous on $[0, 1]$?

(c) Is the sequence uniformly convergent on $[0, 1]$?

(d) What conclusion do you draw?

7.12. Let $f(\mathbf{p})$ be real-valued and continuous on the set $\mathbf{D} = \{\mathbf{p} : \mathbf{p} = (x, y), a \le x \le b, c \le y \le d\}$. Show that the function $F(y) = \int_a^b f(\mathbf{p})dx$ is continuous on $[c, d]$. In other words, verify the limit commutation

$$\int_a^b \left[\lim_{y\to y_0} f(\mathbf{p})\right] dx = \lim_{y\to y_0} \int_a^b f(\mathbf{p})dx$$

for any $y_0 \in [c, d]$.

Section 7.3

7.13. Construct an example to show that the converse of the first half of Lemma 7.3.1 is false.

7.14. Give an example of a sequence of functions $\{f_k\}_{k=1}^{\infty}$ that converges pointwise but not uniformly on $[0, 1]$, and yet

$$\int_0^1 \left[\lim_{k\to\infty} f_k(x)\right] dx = \lim_{k\to\infty} \int_0^1 f_k(x) dx.$$

What do you conclude?

7.15. Define the sequence $\{f_k\}_{k=1}^{\infty}$ by $f_k(x) = \begin{cases} 2k^{-1} & 0 \le x \le k \\ x^{-2} & x > k. \end{cases}$

(a) Show that the sequence is uniformly convergent on $[0, \infty)$.

(b) However, also show that

$$\int_0^{\infty} \left[\lim_{k\to\infty} f_k(x)\right] dx \neq \lim_{k\to\infty} \int_0^{\infty} f_k(x) dx.$$

Is this a violation of Theorem 7.3?

7.16. Write out the proof for Theorem 7.1′.

7.17. Write out the proof of either Theorem 7.2′ or Theorem 7.3′.

7.18. In each case prove that the given series converges uniformly on **D**:

(a) $\sum_{k=1}^{\infty} k^2 x^k$, $\mathbf{D} = \left[-\frac{2}{3}, \frac{2}{3}\right]$;

(b) $\sum_{k=1}^{\infty} (x \ln x)^k$, $\mathbf{D} = [0, 1]$;

(c) $\sum_{k=0}^{\infty} \frac{\sin(2kx)}{k^2+2}$, $\mathbf{D} = (-\infty, \infty)$;

(d) $1 + \sum_{k=1}^{\infty} \frac{(2k-1)!!}{(2k)!!} c^{2k} (\sin\theta)^{2k}$, $\mathbf{D} = [0, \pi/2]$, $0 < c < 1$
$(2k-1)!! = 1 \cdot 3 \cdot 5 \cdots (2k-1)$, $(2k)!! = 2 \cdot 4 \cdot 6 \cdots (2k)$;

(e) $\sum_{k=0}^{\infty} \frac{x^k}{k!}$, $\mathbf{D} = [-10, 10]$;

(f) $\sum_{k=1}^{\infty} \frac{k^{5/2}}{\sqrt{x}(k^4+2)}$, $\mathbf{D} = [c, \infty)$, $c > 0$;

(g) $\sum_{k=1}^{\infty} (-1)^k (1-x) x^{k^2}$, $\mathbf{D} = [0, 1]$.

7.19. Regarding Example 7.11,

(a) Verify the results in the last two lines.

(b) If the series in the integrand is approximated by the partial sum $s_9(x)$, what is the estimated value of the integral? Compare with the stated "exact" value.

7.20. Prove that $\int_0^1 x^x dx = \sum_{k=1}^{\infty} \frac{(-1)^{k-1}}{k^k}$. (Missouri MAA Examination, 2000)

7.21. **(Abel's Identity)** This exercise provides a preliminary result needed for the proof, in the next exercise, of a test for uniform convergence of a series that is more delicate than the Weierstrass M-test. Let $\{u_k\}_{k=1}^{\infty}$, $\{v_k\}_{k=1}^{\infty}$ be two sequences and let the sequence $\{W_k\}_{k=0}^{\infty}$ be defined by

$$W_k = \begin{cases} 0 & k = 0 \\ W_{k-1} + u_k & k > 0. \end{cases}$$

Show that if n, m are natural numbers and $n > m > 1$, then

$$\sum_{k=m+1}^{n} u_k v_k = W_n v_{n+1} - W_m v_{m+1} - \sum_{k=m+1}^{n} W_k (v_{k+1} - v_k).$$

The form of the result is reminiscent of the formula for integration by parts.

7.22. (Abel's Test for Uniform Convergence) If $\{u_k(x)\}_{k=1}^{\infty}, \{v_k(x)\}_{k=1}^{\infty}$ are two sequences of functions[9] defined on a common interval $\mathbf{I}$, and if the following hold:

(a) the partial sums of $\sum_{k=1}^{\infty} u_k(x)$ are uniformly bounded; that is, there is an $M > 0$ such that for all $n \in \mathbf{N}$ and for all $x \in \mathbf{I}$, we have $\left|\sum_{k=1}^{n} u_k(x)\right| < M$;

(b) $\{v_k(x)\}_{k=1}^{\infty}$ converges uniformly on $\mathbf{I}$ to 0;

(c) $\sum_{k=1}^{\infty} |v_k(x) - v_{k-1}(x)|$ converges uniformly on $\mathbf{I}$,

then the series $\sum_{k=1}^{\infty} u_k(x)v_k(x)$ converges uniformly on $\mathbf{I}$.

Proof. Let $\varepsilon > 0$ be given. Then interpret, in succession, Hypothesis (b), Hypothesis (c) (use Theorem 7.1′), and Hypothesis (a) in conjunction with Abel's Identity (what should you denote by W_n?). The completion of the proof is left to you.

7.23. Consider the series $f(x) = \frac{x}{3} + \sum_{k=1}^{\infty} \frac{(-1)^k x^{2k+1}}{2k+3}$, $\mathbf{D} = [-1, 1]$.

(a) Explain why the Weierstrass M-test is not of much use here.[10]

(b) Show, separately, that the series converges pointwise for $-1 < x < 1$, for $x = -1$, and for $x = 1$.

(c) Establish, using Exercise 7.22, that the series is uniformly convergent on $[0, 1]$.

(d) What do you conclude from Theorem 7.2′?

7.24. Prove that each of the following results is valid:

(a) $\int_1^2 \left(\sum_{k=1}^{\infty} k e^{-kx}\right) dx = e/(e^2 - 1)$;

(b) $\int_1^3 \left(\sum_{k=1}^{\infty} \frac{\ln(kx)}{k^2}\right) dx = 2\sum_{k=1}^{\infty} \frac{\ln k}{k^2} + (3\ln 3 - 2)\cdot \frac{\pi^2}{6}$;

(c) $\int_0^{\pi} \left(\sum_{k=1}^{\infty} \frac{\cos(kx/2)}{k^2}\right) dx = 4\sum_{j=0}^{\infty} \frac{48j^2+48j+13}{[16j^2+16j+3]^3} \approx 1.9379$;

(d) $\operatorname{erf}(x) = \frac{2x}{\sqrt{\pi}} \sum_{k=0}^{\infty} \frac{(-1)^k x^{2k}}{k!(2k+1)}$ (refer to Exercise 6.45).

7.25. Bessel, in his investigation (1824) of the solutions of the family of differential equations

---

[9] One, or even both, of these sequences could be sequences of constants. The test will still be valid, but in the latter of these two cases the test becomes just a test for ordinary convergence of an infinite series of constants.

[10] Bromwich says that the French mathematician **René Baire** (1874–1932) designated those series that pass the Weierstrass M-test by the suggestive term **normally convergent**. The series in this exercise is not normally convergent.

$$x^2\frac{d^2y}{dx^2}+x\frac{dy}{dx}+(x^2-n^2)y=0,\qquad \begin{cases} y(0)=1\\ y'(0)=1,\end{cases}\qquad n\in\mathbf{N}\cup\{0\},$$

originally wrote for a solution $J_n(x)$ when $n=0$,

$$J_0(x)=\frac{1}{\pi}\int_0^{\pi}\cos\left[x\sin\theta\right]d\theta.$$

(a) Establish by induction that for any $k\in\mathbf{N}\cup\{0\}$ we have $\int_0^{\pi}(\sin\theta)^{2k}d\theta=\frac{(2k)!\pi}{2^{2k}(k!)^2}$.

(b) Now prove that $J_0(x)=\sum_{k=0}^{\infty}\frac{(-1)^k}{(k!)^2}\left(\frac{x}{2}\right)^{2k}$, and state for which x this expression is equivalent to the integral.

(c) Estimate $J_0(1)$ and compare with the tabulated value of 0.76519769 (Abramowitz and Stegun, 1965).

Section 7.4

7.26. Write out the proof of Theorem 7.5.

7.27. **(a)** Expand e^{-t^2} in powers of t^2. Show that the series is uniformly convergent on $[0,1]$.

(b) Retain terms up to t^{20} in the series of part (a). From this estimate $\int_0^1 e^{-t^2}dt$, and compare with the tabulated value of 0.74682413 (Abramowitz and Stegun, 1965).

7.28. If the power series $\sum_{k=0}^{\infty}c_k(x-a)^k$ has radius of convergence R, $0<R<\infty$, then so does $\sum_{k=0}^{\infty}\frac{c_k}{k+1}(x-a)^{k+1}$. We prove this as follows:

(a) Suppose that a sequence $\{u_k\}_{k=0}^{\infty}$ converges to $U>0$, and that $\{v_k\}_{k=0}^{\infty}$ is any real sequence with nonzero real $\rho=\limsup_{k\to\infty}v_k$. Show that there is a subsequence of k's, $\{k_j\}_{j=1}^{\infty}$, such that $\lim_{j\to\infty}v_{k_j}=\rho$ and $\lim_{j\to\infty}u_{k_j}=U$.

(b) Then why does $U\cdot\rho\le\limsup_{k\to\infty}(u_kv_k)$ follow?

(c) There exists an $N\in\mathbf{N}$ such that for all $k>N$ we have $u_k\neq0$. Explain this. Then we can write $\lim_{k\to\infty}u_k^{-1}=U^{-1}$; why?

(d) Then why does $U^{-1}\cdot\limsup_{k\to\infty}(u_kv_k)\le\rho$? Combine this with the result in (b) and draw the expected conclusion.

(e) Apply the result in (d) and Theorem 5.13 to the series $\sum_{k=0}^{\infty}\frac{c_k}{k+1}(x-a)^{k+1}=(x-a)\sum_{k=0}^{\infty}\frac{c_k}{k+1}(x-a)^k$, and deduce the theorem stated at the start of this Exercise. Example 5.17 may be useful.

7.29. Write out the proof of Corollary 7.6.1.

7.30. Start with the Taylor series expansion

$$\ln(1+t)=\sum_{k=0}^{\infty}(-1)^k\frac{t^{k+1}}{k+1},\qquad -1<t\le1.$$

Integrate both sides of this equation twice, letting the lower limit be 0 and the upper limit be x in each case. Show that if x is set equal to $-1/2$, then the following expression for $\ln 2$ is obtained:

$$\ln 2=\frac{1}{2}+\sum_{k=0}^{\infty}\frac{1}{2^k(k+1)(k+2)(k+3)}.$$

How well does this estimate $\ln 2$ if just 12 terms of the series are taken? Is this procedure better than that in Example 7.13?

7.31. Abel's Limit Theorem and Corollary 7.6.1 have analogs for series $\sum_{k=0}^{\infty} c_k(x-a)^k$ (of radius of convergence R), where we are interested in $x-a \to -R^+$. In view of this, show that for $\sum_{k=1}^{\infty}(-1)^{k+1}\frac{x^k}{k}$, which is the series for $\ln(1+x)$ in Examples 7.12 and 7.13, we have

$$\int_{-1}^{1}\left[\sum_{k=1}^{\infty}(-1)^{k+1}\frac{x^k}{k}\right]dx = 2(\ln 2 - 1).$$

7.32. Prove that $\int_0^2 \left(\sum_{k=1}^{\infty}(-1)^{k+1}\frac{x^k}{2^k \cdot k}\right)dx = 2\sum_{k=1}^{\infty}\frac{(-1)^{k+1}}{k(k+1)} = \ln 16 - 2.$

7.33. Refer to Young's result in Section 3.4. What would be the approximate error (according to Young) in Example 7.13 if 100 terms in the second summation were taken?

7.34. **(a)** Use Newton's Binomial Theorem to show that the derivative of $\mathrm{Sin}^{-1}x, x^2 < 1$, is given by

$$\frac{d\mathrm{Sin}^{-1}x}{dx} = 1 + \sum_{k=1}^{\infty}\frac{(2k-1)!!}{2^k k!}x^{2k} \quad \text{(see Exercise 7.18(d))}.$$

(b) Now prove that

$$\frac{\pi}{6} = \frac{1}{2} + \sum_{k=1}^{\infty}\frac{(2k-1)!!}{2^{3k+1}k!(2k+1)}.$$

(c) Use the result in part (b) to estimate π. How many terms of the infinite series will give a value for π that is correct (after rounding) to 7 decimal places?

7.35. Refer to Exercise 7.34.

(a) Prove that the series in part (a) diverges at $x = 1$.

(b) Next, prove that $\frac{\pi}{2} = 1 + \sum_{k=1}^{\infty}\frac{(2k-1)!!}{2^k k!(2k+1)}$. Is this relation useful for the estimation of π?

(c) On the other hand, obtain the relation

$$\frac{\pi}{12} = \frac{\sqrt{2-\sqrt{3}}}{2}\left[1 + \sum_{k=1}^{\infty}\frac{(2k-1)!!}{k!(2k+1)}\left(\frac{2-\sqrt{3}}{8}\right)^k\right].$$

How many correct decimal places in π are obtained (after rounding) if only the first six terms of the series are taken?

7.36. Refer to Exercise 6.15(c). The substitution $t = \sin\varphi$ converts the elliptic integral into the equivalent integral

$$E(1,m) = \int_0^{\pi/2}\sqrt{1 - m\sin^2\varphi}\,d\varphi,$$

in which form the integral is proper for $m \in [-1, 1]$.

(a) Expand the integrand by means of Newton's Binomial Theorem and show that this series converges on $\mathbf{D} = [0, \pi/2]$ if $-1 < m < 1$.

(b) Prove that the series is uniformly convergent on **D** if, in fact, $-1 \le m \le 1$.

(c) Derive the recursive relationship

$$\int_0^{\pi/2} (\sin\varphi)^{2k+2}\, d\varphi = \frac{2k+1}{2k+2} \int_0^{\pi/2} (\sin\varphi)^{2k}\, d\varphi.$$

(d) Obtain the first seven terms in the integration of the series in part (a), and use them to estimate $E(1, 1/4)$. The tabulated value is 1.4674622 (Abramowitz and Stegun, 1965).

7.37. Let $y(x) = \left(\mathrm{Sin}^{-1}x\right)/\sqrt{1-x^2}$.

(a) Find a first-order, linear, nonhomogeneous differential equation that $y(x)$ satisfies.

(b) Let $y(x) = \sum_{n=0}^{\infty} a_n x^n$. Determine the explicit form of the a_n's in order that the series shall be a formal solution of the differential equation.

(c) Determine R, the radius of convergence of the series.

(d) Estimate $\pi/4$.

(e) Prove that $\pi^2/8 = \sum_{n=0}^{\infty} \frac{2^{2n}(n!)^2}{(2n+2)!}$. This expansion was given by Euler in 1737.

Section 7.5

7.38. Prove that each of the following results is valid:

(a) $\frac{d}{dx}\sum_{k=1}^{\infty} \frac{\sin(kx)}{k^3} = \sum_{k=1}^{\infty} \frac{\cos(kx)}{k^2}$, $\mathbf{D} = \mathbf{R}^1$;

(b) $\frac{d}{dx}\sum_{k=1}^{\infty} \frac{x^k}{2k(k+1)} = \frac{1}{2}\sum_{k=0}^{\infty} \frac{x^k}{k+2}$, $\mathbf{D} = [-r, r]$, $0 < r < 1$;

(c) $\frac{d}{dx}\sum_{k=1}^{\infty} \frac{\sin(kx)}{k^3 x} = \sum_{k=1}^{\infty} \left[\frac{kx\cos(kx) - \sin(kx)}{k^3x^2}\right]$, $\mathbf{D} = [r, \infty)$, $r > 0$;

(d) $\frac{d}{dx}\sum_{k=1}^{\infty} e^{-kx}\cos(2kx) = -\sum_{k=1}^{\infty} ke^{-kx}[2\sin(2kx) + \cos(2kx)]$, $\mathbf{D} = [r, \infty)$, $r > 0$.

7.39. An alternative to Theorem 7.7 appears in (Dubins, 1960): If $\{f_k(x)\}_{k=0}^{\infty}$ is a sequence of real-valued functions defined on a closed, bounded interval **I**, and if (i) $\{f_k\}_{k=0}^{\infty}$ converges uniformly on **I**, (ii) each f_k is twice differentiable on **I** (sided derivatives at the endpoints), (iii) there exists an $M > 0$ such that $\left|f_k''(x)\right| < M$ uniformly in k and x, then $\frac{d}{dx}\left[\lim_{k\to\infty} f_k(x)\right] = \lim_{k\to\infty} f_k'(x)$ for all $x \in \mathbf{I}$. Read the paper, write out the proof in your own words, state and prove an alternative to Theorem 7.7′, and apply this alternative to Example 7.17.

7.40. Complete the proof of Theorem 7.7′.

7.41. In this exercise we work out the sum of this series (which, by Theorems 3.4 and 3.8, clearly converges) from Example 7.17:

$$1 + \sum_{k=1}^{\infty} f_k(\theta) \cong 1 + \sum_{k=1}^{\infty} \frac{\cos(k\theta)}{2^{2k-1}}.$$

The approach for doing this is by a series expansion of a particular rational algebraic function of $\cos\theta$.

(a) Let the parameter m be nonzero ($m = 0$ is uninteresting) and define

$$F(\theta) = \frac{1 - m\cos\theta}{1 - 2m\cos\theta + m^2}, \qquad -\pi \le \theta \le \pi.$$

The function $(1-2m\cos\theta+m^2)^{-1}$ can be expressed as $\sum_{k=0}^{\infty}(2m\cos\theta-m^2)^k$, provided that $|2m\cos\theta-m^2|<1$. But in order to guarantee that $F(\theta)$ can be rearranged as a convergent power series in m, we can require that $2|m\cos\theta|+m^2\le 2|m|+m^2<1$ hold for all $\theta\in[-\pi,\pi]$. Why will this suffice? Show that $0<|m|\le\frac{2}{5}$ will work.

(b) Hence, with m restricted as above, we can write $F(\theta)=\sum_{k=0}^{\infty}C_k m^k$, where each C_k is a function of $\cos\theta$. Prove first the following lemma: If $n\in\mathbf{N}$, then for any θ we have $\cos[(n+1)\theta]+\cos[(n-1)\theta]=2\cos(n\theta)\cos\theta$.

(c) Use the lemma to next prove that $C_k=\cos(k\theta)$ for each $k\in\mathbf{N}\cup\{0\}$.

(d) From the result in (c), deduce that

$$\frac{\cos\theta-m}{1-2m\cos\theta+m^2}=\sum_{k=1}^{\infty}m^{k-1}\cos(k\theta).$$

Finally, by making a suitable choice for m, consistent with the restriction in part (a), deduce that for any $\theta\in[-\pi,\pi]$ we have

$$1+\sum_{k=1}^{\infty}f_k(\theta)=\frac{15}{17-8\cos\theta}.$$

7.42. This exercise continues the analysis of Example 7.17. Prove that the series of derivatives

$$\sum_{k=1}^{\infty}f_k'(\theta)=2\sum_{k=1}^{\infty}\frac{-k\sin(k\theta)}{4^k}$$

converges uniformly on $[-\pi,\pi]$ by showing that $\sum_{k=1}^{\infty}\frac{k}{4^k}=\frac{4}{9}$.

7.43. Refer to Exercise 7.32. Use the method there to prove that if $\sum_{k=0}^{\infty}c_k(x-a)^k$ has radius of convergence R, $0<R<\infty$, then so does $\sum_{k=1}^{\infty}kc_k(x-a)^{k-1}$.

7.44. Use Example 7.22 to obtain a good estimate of $\sqrt{73}$.

7.45. Consider the generating function $g(x)$ in Exercise 3.42. Prove that $\sum_{n=0}^{\infty}\frac{a_n}{n!}x^n$ is differentiable term-by-term at $x=9/8$, and determine the value of the derivative there. How well do you do if you estimate the derivative at $x=9/8$ by adding only the first six terms of the differentiated series?

7.46. Newton, in his first letter to Leibniz in 1676, gave the expansion[11]

$$\sin(\alpha\sin^{-1}x)=\alpha x+\alpha\sum_{k=1}^{\infty}\frac{(-1)^k}{(2k+1)!}\left\{\prod_{j=1}^{k}\left[\alpha^2-(2j-1)^2\right]\right\}x^{2k+1},\ \alpha\in\mathbf{R}^1.$$

(a) Determine R, the radius of convergence of the series.
(b) Does the series converge at the endpoints?
(c) It is convenient to let $x=\sin\theta$. Obtain a series expansion for $\cos(\alpha\theta)$, $\alpha\neq 0$.
(d) In the series obtained in (c), let $\alpha=3$ and deduce an elementary trigonometric identity.
(e) In the series obtained in (c), let $\alpha=1/2$ and $\theta=\pi/4$. Estimate $\cos(\pi/8)$.

[11] Private communication to one of the authors (JBD) from R. Roy (Beloit College).

Section 7.6

7.47. Write a short program and plot the first few partial sums of $f(x) = \sum_{k=0}^{\infty} \left(\frac{6}{11}\right)^k \cos(11^k \pi x)$ for $x \in [0, 1/2]$. How might you describe qualitatively the feature that makes this function nondifferentiable everywhere on $[0, 1/2]$?

7.48. Consult (Hildebrandt, 1933), which gives another example of a function f continuous on $\mathbf{R}^1$ and differentiable nowhere. Read this short paper, write up the analysis in your own words, and include graphs and calculations to enrich your report.

7.49. Wiener is right; direct integration of $\int_0^{\pi/2} x \cot x \, dx$ (in Example 7.24) is not easy. However, we can do this:

(a) Show that $\mathrm{I} = \int_0^{\pi/2} x \cot x \, dx$ is equivalent to $-\int_0^{\pi/2} [\ln(\sin x)] dx$.

(b) Replace $\sin x$ by its Taylor series; show that this leads to

$$I = -\int_0^{\pi/2} \left[\ln x - \frac{x^2}{6} - \frac{x^4}{180} - \frac{x^6}{2835} - \frac{x^8}{37800} - \cdots \right] dx.$$

(c) With the series truncated as shown, obtain an estimate for I, and compare with the exact value.

7.50. Use Theorem 7.9 to prove each of the following assertions:

(a) If $F(x) = \int_0^{\ln 2} \sin(xe^y) \, dy$, then $\lim_{x \to 0} F'(x) = 1$.

(b) If $F(x) = \int_0^1 \sin(xy) dy, x \neq 0$, then $\left| \frac{d^{99}}{dx^{99}} \left(\frac{1 - \cos x}{x} \right) \right| \leq \frac{1}{100}$.

(c) If $F(x) = -\ln x + \int_1^{x^2} \frac{e^{-t/x}}{t} dt, x > 0$, then $F(x)$ is a decreasing function.

7.51. **(Leibniz's Theorem)** Assume that the following hold: (i) f is defined and continuous on the closed, bounded rectangle $\mathbf{D} = \{(x, y) : a_1 \leq x \leq a_2, b_1 \leq y \leq b_2\}$; (ii) $D_2 f$ is defined and continuous on $\mathbf{D}$, (iii) for any function $g(x, y)$ continuous on $\mathbf{D}$, $\int_{a_1}^{a_2} g(x, y) dx$ is continuous in y, $\int_{b_1}^{b_2} g(x, y) dy$ is continuous in x, and the equality

$$\int_{b_1}^{y} dt \int_{a_1}^{a_2} g(x, t) dx = \int_{a_1}^{a_2} dx \int_{b_1}^{y} g(x, t) dt$$

holds for any $y \in [b_1, b_2]$.[12]

(a) How does it follow that

$$\int_{a_1}^{a_2} f(x, y) dx = \int_{a_1}^{a_2} dx \left[\int_{b_1}^{y} D_2 f(x, t) dt + f(x, b_1) \right]$$

holds for any $y \in [b_1, b_2]$?

---

[12] This is a weak version of **Fubini's Theorem** (after the Italian mathematician **Guido Fubini** (1879–1943). It is not hard to prove it for Riemann integrals; our Section 6.8 broke off just before a presentation of this result. We, therefore, take this version of Fubini's Theorem for granted. Naturally, the Theorem can be strengthened by weakening the hypotheses somewhat.

(b) Obtain the relation

$$\int_{a_1}^{a_2} f(x,y)dx = \int_{b_1}^{y} dt \left[\int_{a_1}^{a_2} D_2 f(x,t)dx \right] + \int_{a_1}^{a_2} f(x,b_1)dx.$$

(c) Finally, how does it follow that

$$\frac{d}{dy} \int_{a_1}^{a_2} f(x,y)dx = \int_{a_1}^{a_2} D_2 f(x,y)dx?$$

(Seeley, 1961).

7.52. Leibniz's Theorem can also be obtained as a consequence of the Theorem of the Equality of Mixed Second-Order Derivatives (our Exercise 5.77). See if you can prove this on your own. If you get stuck, then consult Fisher and Shilleto (1986) and write up the brief proof there more completely and in your own words.

7.53. This exercise and the next lay the groundwork for the proof of the extension of Leibniz's Theorem to improper integrals of the first kind. The following definition is standard.

UNIFORM CONVERGENCE OF AN INTEGRAL

Suppose that for all $x \in [a,b]$ the integral $\int_0^\infty f(x,y)dy$ converges. We say that the convergence is **uniform** on $[a,b]$ iff for each $\varepsilon > 0$ there is a number $\delta > 0$ such that for all $x \in [a,b]$ and for all $c \geq \delta$ we have

$$\left| \int_c^\infty f(x,y)dy \right| < \varepsilon.$$

(a) Prove the following simple test for uniform convergence of an integral: $\int_0^\infty f(x,y)dy$ is uniformly convergent on $[a,b]$ if for all sufficiently large y the inequality $|f(x,y)| < M/y^k$ holds, where $M > 0$ and $k > 1$.

(b) Let $\mathbf{D} = [a,b]$, where $0 < a < b$; apply the test in part (a) to the integral $\int_0^\infty e^{-xy}dy, x \in \mathbf{D}$.

(c) Use Theorem 6.22 to establish that, if $\varepsilon > 0$ is given, then there is a number $r > 0$ such that for any $u^* \geq r$ we have $\left|\int_{u^*}^\infty \frac{\sin u}{u} du\right| < \varepsilon$. From this, show that if $\mathbf{D} = [a,b], 0 < a < b$, then $\int_0^\infty \frac{\sin xy}{y} dy$ is seen to be uniformly convergent on $\mathbf{D}$ if we take (in the definition) $\delta = u^*/a$.

7.54. Theorem 7.10 (Fubini's Theorem). If $f(x,y)$ is continuous on $\mathbf{D} = \{(x,y): x \geq 0, b_1 \leq y \leq b_2\}$ and the integral $\int_0^\infty f(x,y)dx$ converges uniformly on $[b_1,b_2]$, then

$$\int_{b_1}^{b_2} dy \int_0^\infty f(x,y)dx = \int_0^\infty dx \int_{b_1}^{b_2} f(x,y)dy.$$

(a) Assume that $\int_0^\infty f(x,y)dx$ is integrable on $[b_1, b_2]$, and let c satisfy $0 < c < \infty$. Establish that

$$\left| \int_{b_1}^{b_2} dy \int_0^\infty f(x,y)dx - \int_0^c dx \int_{b_1}^{b_2} f(x,y)dy \right| = \left| \int_{b_1}^{b_2} dy \int_c^\infty f(x,y)dx \right|.$$

(b) Show how to choose c in part (a) so that there results

$$\int_{b_1}^{b_2} dy \int_0^\infty f(x,y)dx = \lim_{c\to\infty} \int_0^c dx \int_{b_1}^{b_2} f(x,y)dy$$
$$= \int_0^\infty dx \int_{b_1}^{b_2} f(x,y)dy.$$

(c) (Optional) For extra credit, establish that $G(y) \cong \int_0^\infty f(x,y)dx$ is continuous on $[b_1, b_2]$ and, hence, is integrable there.

7.55. Theorem 7.10 (Leibniz's Theorem for Improper Integrals). If $f(x,y)$ satisfies the hypotheses in Theorem 7.10 and, additionally, $D_2 f$ is continuous on $\mathbf{D}$ and $\int_0^\infty D_2 f(x,y)dx$ converges uniformly on $[b_1, b_2]$, then

$$\frac{d}{dy} \int_0^\infty f(x,y)dx = \int_0^\infty D_2 f(x,y)dx.$$

(a) How does it follow immediately that for any $y \in [b_1, b_2]$

$$\int_{b_1}^{y} dt \int_0^\infty D_2 f(x,t)dx = \int_0^\infty dx \int_{b_1}^{y} D_2 f(x,t)dt$$

holds?

(b) Then how does it follow that the equation in (a) can be rewritten as

$$\int_{b_1}^{y} dt \int_0^\infty D_2 f(x,t)dx = \int_0^\infty f(x,y)dx - \int_0^\infty f(x,b_1)dx?$$

(c) Finally, differentiate both sides, justifying what you do.

7.56. Refer to Exercise 6.56. Let $\mathbf{D} = [a,b]$, $0 < a < b$; prove that $\Gamma(\alpha)$ is infinitely differentiable at any $\alpha \in \mathbf{D}$.

7.57. The difficult integral $\int_0^\infty e^{-x^2} dx$ has been evaluated by several methods. A recent, interesting method that uses Leibniz's Theorem appeared in Weinstock (1990). Read this and write up the proof in your own words.

REFERENCES

Cited Literature

Abramowitz, M. and Stegun, I.A. (eds.), *Handbook of Mathematical Functions*, Dover Publications, NY, 1965. The source of data referred to in Exercises 7.25, 7.27, 7.36.

Apostol, T.M., "Term-Wise Differentiation of Power Series," *Amer. Math. Monthly*, **59**, 323–326 (1952). An elementary proof, not making use of uniform convergence, of the theorem that a power series can be differentiated everywhere in the interior of its interval of convergence.

Apostol, T.M., "Formulas for Higher Derivatives of the Riemann Zeta Function," *Math. Comp.*, **44**, 223–232 (1985). Very interesting paper. Values of $\zeta(s) = \sum_{k=1}^{\infty} \frac{1}{k^s}$ can be estimated from the series expansion

$$\zeta(s) = \frac{1}{s-1} + \sum_{n=0}^{\infty} A_n (s-1)^n, \quad s > 1.$$

Apostol gives a table of values of the first 19 A_n's.

Buck, R.C., *Advanced Calculus*, 3rd ed., McGraw-Hill, NY, 1978, pp. 279–280. This proof of Abel's Limit Theorem (our Theorem 7.6) is short and sweet.

Dubins, L.E., "On Differentiation of Series Term-by-Term," *Amer. Math. Monthly*, **67**, 771–772 (1960). An alternative to our Theorem 7.7 that replaces the uniform convergence of the series of (first) derivatives by the condition of uniform bounding of the second derivatives. Cited in our Exercise 7.39.

Fisher, J.C. and Shilleto, J., "Three Aspects of Fubini's Theorem," *Math. Mag.*, **59**, 40–42 (1982). The motivating source for our Exercise 7.52. This brief paper would make a nice reading assignment for students.

Hardy, G.H., "Sir George Stokes and the Concept of Uniform Convergence," *Proc. Camb. Phil. Soc.*, **19**, 148–156 (1916/1919). The concept of uniform convergence is solidified today, but Hardy points out that in the past there have been several variants in use. For the scholar, there are some interesting references in this paper.

Hawkins, T., *Lebesgue's Theory of Integration*, Chelsea Publ. Co., NY, 1975, pp. 42–54. See these pages for a historical discussion of nondifferentiable continuous functions.

Hildebrandt, T.H., "A Simple Continuous Function with a Finite Derivative at No Point," *Amer. Math. Monthly*, **40**, 547–548 (1933). The source of our Exercise 7.48. On the other hand, if you'd like to exercise your French, you may consult pp. 107–108 of Darboux, J.G., "Mémoire sur les fonctions discontinues," *Ann. de l'École Norm.* (2), **4**, 57–112 (1875), in which Darboux discusses the example $f(x) = \sum_{n=1}^{\infty} \frac{\sin[(n+1)!x]}{n!}$.

Ore, O., "Niels Henrik Abel," in Gillispie, C.C. (ed.), *Dictionary of Scientific Biography*, Vol. 1, Charles Scribner's Sons, NY, 1970, pp. 12–17. Nice summary by an authority on the accomplishments of the brilliant Abel.

Ore, O., *Niels Henrik Abel*, Chelsea Publ. Co., NY, 1974. The life of the brilliant Norwegian mathematician (he died at age 26) was tragic; read about it here.

Osler, T.J., "Euler's Little Summation Formula and Special Values of the Zeta Function," *Math. Gaz.*, **92**, 295–299 (2008). The main, neat result is the following: $\zeta(0) = -\frac{1}{2}, \zeta(-n) = -B_{n+1}/(n+1), n \in \mathbf{N}$.

Phillips, E.R., *An Introduction to Analysis and Integration Theory*, Dover Publications, NY, 1984, pp. 104–112. General discussion of spaces of continuous mappings from one metric space into another and of the **Arzelà-Ascoli Theorem** as the analog of the Bolzano-Weierstrass Theorem.

Roy, R., "The Discovery of the Series Formula for π by Leibniz, Gregory and Nilakantha," *Math. Mag.*, **63**, 291–306 (1990). A fine, historical article with some good mathematics in it. One wishes that Gregory and his all-too-brief output were better known. Cited in our Example 7.14.

Seeley, R.T., "Fubini Implies Leibniz Implies $F_{yx} = F_{xy}$," *Amer. Math. Monthly*, **68**, 56–57 (1961). The motivating source for our Exercise 7.51; the content of this paper overlaps that in Fisher and Shilleto (1986) (see above).

Sprecher, D.A., *Elements of Real Analysis*, Dover Publications, NY, 1987, pp. 213–214, 214–218, 238–243. An excellent book that is worth owning; the citations are to Dini's Theorem, to a weakened set of conditions for the differentiation of a convergent sequence of functions, and to general spaces of continuous mappings between metric spaces, respectively.

Titchmarsh, E.C., *The Theory of Functions*, 2nd ed., Oxford University Press, Oxford, 1939, pp. 351–353. Our outline in Section 7.6 of Weierstrass's nondifferentiable, continuous function is taken from these pages. This superb book has the stamp of Hardy on it; Titchmarsh was, in fact, a research student under Hardy (see Kanigel, R., *The Man Who Knew Infinity: A Life of the Genius Ramanujan*, Charles Scribner's Sons, NY, 1991, p. 151).

Vernescu, A., "The Summation of a Family of Series," *Amer. Math. Monthly*, **115**, 939–943 (2008). As the author says: "The summation of a convergent series is often an interesting problem, which can involve various techniques." Cited in our Example 7.15.

Weinstock, R., "Elementary Evaluations of $\int_0^\infty e^{-x^2} dx$, $\int_0^\infty \cos(x^2) dx$, and $\int_0^\infty \sin(x^2) dx$," *Amer. Math. Monthly*, **97**, 39–42 (1990). The source of our Exercise 7.57; the author also does an evaluation of the two Fresnel integrals (see Example 6.15).

Wiener, J., "Differentiation with Respect to a Parameter," *Coll. Math. J.*, **32**, 180–184 (2001). A very good paper for discussion in senior seminar; the source of our Example 7.24. See also Exercise 7.49, which replaces a differentiation of an integral by integration of a series.

Additional Literature

Bridger, M., *Real Analysis: A Constructive Approach*, John Wiley & Sons, NY, 2006.

Bromwich, T.J.I'a., *Introduction to the Theory of Infinite Series*, 2nd ed. rev., Macmillan & Co., London, 1926, pp. 119–155 (topics in uniform convergence).

Carslaw, H.S., *An Introduction to the Theory of Fourier's Series and Integrals*, 3rd ed. rev., Dover Publications, NY, 1950, pp. 172–181 (integration of series).

Clarke, L.E., "Integration Under the Integral Sign Again," *Math. Gaz.*, **69**, 131–134 (1985).

Courant, R. and John, F., *Introduction to Calculus and Analysis*, Vol. II, John Wiley & Sons, NY, 1974, pp. 71–82, 462–474 (commutation of limit operations).

Flanders, H., "Differentiation Under the Integral Sign," *Amer. Math. Monthly*, **80**, 615–627 (1973).

Gordon, R.A., "A Convergence Theorem for the Riemann Integral," *Math. Mag.*, **73**, 141–147 (2000).

Hallenbeck, D.J. and Tkaczynska, K., "The Absolute and Uniform Convergence of Infinite Improper Integrals," *Amer. Math. Monthly*, **95**, 124–126 (1988).

Hunt, J.N., "An Extension of Leibnitz's Rule for Integrals," *Amer. Math. Monthly*, **75**, 172–173 (1968).

Kestelman, H., "Riemann Integration of Limit Functions," *Amer. Math. Monthly*, **77**, 182–187 (1970).

Osgood, W.F., "On the Differentiation of Definite Integrals," *Ann. Math.*, **9**(2), 119–122 (1907–1909).

Parkinson, E.M., "George Gabriel Stokes," in Gillispie, C.C. (ed.), *Dictionary of Scientific Biography*, Vol. XIII, Charles Scribner's Sons, NY, 1976, pp. 74–79 (one of the discoverers of uniform convergence).

Rosenlicht, M., *Introduction to Analysis*, Dover Publications, NY, 1986, pp. 137–167 (Chap. VII. "Interchange of Limit Operations").

APPENDIX A

Hints and Answers to Selected Exercises

Chapter 1

1.4. **(a)**, **(b)**, **(f)** hold.

1.6. **(b)** Suppose that the set **P** of positive elements is nonempty; let $x \in \mathbf{P}$. Now look at x added to itself.

1.12. Suppose $x > x$ were to hold for some $x \in \mathbf{R}$; now look at $x + (-x)$.

1.15. Suppose that there were two (distinct) suprema of **S**.

1.19. **(a)** $x > y$ and $z < 0$ mean $[x + (-y)] \in \mathbf{P}, -z \in \mathbf{P}$.

1.25. Suppose that $a, b \in \mathbf{N}$ exist such that $a^2/b^2 = 5$. Now look at the factorizations of both sides of $a^2 = 5b^2$.

1.32. $z = \frac{\sqrt{53}}{53} \cdot (-6, 1, -4)$.

1.33. **(a)** Begin two lines of development separately with $x = (x - y) + y$ and $y = (y - x) + x$.

(b) Do this in cases.

1.35. **(b)** $||\mathbf{x}||^2 + 2c(\mathbf{x} * \mathbf{y}) + c^2||\mathbf{y}||^2 > 0$.

1.39. **(c)** $f = \{(x, y) : y = x^2 + 1, x \in \mathbf{R}^1\}, \mathbf{I} = [0, 1]$.

1.40. **(a)** Start with $y \in f(f^{-1}(\mathbf{H})) \subseteq \mathbf{S}$; deduce that $f(f^{-1}(\mathbf{H})) \subseteq \mathbf{H}$. Then reason why proper set inclusion, $f(f^{-1}(\mathbf{H})) \subset \mathbf{H}$, can be rejected.

1.41. Assume that $f(\mathbf{I}), f(\mathbf{J})$ are defined for any x in either **I** or **J** and, therefore, f may be taken as onto $f(\mathbf{I} \cup \mathbf{J})$.

1.42. **(c)** Neither makes sense.

1.43. **(a)** Let $y \in \mathbf{R}(f)$ be arbitrary; then look at $(x_1, y), (x_2, y) \in f$.

1.45. Construct an indexing scheme for the elements of $\mathbf{Q}^- \cup \{0\} \cup \mathbf{Q}^+$.

1.46. **(b)** 53,615.

1.48. Explore the geometry of a right triangle inscribed in a semicircle.

1.49. Let $\mathbf{S}'$ denote the set of all cluster points of a set **S**;

(a) $\mathbf{S}' = [0, 1]$

(d) $\mathbf{S}' = \{0\}$.

1.51. Suppose that $\mathbf{L}_1, \mathbf{L}_2$ are distinct values of $\lim_{\mathbf{x}\to\mathbf{a}} \mathbf{f}$. Choose $\varepsilon = ||\mathbf{L}_1 - \mathbf{L}_2||$, and look at $||\mathbf{L}_1 - \mathbf{L}_2|| = ||(\mathbf{L}_1 - \mathbf{y}) \oplus (\mathbf{y} - \mathbf{L}_2)||$, where $\mathbf{y} = \mathbf{f}(\mathbf{x})$.

1.53. (b) In the implementation of the Triangle Inequality, use $|f(\mathbf{x})| < 1 + |\mathbf{F}|$.

1.60. (a) $\lim_{x\to 1^-} \sqrt{1-x^3} = 0$;

(e) $\lim_{x\to\infty} \left\lfloor \frac{3x}{x+2} - \frac{x^2}{x+2} \right\rfloor = -\infty$.

Chapter 2

2.2. (a) Use differentiation.

(c) Mathematical induction.

2.5. (b) Bounded above by x_6; bounded below by 0.

2.6. (b) Mathematical induction.

2.7. (b) Becomes decreasing.

2.10. Look at b_{n+1}/b_n.

2.11. Look at $x_n - x_{n+1}$, and make use of $\ln\left(1+\frac{1}{n}\right) = \sum_{k=1}^{\infty}(-1)^{k+1}\frac{1}{kn^k}$.

2.13. (b) Obtain $x_{n+1} = \frac{12+2x_n^3}{3x_n^2}$, and from this, $\sqrt[3]{12} - x_{n+1} = \frac{-2x_n^3+3\sqrt[3]{12}x_n^2-12}{3x_n^2}$. Multiply by $3x_n^2$, factor the cubic, and look at the minimization of the quadratic factor.

2.14. (b) Show that $x_{k+2} - x_{k+1} = x_k \geq 1$ for $k \geq 2$.

2.15. (a) Use differentiation.

2.23. (b) Increase half of the factors in the numerator by 1, and increase half of the factors in the denominator by 1.

2.24. (b) (ii) $\lim_{n\to\infty} x_n = 3$.

2.27. $\lim_{n\to\infty} \mathbf{p}_n = (2/5, 1/5)$. An artistic hint is to note, after drawing several of the auxiliary lines on a full sheet of paper, that there is a kind of periodicity (period $= 8$) of the nature of the points $\mathbf{p}_n$.

2.28. (a) Start with the MacLaurin series for cosine:

$$\cos(1/n) = 1 - (1/2n^2) + (1/24n^4) - \cdots.$$

2.31. (d) Given any $x_0 \in (-1, 1)$, there is a sequence $\{n_k\}_{k=1}^{\infty}$ of natural numbers such that x_0 is a cluster point of $\{\cos n_k\}_{k=1}^{\infty}$.

2.32. Construct nested open intervals $\mathbf{I}_1 \supset \mathbf{I}_2 \supset \mathbf{I}_3 \supset \cdots$ but with $x_0 \in (-1, 1]$ as the common right-hand endpoint of them.

2.36. Let $\mathbf{s} = \left\{x_{n_1}, x_{n_2}, x_{n_3}, \ldots\right\}$ be the set of terms of an arbitrary convergent subsequence of $\{x_n\}_{n=1}^{\infty}$. Then $\inf \mathbf{S} \leq \inf \mathbf{s} \leq \sup \mathbf{s} \leq \sup \mathbf{S}$ (why?). Now use the facts that $\lim_{k\to\infty} x_{n_k}$ is bracketed by $\inf \mathbf{s}$ and $\sup \mathbf{s}$, and that $\left\{x_{n_k}\right\}_{k=1}^{\infty}$ is arbitrary.

2.41. **(f)** Write $n^{-1}\csc(n^{-1})$ as $\frac{n^{-1}}{\sin(n^{-1})}$; then make use of the first two terms in the MacLaurin series for sine:

$$\sin(1/n) = n^{-1} - (1/6n^3) + (1/120n^5) - \cdots .$$

2.42. Consider odd m and even n.

2.47. **(a)** Choose $\varepsilon = 1/2$. There is an $N \in \mathbf{N}$ such that $n > N$ implies $|x_n - x_{N+1}| < 1/2$; arrive at $|x_n| < \frac{1}{2} + |x_{N+1}|$.

(b) Use, actually, Corollary 2.11.1.

Chapter 3

3.8. **(b)** Examine $n^{n/10}$ versus 2^n.

(c) Examine $\sqrt{n!}$ versus 2^n.

3.11. **(a)** Use mathematical induction to show that $y_n - y_{n-1} = \frac{1}{2}(a_{n-1} + a_n)$.

3.12. **(a)** In the symbolism of Exercise 3.11, show that $\lim_{n\to\infty} y_{3n-1} = 1$ and $\lim_{n\to\infty} y_{3n} = 1/2$, so $\lim_{n\to\infty} y_n$ does not exist and $\sum_{n=1}^{\infty} b_n$ is not summable Y.

3.14. Start with $\sum\limits_{n=1}^{\infty} \frac{1}{X_n(9)} = \sum\limits_{k=1}^{8} \frac{1}{k} + \sum\limits_{k=2}^{\infty} \sum\limits_{10^{k-1} \le X_n < 10^k} \left(\frac{1}{X_n(9)}\right)$.

3.17. Use mathematical induction to show that $F_n > \frac{1}{2}\left(\frac{8}{5}\right)^n$ for $n \ge 10$.

3.26. **(a)** Look at $|a_n| - |a_{n+1}|$.

3.27. 49 terms.

3.28. 1.8×10^{-12}.

3.31. **(f)** Converges absolutely;

(k) Diverges;

(l) Diverges.

3.35. **(a)**

$$\left(\sum_{n=0}^{\infty} \frac{n+1}{n!}\right) \otimes \left(\sum_{n=0}^{\infty} \frac{2}{n!}\right) = \sum_{n=0}^{\infty} \frac{2}{n!} \sum_{k=0}^{n} (k+1)\binom{n}{k}$$

$$= \sum_{n=0}^{\infty} \frac{2}{n!}(n \cdot 2^{n-1}) + \sum_{n=0}^{\infty} \frac{2}{n!}(2^n)$$

$$= 4\sum_{n=0}^{\infty} \frac{2^n}{n!}.$$

3.39. $s_7 = \sum\limits_{k=0}^{7} c_k = 87947/279936$; % error $= 21.5\%$.

$\sum\limits_{k=0}^{7} |c_k| = 1555105/279936$; % error $= 7.41\%$.

3.40. **(b)** $\sum_{n=1}^{5} \frac{1}{1+2^n} < f(2) < \sum_{n=1}^{\infty} \frac{1}{2^n}$.

3.42. For the second part of the exercise, show by mathematical induction that

$$a_n = (-1)^{n-1} 1 \cdot 3 \cdot 5 \cdots (2n-3)/2^{2n};$$

then look at Exercise 2.10.

3.44. $\zeta(2) = \zeta(3) \sum_{n=1}^{\infty} \frac{\Phi(n)}{n^3}$ becomes

$$\sum_{k=1}^{\infty} \frac{1}{k^2} = \left(\sum_{k=1}^{\infty} \frac{1}{k^3}\right) \otimes \left(\sum_{n=1}^{\infty} \frac{\Phi(n)}{n^3}\right) = \left(\sum_{k=0}^{\infty} \frac{1}{(k+1)^3}\right) \otimes \left(\sum_{n=0}^{\infty} \frac{\Phi(n+1)}{(n+1)^3}\right)$$

$$= \sum_{n=0}^{\infty} \sum_{k=0}^{n} \frac{1}{(k+1)^3} \frac{\Phi(n-k+1)}{(n-k+1)^3}$$

$$= \frac{1}{1^3}\frac{\Phi(1)}{1^3} + \left[\frac{1}{1^3}\frac{\Phi(2)}{2^3} + \frac{1}{2^3}\frac{\Phi(1)}{1^3}\right]$$

$$+ \left[\frac{1}{1^3}\frac{\Phi(3)}{3^3} + \frac{1}{2^3}\frac{\Phi(2)}{2^3} + \frac{1}{3^3}\frac{\Phi(1)}{1^3}\right]$$

$$+ \left[\frac{1}{1^3}\frac{\Phi(4)}{4^3} + \frac{1}{2^3}\frac{\Phi(3)}{3^3} + \frac{1}{3^3}\frac{\Phi(2)}{2^3} + \frac{1}{4^3}\frac{\Phi(1)}{1^3}\right] + \cdots.$$

Continue the pattern for $n = 5, 6$, and then rearrange terms.

3.45. The transposition produces

$$\frac{x}{e^x - 1} + \frac{x}{2} = 1 + \sum_{n=2}^{\infty} \frac{B_n}{n!} x^n.$$

Examine the left-hand side when x is replaced by $-x$.

3.46. Begin with $\coth x = \frac{e^x + e^{-x}}{e^x - e^{-x}} \frac{e^x}{e^x} = 1 + \frac{1}{x} \frac{2x}{e^{2x}-1}$.

3.47. Using terms out to B_{10}, we obtain $\cot(\pi/4) \approx 1.000000164$.

3.48. **(c)** $\sum_{k=1}^{100} k^5 = 171{,}708{,}332{,}500$.

3.49. **(c)** Differentiate both sides of the defining equation with respect to x.

3.50. Rearrange the defining equations in Exercise 3.49(a) to give

$$ze^{xz} = (e^z - 1) \sum_{n=0}^{\infty} \frac{B_n(x)}{n!} z^n.$$

Chapter 4

4.2. Recall Lemma 2.3.1.

4.4. $(a) \to (b)$: Since $\mathbf{a} \in \mathrm{Int}(\mathbf{D}(\mathbf{f}))$, there is an n-ball $\mathbf{B}_n(\mathbf{a}; \delta)$ of small enough $\delta > 0$ such that $\mathbf{B}_n(\mathbf{a}; \delta) \subset \mathbf{D}(\mathbf{f})$.

4.10. Continuous only at $x = 0$.

4.12. **(d)** $x = 0$; removable.

4.16. Use Lemma 2.3.1.

4.18. Use Theorems 4.2 and 4.3.

4.20. **(a)** Continuous on $\mathbf{R}$.

4.22. Make use of the identity $\sin(x_1) - \sin(x_2) = \left(2\cos\frac{x_1+x_2}{2}\right)\left(\sin\frac{x_1-x_2}{2}\right)$, and of the MacLaurin series for $\sin x$. Finally, call on Exercise 4.21.

4.26. **(b)** Consider an adaptation of Corollary 2.9.2.

4.30. **(a)** For the set inclusion $\mathbf{B}(\mathbf{p};\delta) \subseteq f^{-1}[f(\mathbf{B}(\mathbf{p};\delta))]$, begin by letting $\mathbf{x} \in \mathbf{B}(\mathbf{p};\delta)$ be arbitrary.

(b) For the set inclusion $f[f^{-1}(\mathbf{B}'(f(\mathbf{p});\varepsilon))] \subseteq \mathbf{B}'(f(\mathbf{p});\varepsilon)$, begin by letting $\mathbf{y} \in \mathbf{B}'(f(\mathbf{p});\varepsilon)$ be arbitrary.

4.31. **(a)** Let $x, x' \in \mathbf{R}^1 \backslash \mathbf{S}_i$; show that $g_i(x) \le |x - x'| + g_i(x')$.

(b) If $f(x) = g_1(x) - g_2(x)$, then $\mathbf{T} = f^{-1}((-\infty, 0))$.

4.33. **(a)** Use Theorem 4.5.

4.34. If n is replaced by ∞ in $\bigcup_{k=1}^{n} \mathbf{F}_k = M \backslash \bigcap_{k=1}^{n} \mathbf{F}_k^c$, both sides may turn out to be neither open nor closed.

4.37. **(a)** It is useful to show that for $n \ge 4$ we have $n! > 2^n$. Then $\lim_{n\to\infty} x_n < 67/24$.

(b) $\sum_{k=q+1}^{\infty} \frac{q!}{k!} = \frac{1}{q}$, which is less than 1.

4.39. Begin by letting $\mathbf{\Theta} = \{\mathbf{B}_n(\mathbf{p};k) : k \in \mathbf{N}\}$ be an open cover of $\mathbf{S}$.

4.40. A reasonable conjecture is that $\text{diam}(\mathbf{S}) \approx 4.5363$.

4.43. ABSOLUTE MINIMUM $= -\frac{1}{2}\left(\sqrt{5} - 1\right)\exp\left[-(7 + \sqrt{5})/4\right]$;
ABSOLUTE MAXIMUM $= 2$.

4.45. **Lemma 4.8.2**: $(\rightarrow)$ Suppose that there is a nonempty, proper subset $\mathbf{S}_1 \subset \mathbf{M}$ that is both open and closed; let $\mathbf{S}_2 = \mathbf{S}_1^c = \mathbf{M} \backslash \mathbf{S}_1$.

4.49. Use Theorem 4.13.

4.50. **(a)** $x \approx 4.720358$.

4.52. **(c)** Use Theorem 4.15.

(d) Assume f^{-1} is discontinuous at some $y_0 \in \mathbf{R}(f)$. As f^{-1} is, nevertheless, defined at $y_0 \in \mathbf{I}$, the discontinuity must be a jump discontinuity. Assume from part (c) that $\lim_{y\to y_0^-} f^{-1}(y) < f^{-1}(y_0) < \lim_{y\to y_0^+} f^{-1}(y)$, and choose a such that $\lim_{y\to y_0^-} f^{-1}(y) < a < f^{-1}(y_0)$. Is there a $y \in \mathbf{R}(f)$ such that $a = f^{-1}(y)$?

4.57. Consider $f(x) \cong x$ and $g(x) \cong \cos x$, $\mathbf{D} = [0, \infty)$. First show that g is, in fact, uniformly continuous on $\mathbf{D}$ (a hint is contained in Exercise 4.22). Next, consider $h(x) \cong f(x)g(x)$, and choose $\varepsilon = 1/2$, $x_0 = (2n+1)\pi/2$, $n \in \mathbf{N}$, and $x = x_0 + (\delta/3)$, $\delta \in \mathbf{Q}$.

4.58. $\mathbf{S}$ can be covered by finitely many n-balls of radius $\frac{1}{2}\delta(\varepsilon)$ (why?). Apply the Triangle Inequality to an arbitrary pair of points in each n-ball, and make use of the uniform continuity.

4.59. **(b)** First prove the general theorem that function composition preserves uniform continuity.

4.63. $\lambda = 1/3$; limiting result ≈ 1.324718.

4.65. **(a)** $\lambda = 3/5$; $x \approx 0.6566204$.

4.66. Suppose that $\mathbf{x}, \mathbf{x}'$ are two distinct fixed points of the contraction f; then $d(\mathbf{x}, \mathbf{x}') = d(f(\mathbf{x}), f(\mathbf{x}'))$.

Chapter 5

5.4. Half of the analysis: suppose $a \in \mathbf{Q}$ and assume that $f'(a)$ exists; let x's be irrational.

5.10. The identity $\cos(a) - \cos(b) = -2\sin\frac{a+b}{2}\sin\frac{a-b}{2}$ is useful.

5.12. **(a)** To get started, let $y \neq f(a)$ be in $\mathbf{U}$; then $y - b = f(g(y)) - f(a) = \phi_a[g(y)]\,(g(y) - g(b))$.

5.14. Use Exercise 5.13(a).

5.19. **(c)** Yes.

5.21. One.

5.22. **(b)** Arrive at $f(b) - f(a) = f'(c)(b-a), c \in (a, b)$.

5.24. **(a)** Write the Mean-Value Theorem separately for the intervals $[0, \theta]$, $[\theta, \pi/4]$.

5.25. **(e)** Do integrations of both sides of the inequality in part (b).

5.26. Begin by writing the equation of the line through B, C in Figure 5.5(a).

5.31. By Corollary 5.7.2, for any nonnegative numbers a, b and any natural numbers m, n,

$$a^n b^m \leq \left(\frac{ma + nb}{m+n}\right)^{m+n}.$$

The natural numbers are continuous with the nonintegral, positive reals.

5.32. Examine $\left(1 - \frac{2}{6r+1}\right)^r > \frac{2}{3}$.

5.34. If f is assumed not to have a relative minimum at c, then there is an $x_0 \in [a, c) \cup (c, b]$ such that $f(x_0) < f(c)$. Consider separately $x_0 \in [a, c), x_0 \in (c, b]$.

5.35. Think of the operator $D \cong \frac{d}{dx}$ as the sum of two operators D_1 and D_2, where D_1 operates only on f and D_2 operates only on g. Endow D_1, D_2 with desired properties.

5.36. **(b)** For all n, $L_n(0) = n!$.

5.38. Working with logarithms will be useful.

5.40. **(a)** $a = ½$

(b) Use the theorem that if f is integrable on some interval $\mathbf{I}$, if $g(x), h(x) \in \mathbf{I}$ are such that $f(g(x)), f(h(x))$ are continuous at x, and if $g'(x), h'(x)$ exist, then

$$\frac{d}{dx}\int_{g(x)}^{h(x)} f(t)dt = f(h(x))h'(x) - f(g(x))g'(x).$$

5.42. Interpret s_6 from Table 3.2 to mean the sum of the first six nonzero terms; $R_{11}(1) < 2.1 \times 10^{-9}$.

5.44. $T_7(\theta) = \theta + \frac{\theta^3}{3} + \frac{2\theta^5}{15} + \frac{17\theta^7}{315}$; $|\tan(1/2) - T_7(1/2)| \approx 4.7 \times 10^{-5}$.

5.45. Mathematical induction.

5.46. Make use of the Ratio Test, Theorem 3.3, Exercise 2.22, and Corollary 5.11.1.

5.47. **(b)** Yes.

5.49. **(b)** Two real roots: $x = \pm 4.2426406$.

5.50. Take the kth root of both sides.

5.51. **(a)** Second part of Property (7): let $x_2 > x_1$ be arbitrary and use Theorem 2.8.

(c) Use Property (5).

5.52. **(b)** Use Property (6) for exponential functions.

(d) The analysis is similar in spirit to that used in Exercise 5.38.

(e) Make use of the Mean-Value Theorem (Theorem 5.7). Consider, separately, the cases $x > 0$ and $-1 < x < 0$.

5.53. See the hint for Exercise 5.38.

5.55. **(a)** Arrive at $xf'(x) = \alpha x + \alpha x\,[f(x) - 1] - x^2 f'(x)$, where $f(x) = 1 + \sum_{k=1}^{\infty}\binom{\alpha}{k}x^k$.

5.56. Arrive at $\frac{1}{n(n+1)} > x_n - x_{n+1} > \frac{1-n^{-1}}{2n(n+1)} \geq 0$.

5.58. Show that $0 < \ln n < \sqrt{n}$ iff $0 < n < \sqrt{n} + \frac{n}{2} + \frac{n\sqrt{n}}{6} < e^{\sqrt{n}}$.

5.59. $|R_6(3/4)| < 0.0201$, but the actual $|R_6(3/4)| = \left|f(3/4) - T_6(3/4)\right| \approx 0.00720$.

5.60. $\sum_{k=0}^{\infty} c_k = -3/4$.

5.61. **(g)** Write $13 = (9/4)^3 + [13 - (9/4)^3]$; apply the Binomial Theorem to this.

5.62. **(e)** 4.

5.63. Use the Mean-Value Theorem.

5.65. **(a)** Arrive at $\frac{F'(z)}{G'(z)} = \frac{(-z^{-2})f'(z^{-1})}{(-z^{-2})g'(z^{-1})} = \frac{f'(x)}{g'(x)}$.

(b) 1.

5.66. The limit is 1.

5.68. **(a)** Make use of the Squeeze Theorem; the substitution $x = u^{-1}$ is useful.

(b) Establish by induction that $g^{(n)}(x) = g(x)\sum_{k=1}^{3n} c(k,n)\left(\frac{1}{x}\right)^k$, where the $c(k,n)$'s are real.

(d) $R = \infty$.

5.70. If $f(x_1, x_2) = x_1 + x_2^2 + 2x_1x_2$, $E(\mathbf{h}) = \frac{2h_1h_2+h_2^2}{\sqrt{h_1^2+h_2^2}}$.

5.73. (b) $\left(\nabla^2 F\right)(\mathbf{p}) = \frac{-1}{\|\mathbf{p}\|^2}$, $\mathbf{p} \neq \mathbf{0}$.

5.74. (a) Assume that $\mathbf{L}_1, \mathbf{L}_2$ are two distinct linear mappings, and that $E_1(\mathbf{h}), E_2(\mathbf{h})$ are the corresponding functions.

(b) For any natural number $k \in [1, n]$, let $\mathbf{h} = t \cdot \mathbf{u}_k$, where $t \neq 0$ is small and $\mathbf{u}_k = (0, 0, \ldots, 1_k, 0, \ldots)$ is a unit basis vector in $\mathbf{R}^n$.

(c) What are $\lim_{\mathbf{h}\to\mathbf{0}} \mathbf{L} * \mathbf{h}$ and $\lim_{\mathbf{h}\to\mathbf{0}} E(\mathbf{h})\,\|\mathbf{h}\|$?

5.77. (a), (b) Use the Mean-Value Theorem.

5.78. Begin by writing Carathéodory's definition of f being differentiable at **a**.

Chapter 6

6.2. Use Exercise 6.1.

6.3. (a) $D^{-1}f = \frac{f(t)}{\ln 4}\left\lfloor \sqrt{2t+1} - \frac{1}{\ln 4}\right\rfloor + C$.

6.4. (a) Use mathematical induction and the trigonometric identity for $\sin(\theta + \phi)$.

(c) l'Hôpital's Rule will be useful.

6.6. (a) $S \approx 1.0338$; $\int_0^{\pi/2} \cos x dx = 1$.

6.8. Make use of the Taylor series for $\sin t$.

6.9. (a) Refer to Example 5.18.

6.12. The integral is less than 1.

6.15. (a) $\prod(x, n, m) = \int_0^{\infty} \frac{\sec\theta d\theta}{(1-n\sin^2\theta)\sqrt{1-m\sin^2\theta}}$, $\phi = \text{Sin}^{-1} x$.

(b) $C \approx 4.85120$.

6.16. (d) $\int_0^x \left(mt\sqrt{\frac{1-t^2}{1-mt^2}}\right)dt = E(x,m) - (1-m)\prod(x, m, m)$.

(f) $F\left(\frac{\sqrt{3}}{2}, \frac{1}{2}\right) \approx 1.1424$; $E\left(\frac{\sqrt{3}}{2}, \frac{1}{2}\right) \approx 0.9650$.

6.18. (a) If f is monotonic increasing, choose $\delta = \varepsilon/[f(b) - f(a)]$.

6.19. (b) Example 6.7(d): Show that on $[0, 4]$ $f(y) = \frac{\ln(1+y)}{y+2}$ is increasing on $[0, 2.59112]$ and decreasing on $[2.59112, 4]$.

6.23. Use Lemma 6.4.3 and the Squeeze Theorem.

6.25. (b) $k \approx 6.426907$.

6.28. Express $H_a(x)$ in two different ways: (1) by using Exercise 6.27(a) and l'Hôpital's Rule, and (2) by applying Corollary 6.13.1 to the numerator of $H_a(x)$.

6.29. Use Theorems 4.10 and 6.2.

6.33. The limit is $\frac{\pi}{4} + \frac{1}{2}\ln 2$.

6.34. The limit is 25/4.

6.36. In the theorem of Exercise 6.35, let $t(x)$ there be defined by $\frac{x}{\sqrt{1-x^2}}$, where $0 \le x \le \frac{c}{\sqrt{1+c^2}}, c > 0$.

6.37. Begin by applying integration by parts to $\int_a^b f(t)g(t)dt(u = g(t), v = \int_a^t f(x)dx = F(t))$.

6.38. **(c)** Use Corollary 6.16.1 and Exercise 4.52.

6.39. **(c)** In the integral form of the remainder in part (a), let $1/n!$ play the role of the $g(x)$ in Exercise 6.29.

6.41. Use the definition in Section 6.2 and Theorem 6.16.

6.44. **(a)** Begin by rewriting $\frac{t^2}{e^t-1}$ as $\frac{t}{\left(\frac{e^t-1}{t}\right)}$; eventually, use Theorem 6.2.

6.45. **(a)** For all $t \in (1, x], x > 1, e^{-t^2} < e^{-t}$ holds. Show that $\operatorname{erf} x$ increases, and use Theorem 2.2 (as adapted for real-valued functions, generally).

6.48. **(b)** Example 6.20(c): converges to $\sqrt[4]{8}$.

6.49. Begin by simplifying the problem to $I = 2\int_0^\infty \frac{u^2du}{1+u^4} = 2\int_0^1 \frac{u^2du}{1+u^4} + 2\int_1^\infty \frac{u^2du}{1+u^4}$. Be prepared to factor $1 + u^4$ in $\mathbf{R}^1$.

6.50. **(b)** Begin with $\pi = 4\int_0^1 \frac{dt}{1+t^2}$, from part (a). Simpson's Rule gives $\pi \approx 3.14159264$.

6.51. **(d)** You are to show that $\lim\limits_{y\to(\pi/2)^-} \sin y = \lim\limits_{y\to(\pi/2)^+} \sin y = \sin(\pi/2)$.

6.52. Two integrations by parts and use of mathematical induction will give explicit formulas for $F_n(s)$ when n is even and n is odd.

6.53. **(c)** Diverges, but has a Cauchy principal value.
(i) Converges.

6.54. **(c)** Factor $1 + t^4$ as the difference of two squares.
(f) Let $t = u - 3$.

6.55. **(a)** $\alpha > 0$.

6.56. **(d)** Begin with $\Gamma'(\alpha) = \int_0^1 t^{\alpha-1}e^{-t}\ln t\,dt + \int_1^\infty u^{\alpha-1}e^{-u}\ln u\,du$.

6.57. **(a)** In $\int_1^\infty \frac{xdx}{x^3+1}$ let $x = t^{-1}$.

6.58. **(c)** Use the simplified version of Stirling's Approximation in Exercise 3.19.

6.59. Pattern the proof after that of Theorem 6.1.

6.60–6.61. Ditto, analogously.

6.62. $S(\mathbf{D}, \overline{P}, f) = 9.60$.

6.64. Establish that $\text{Bd}(\mathfrak{D})$ is actually $\mathbf{T}$.

6.65. First prove that a function $\sigma(\mathbf{p})$ that is defined on a closed, bounded rectangle $\mathbf{R}$ and has a constant value on $\text{Int}(\mathbf{R})$ is Riemann-integrable.

Next, use Theorem 6.5′ to show that $\iint_{\mathbf{R}} \sigma(\mathbf{p}) = \iint_{\text{Int}(\mathbf{R})} \sigma(\mathbf{p})$. Proceed to take $\sigma(\mathbf{p}) = 1$ everywhere in $\mathbf{R}$. Finally, make use of Theorems 6.4′(i), 6.4′(ii).

6.66. Assume that Theorem 4.10 can be extended to real-valued functions defined on compact subsets of $\mathbf{R}^2$.

Chapter 7

7.2. Use the concept of uniformly Cauchy.

7.5. **(b)** Yes; yes.

7.11. **(c)** No.

7.14. See Exercise 7.11.

7.18. **(e)** Establish that $k \geq 25$ implies $\frac{10^k}{k!} < \frac{1}{(k-24)^2}$.

(f) Let $M_k = \frac{1}{\sqrt{c}} k^{-3/2}$.

7.19. **(b)** Estimated value is ≈ 2.1948.

7.20. Begin by writing $x^x = e^{x \ln x}$.

7.24. **(b)** For $k \geq 16$, use $M_k = k^{-3/2}$ in order to show uniform convergence.

7.25. **(b)** All real x.

(c) Agreement is excellent.

7.28. **(a)** Use Corollary 2.13.1.

(d) $\{u_k\}_{k=0}^{\infty}$, $\{v_k\}_{k=0}^{\infty}$ are bounded. Then let $x_k \cong \frac{1}{u_k}$ and $w_k \cong u_k v_k$, so $\lim_{k\to\infty} \sup v_k = \lim_{k\to\infty} (x_k w_k)$.

7.29. Use Exercise 7.28.

7.30. $\ln 2 \approx \frac{1}{2} + s_9 \approx 0.693147024$.

7.33. $\left|\ln 2 - \left(\frac{5}{8} + \frac{1}{2}\sum_{k=0}^{100} (-1)^k \frac{1}{(k+1)(k+2)(k+3)}\right)\right| < \frac{b_{99}}{2} = \frac{1}{2}\left[\frac{1}{2(100)(101)(102)}\right] \approx 2.43 \times 10^{-7}$.

7.34. **(c)** $\pi \approx 3\left[1 + \sum_{k=1}^{n} \frac{(2k-1)!!}{8^k k!(2k+1)}\right] = 3s_n \approx 3.14159198$.

If we require $|\pi - 3s_n| < \frac{3b_n}{2} < 1 \times 10^{-8}$, then $\frac{(2n-1)!!}{8^n n!(2n+1)} < 0.67 \times 10^{-8}$, and $n = 11$ will suffice.

7.35. (a) Use Raabe's Test (Exercise 3.22).

(b) Not useful.

(c) Let $x = \frac{1}{2}\sqrt{2-\sqrt{3}}$ in the series for $\mathrm{Sin}^{-1}x$ from Exercise 7.34(a) (upon integration). Six terms yield $\pi \approx 3.141592654$.

7.36. (b) Establish by induction that $\frac{(2k-3)!!}{2^k k!} < \frac{1}{k^{3/2}}$ if $k \geq 2$, and take $M_k = k^{-3/2}$.

(c) Prove the recursive relationship as in Exercise 7.25(a). The explicit formula is $\int_0^{\pi/2} \sin^{2k}\varphi d\phi = \frac{\pi}{2}\frac{(2k)!}{2^{2k}(k!)^2}$.

(d) The estimated value is $E(1, 1/4) \approx 1.46746261$.

7.37. (a) $(1-x^2)y'(x) - xy(x) - 1 = 0$.

(b) In (a), let $y(x) = \sum_{n=0}^{\infty} a_n x^n$. Use the facts that $\mathrm{Sin}^{-1}x$ is an odd function of x, and $\sqrt{1-x^2}$ is an even function of x.

(c) $R = 1$.

(d) $\frac{\pi}{4} \approx 0.785277$, if the pertinent series is truncated to 10 terms.

(e) Use Raabe's Test to show that the series for $y(x)$ diverges at $x = 1$. However, Corollary 7.6.1 applies.

7.38. (b) Abel's Test (Exercise 7.22) is useful.

7.41. (a) Use Theorem 3.11.

(c) Mathematical induction.

7.42. Each term of the indicated series satisfies $\left|\frac{k\sin(k\theta)}{4^k}\right| \leq \frac{k}{4^k} = M_k$.

7.43. The Ratio Test shows that $\sum_{k=1}^{\infty} kc_k(x-a)^{k-1}$ has the same radius of convergence as does $\sum_{k=1}^{\infty} kc_k(x-a)^k$. Apply Exercise 7.28(d) to this latter series.

7.45. $g'(x) = \frac{1}{2}\sum_{n=1}^{\infty} n\binom{1/2}{n}\left(\frac{x}{2}\right)^{n-1}, x \in [-r, r], r \in (0, 2)$.

7.46. (a) $R = 1$, whenever α is not an odd integer.

(b) Use Raabe's Test (Exercise 3.22).

(c) $\cos(\alpha\theta) = \cos(\theta)\left[1 + \sum_{k=1}^{\infty} \frac{(-1)^k}{(2k)!}\left\{\prod_{j=1}^{k}[\alpha^2 - (2j-1)^2]\right\}(\sin^2\theta)^k\right]$.

7.49. (b) Begin with $I = -\int_0^{\pi/2}\left[\ln x + \ln\left(1 + \sum_{k=1}^{\infty}(-1)^k\frac{x^{2k}}{(2k+1)!}\right)\right]dx$, and then use the series for $\ln(1+u)$, where $u \cong \sum_{k=1}^{\infty}(-1)^k\frac{x^{2k}}{(2k+1)!}$. Obtain, finally, $I \approx 1.088759$, to be compared with 1.088793.

7.50. (a) Use the analog of Theorem 7.9 in which D_1 replaces D_2.

(c) If $F(x) = -\ln x + \int_1^{x^2}\frac{e^{-t/x}}{t}dt, x > 0$, differentiation of the integral gives two terms (see Example 7.25).

7.51. (b) Use Fubini's Theorem here.

(c) Use the Fundamental Theorem of the Calculus—B in this step.

7.52. Get to $\int_{a_1}^{a_2} D_2 f(x,y)dx = \int_{a_1}^{a_2} (D_{2,1}g)(\mathbf{p})dx$, where $\mathbf{p} \in \mathbf{D} = \{(x,y) : a_1 \leq x \leq a_2, b_1 \leq y \leq b_2\}$ and $g(\mathbf{p}) \cong \int_{a_1}^{x} f(t,y)dt$.

7.53. (b) Use the test in part (a); the integral $\int_0^\infty e^{-xy}dy$ is uniformly convergent on $\mathbf{D} = [a,b], 0 < a < b < \infty$.

(c) Write $\int_0^\infty \frac{\sin u}{u}du = \int_0^\pi \frac{\sin u}{u}du + \sum_{k=1}^{\infty} \int_{k\pi}^{(k+1)\pi} \frac{\sin u}{u}du$, and use Theorem 6.22. Finally, for the desired uniform continuity, if $\varepsilon > 0$ is given, then require a $\delta > 0$ such that for all $x \in \mathbf{D}$ and all $c \geq \delta$,

$$\left| \int_c^\infty \frac{\sin(xy)}{y}dy \right| < \varepsilon.$$

In this, let $u = xy$.

7.54. (a) Apply (weak) Fubini to $\int_0^c dx \int_{b_1}^{b_2} f(x,y)dy$.

(b) Assume that $F_c(y) \cong \int_c^\infty f(x,y)dx$ is integrable on $b_1 \leq y \leq b_2$. Use Lemma 7.3.1 and the uniform convergence (by hypothesis) of $\int_0^\infty f(x,y)dx$.

(c) Let $y_0 \in [b_1, b_2]$ be fixed (momentarily) and $y_1 \in [b_1, b_2]$ be variable, and let $G(y) \cong \int_0^\infty f(x,y)dx$; let $\varepsilon > 0$ be given. Then look at $|G(y_1) - G(y_0)|$, and use the uniform continuity of f on the rectangle $\mathbf{R}_0 = \{(x,y) : 0 \leq x \leq c, b_1 \leq y \leq b_2\}$, where c is such that $|f_c(y)| < \min\left\{\frac{\varepsilon}{4}, \frac{\varepsilon}{4(b_2 - b_1)}\right\}$.

7.55. (a) Theorem 7.10.

(b) Fundamental Theorem of the Calculus—A.

(c) Fundamental Theorem of the Calculus—B, assuming (or if we prove) that $\int_0^\infty D_2 f(x,t)dx$ is continuous on $b_1 \leq t \leq b_2$.

To prove this latter, argue from the uniform continuity of $D_2 f(x,y)$ on the rectangle $\mathbf{R}_0 = \{(x,y) : 0 \leq x \leq c, b_1 \leq y \leq b_2\}$.

7.56. Begin by writing $\Gamma(\alpha) = \int_1^\infty t^{\alpha-1}e^{-t}dt + \int_0^1 t^{\alpha-1}e^{-t}dt = I_1 + I_2$. Show that I_1 converges uniformly on $[a,b], 0 < a \leq \alpha \leq b < \infty$. For $t \in (0,1]$, use $t^{\alpha-1}e^{-t} \leq t^{\alpha-1}$, and establish that I_2 is uniformly convergent on $[a,b]$. Use similar reasoning to establish the uniform convergence of $\int_0^\infty t^{\alpha-1}e^{-t}(\ln t)dt$ on $[a,b]$. Generalization to derivatives of $\Gamma(\alpha)$ of any order is immediate.

Sample Final Examinations

* * * FINAL EXAMINATION 1 * * *

(closed book; 120 min)

1. (16 pts) **DEFINITIONS**

Provide brief but accurate definitions or statements of the following concepts:

(a) limit of $f : \mathbf{R}^n \to \mathbf{R}^1$ is ∞ as $\mathbf{x}$ approaches $\mathbf{x}_0 \in \mathbf{R}^n$;

(b) open set;

(c) Mean-Value Theorem;

(d) the real-valued function f is uniformly continuous on $\mathbf{S} \subset \mathbf{R}^1$.

2. (16 pts) **EXAMPLES**

Give and explain examples that illustrate each of the following observations:

(a) an interval $\mathbf{I}$ that is not compact and a nonconstant function f that is continuous on $\mathbf{I}$ and attains its supremum on $\mathbf{I}$;

(b) a collection $\mathfrak{C}$ of open sets whose intersection is not open;

(c) a function f that is unbounded on $[0, 1]$ but is integrable there;

(d) a function $f : \mathbf{R}^1 \to \mathbf{R}^1$ that is continuous on all of $\mathbf{R}^1$ and is differentiable everywhere except at two points.

3. (12 pts) **LIMITS**

Determine the following limits and show all work:

(a) $\lim_{n\to\infty} a_n$, if $\{a_n\}_{n=1}^{\infty}$ is defined by $a_n = \left(1 + \frac{1}{3n}\right)^{3n/5}$;

(b) $\lim_{n\to\infty} b_n$, if $\{b_n\}_{n=1}^{\infty}$ is defined by $b_n = (-1)^n \binom{-1/2}{n}$, where $\binom{c}{n} = \frac{c(c-1)(c-2)\cdots(c-n+1)}{n!}$;

(c) $\lim_{n\to\infty} \sum_{k=3}^{n} \frac{1}{k(k-2)}$.

4. (12 pts) **INTEGRALS**

Ascertain the following facts about the indicated integrals:

(a) $\int_{1/3}^{13/4} \frac{\ln t}{1+t^2} dt > 0$;

(b) $\int_0^1 \sqrt{\frac{1+t}{1-t}} dt = \frac{\pi}{2} + 1$;

(c) If $D(n, x)$ is defined by $D(n, x) = \int_0^x \frac{t^n}{e^t - 1} dt$, $n \in \mathbf{N}$ and $|x| < 2\pi$, then $D(2, 3/2) < 3 - 4\ln(7/4)$.

5. (14 pts) **PROOF**

Let $f : (-2, 2) \to \mathbf{R}^1$ be defined by $f(x) = 2x^3 + 3x^2 - 36x + 5$. Prove (a) that f has an inverse function f^{-1}, and (b) determine and justify the value of $(f^{-1})'(-26)$.

6. (14 pts) **PROOF**

Prove the Mean-Value Theorem for Integrals: If f is continuous on $[a, b]$, then there is a point $c \in (a, b)$ such that $\int_a^b f(x)dx = f(c)(b - a)$.

7. (16 pts) **PROOF**

Let $f : [0, 2] \to \mathbf{R}^1$ be defined by $f(x) = \sum_{k=0}^{\infty} (-1)^k \frac{x^{2k+1}}{(2k+2)!}$. Prove that $\int_0^2 f(x)dx = \sum_{k=0}^{\infty} (-1)^k \left(1 + \frac{1}{2k+2}\right) \frac{4^{k+1}}{(2k+3)!}$.

8. (5 pts; not required; no partial credit) **BONUS**

State the Continuum Hypothesis. What is the status of this proposition?

* * * FINAL EXAMINATION 2 * * *

(closed book; 120 min)

1. (16 pts) **DEFINITIONS**

Provide brief but accurate definitions or statements of the following concepts:

(a) Cauchy sequence;

(b) the series $\sum_{k=0}^{\infty} \mathbf{x}_k$ is absolutely convergent in $\mathbf{R}^n$;

(c) Heine-Borel Theorem;

(d) Fundamental Theorem of the Calculus.

2. (16 pts) **SHORT ANSWER**

(a) If a sequence is monotonic and bounded, then what do we know for sure about it? Explain.

(b) If $\mathbf{S} = \mathbf{R} \backslash \mathbf{Q}$, then

(i) is $\mathbf{S}$ open, closed, both, or neither?

(ii) is $\mathbf{S}$ compact?

(iii) what is the interior of **S**?

(iv) what are the boundary points of **S**?

(c) If $f : \mathbf{R}^1 \to \mathbf{R}^1$ is defined by

$$f(x) = \begin{cases} x & x \in Q \\ -x & x \notin Q, \end{cases}$$

where (if at all) is f continuous?

(d) If $\Gamma(\alpha) = \int_0^\infty t^{\alpha-1}e^{-t}dt$, for which $\alpha \in \mathrm{R}$ does the integral converge?

3. (12 pts) **SERIES**

(a) A series $\sum_{n=1}^{\infty} a_n$ has the sequence of partial sums $\{s_n\}_{n=1}^{\infty}$, where each $s_n = (-1/3)^n$. Determine the general term a_n.

(b) Determine whether $\sum_{n=1}^{\infty} \frac{n!\, n^2}{(2n)!}$ converges or not; explain.

(c) If $\sum_{n=1}^{\infty} x_n$ converges absolutely, then explain how you know that $\sum_{n=1}^{\infty} \frac{x_n^2}{x_n^2+3}$ converges.

4. (12 pts) **INTEGRALS**

Ascertain the following facts about the indicated integrals:

(a) Determine $\lim_{x\to 0^+} x^8 \left[\int_0^x t(1-\cos t)dt\right]^{-1}$;

(b) Apply the Integral Test to determine if $\sum_{k=1}^{\infty} \frac{\mathrm{Sec}^{-1}(2k)}{k^2}$ converges or diverges;

(c) Verify that $(P) \int_{-\infty}^{\infty} \frac{1+t}{1+t^2}dt = \pi$.

5. (14 pts) **PROOF**

Let **S** be the set of numbers x such that $0 < x < 1/2$. Prove that **S** is uncountably infinite.

6. (14 pts) **PROOF**

The function f is defined by $f(x) = x^3 + x - 3,\ x \in [0, 4]$. Prove that f is uniformly continuous on its indicated domain.

7. (16 pts) **PROOF**

Suppose that $\{f_k(x)\}_{k=0}^{\infty}$ and $\{g_k(x)\}_{k=0}^{\infty}$ are uniformly convergent on a common domain $\mathbf{D} \subseteq \mathbf{R}^1$. Prove that $\{h_k(x)\}_{k=0}^{\infty}$, where $h_k(x) = f_k(x) + g_k(x)$ is also uniformly convergent on **D**.

8. (5 pts; not required; no partial credit) **BONUS**

State the Riemann Hypothesis. What is the status of this proposition?

* * * FINAL EXAMINATION 3 * * *

(closed book; 120 min)

1. (16 pts) **DEFINITIONS**

   Provide brief but accurate definitions or statements of the following concepts:

   (a) limit superior of a sequence $\{x_n\}_{n=1}^{\infty}$ in $\mathbf{R}^1$;

   (b) Completeness Axiom;

   (c) Cauchy principal value of an improper integral of the second kind;

   (d) Weierstrass's M-Test.

2. (16 pts) **EXAMPLES**

   Give and explain examples that illustrate each of the following observations:

   (a) a sequence of irrational numbers that converges to a rational number;

   (b) a power series in x that represents $\ln 3$ when x is assigned a judicious value;

   (c) a fourth-degree polynomial $P(x)$ with leading coefficient 1 and such that 2 is a root of $P(x) = 0$ and also of $P'(x) = 0$, but is not a root of $P''(x) = 0$;

   (d) a simple relationship that connects $I_n = \int_0^{\pi/2} \sin^n x\, dx$ with I_{n-2}, $n \geq 2$.

3. (12 pts) **LIMITS AND SERIES**

   (a) Determine

$$\lim_{n\to\infty} \sum_{k=1}^{n} \left[\sin\left(\frac{\pi/2}{k}\right) - \cos\left(\frac{\pi/2}{k}\right) - \sin\left(\frac{\pi/2}{k+2}\right) + \cos\left(\frac{\pi/2}{k+2}\right) \right];$$

   (b) If $\sum_{k=1}^{\infty} \frac{1}{k^2} = \frac{\pi^2}{6}$, then determine $\sum_{k=1}^{\infty} \frac{1}{(2k-1)^2}$, and provide justification for your procedure;

   (c) If $\theta \in (0, \pi/2]$, then prove that $\left(\frac{\sin\theta}{\theta}\right)^3 > \cos\theta$.

4. (12 pts) **INTEGRALS**

   Ascertain the following facts about the indicated integrals:

   (a) $e < \int_2^3 \frac{x dx}{\ln x} < \frac{23}{8}$;

   (b) If f is integrable on $[a, b]$, $b > a$, then

$$\int_a^b f(x)dx = \int_a^b f(a + b - x)dx;$$

   (c) $\lim_{n\to\infty} \left[n \sum_{k=0}^{n-1} \frac{1}{n^2+k^2} \right] = \pi/4.$

5. (14 pts) **PROOF**
Establish, with justification, that $\int_0^\infty \frac{dt}{(e^t+1)(e^{-t}+1)} = 1/2$.

6. (14 pts) **PROOF**
Suppose that $\mathbf{S}$, $\mathbf{T}$ are bounded subsets of $\mathbf{R}$ and that $\mathbf{S} \subseteq \mathbf{T}$. Prove that $\inf \mathbf{T} \le \inf \mathbf{S}$.

7. (16 pts) **PROOF**
Consider the sequence $\{y_n\}_{n=1}^{\infty}$, where $y_n = \sum_{k=1}^{n} \frac{1}{k} - \ln n$. Prove that this is a decreasing sequence and that it converges.

8. (5 pts; not required; no partial credit) **BONUS**
What is meant by the Fourier transform of a function $f(x)$? Then state two questions about such a transform to which we might wish to have answers.

Supplementary Problems for Further Study[1]

C.1 SEQUENCES

1. The sequence $\{u_n\}_{n=0}^{\infty}$ is defined by

$$u_n = \begin{cases} 1 & n = 0 \\ \frac{u_0}{0!} + \frac{u_1}{1!} + \cdots + \frac{u_{n-1}}{(n-1)!} & n > 0. \end{cases}$$

Show that the sequence converges and that the limit L satisfies $14/4 < L < 15/4$.

2. We let the symbol $\lfloor x \rfloor$ denote (sometimes known as the **greatest integer function**, or the **floor function**) the largest integer that does not exceed x. The sequence $\{P(N)\}_{N=1}^{\infty}$ is defined by the statement that $P(N)$ is the number of partitions of the positive integer N that consist only of 2's and 3's. For example, $P(6) = 2$ because $6 = 3 + 3 = 2 + 2 + 2$. It is known that

$$P(N) = \begin{cases} \lfloor \frac{N}{6} \rfloor & N \equiv 1 (\mathrm{mod} 6) \\ \lfloor \frac{N}{6} \rfloor + 1 & N \equiv 0, 2, 3, 4, 5 (\mathrm{mod} 6). \end{cases}$$

Reexpress $P(N)$ as a single formula.

3. The sequence $\{f(n)\}_{n=1}^{\infty}$ is defined by the statement that $f(n)$ is the product of all the positive divisors of n. When n, m are distinct, can it ever happen that $f(n) = f(m)$?

4. Each side of an equilateral triangle is partitioned into n equal subintervals. All possible straight lines interior to the triangle are drawn parallel to the sides. For example, when $n = 3$, the 3-triangle shown is obtained. Then the sequence $\{u_n\}_{n=1}^{\infty}$ is defined by the statement that u_n is the number

---

[1] Sources, or leading references, appear in the *Instructor's Solution Manual to Accompany Advanced Calculus*.

of triangles present in the n-triangle, including the n-triangle itself. Thus, $u_1 = 1$ and $u_2 = 5$. Compute u_{20}, and find the general formula.

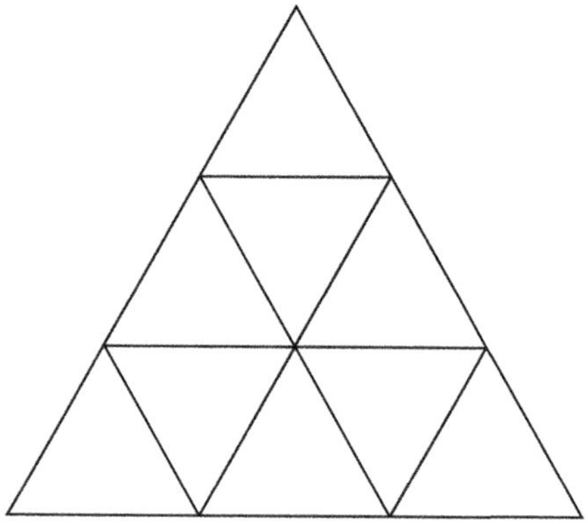

5. A sequence of polynomials, $\{P_n(x)\}_{n=0}^{\infty}$, is defined by the relation

$$\sum_{n=0}^{\infty} P_n(x)\frac{z^n}{n!} = \exp\left(xz - \frac{1}{2}z^2\right).$$

Derive a relationship that connects $P_{n+1}(x)$, $P_n(x)$, and $P_{n-1}(x)$.

6. Two sequences $\{a_n\}_{n=0}^{\infty}$ and $\{b_n\}_{n=0}^{\infty}$ are defined as follows:

$$a_n = \begin{cases} 3\sqrt{3} & n = 0 \\ 2\left(\frac{a_{n-1}b_{n-1}}{a_{n-1}+b_{n-1}}\right) & n > 0 \end{cases} \qquad b_n = \begin{cases} \frac{3}{2}\sqrt{3} & n = 0 \\ \sqrt{a_{n-1}b_{n-1}} & n > 0. \end{cases}$$

Show that both sequences converge, and to the same limit. What is this limit?

7. Show that the interesting sequence $\{u_n\}_{n=1}^{\infty}$, where

$$u_n = \begin{cases} c & n = 1 \\ c^{u_{n-1}} & n > 1, \end{cases}$$

is convergent for $c > 0$ iff $e^{-e} \leq c \leq e^{1/e}$.

8. A sequence $\{a_n\}_{n=1}^{\infty}$ of real numbers is defined by $a_{n+1} = |a_n| - a_{n-1}$, $n > 0$, $a_0, a_1 \in \mathbf{R}$. Show that regardless of the choice of a_0 and a_1, the sequence is periodic with period 9.

9. The sequence of polynomials $\{P_n(x)\}_{n=0}^{\infty}$ is defined by

$$P_n(x) = \begin{cases} 1 & n = 0 \\ x & n = 1 \\ 2xP_{n-1}(x) - P_{n-2}(x) & n > 1. \end{cases}$$

Show that the polynomials are commutative under function composition, that is, for any nonnegative integers m, n we have $P_m[P_n(x)] = P_n[P_m(x)]$.

10. The sequence of rep-units is given by $\{R_n\}_{n=1}^{\infty}$, where $R_n = (10^n - 1)/9$. It is known that for $n \le 10000$, R_n is prime for just five choices of n. Frequently, members of the sequence $\{H_m\}_{m=1}^{\infty}$, where $H_m = 10^{2m} - 10^m + 1$, are divisors of some R_n. Establish these properties:

(a) If $n = 6m$, then H_m divides R_n.

(b) If p is any odd prime, then H_p is composite and, in particular, 13 divides H_p if $p > 3$.

(c) If $m = 3p$, where $p > 3$ is a prime, then 19 divides H_m.

11. Define the sequences $\{x_n\}_{n=0}^{\infty}$ and $\{y_n\}_{n=0}^{\infty}$ by the relation $x_n + y_n\sqrt{2} = (3 + 2\sqrt{2})^n$.

(a) Show that any ordered pair (x_n, y_n) is a solution of $x^2 - 2y^2 = 1$.

(b) A third sequence, $\{q_n\}_{n=0}^{\infty}$, is defined by

$$q_n = \left(2+\sqrt{2}\right)\left(3+2\sqrt{2}\right)^n + \left(2-\sqrt{2}\right)\left(3-2\sqrt{2}\right)^n.$$

Show that for those $n \ge 0$ that make q_n a perfect square, it is also true that $x_n + y_n$ is a perfect square.

(c) Find two n that do as indicated in part (b).

12. Establish whether the sequence $\{a_n\}_{n=1}^{\infty}$, where

$$a_n = \left[(n+1)^{n+1}(n-1)^n\right] \Big/ n^{2n+1},$$

is decreasing, increasing, or neither.

13. A sequence of numbers, $\{C_n\}_{n=0}^{\infty}$, is defined by

$$C_n = \frac{1}{n+1}\binom{2n}{n}.$$

(a) Show that every C_n is a positive integer.

(b) Establish that the C_n's also are given by the generating function

$$\sum_{n=0}^{\infty} C_n x^n = \frac{1-\sqrt{1-4x}}{2x}.$$

14. A primitive Pythagorean triangle is a right triangle, all of whose sides are of integral lengths, and the only common divisor of the three lengths is 1. For example, $(36, 77, 85)$ are the dimensions (x, y, z), $x < y < z$, of a primitive Pythagorean triangle. We note that the smallest leg is even and is a perfect square. Are there infinitely many primitive Pythagorean triangles with this property, or is the sequence of such triangles finite?

15. Let r be an as-yet unspecified positive constant and let $C \geq 0$ be specified. A sequence $\{c_n\}_{n=0}^{\infty}$ is defined recursively as follows:

$$c_n = \begin{cases} C & n = 0 \\ c_{n-1} + r + \left[c_{n-1} \Big/ \sqrt{1 + c_{n-1}^2}\right] & n > 0. \end{cases}$$

For which choices of r does the sequence converge?

16. What might be called the Tanny sequence, $\{T(n)\}_{n=0}^{\infty}$, is defined by

$$T(n) = \begin{cases} 1 & n = 0, 1, 2 \\ T(n - 1 - T(n - 1)) + T(n - 2 - T(n - 2)) & n > 2. \end{cases}$$

Prove that this sequence is "well-behaved" (not erratic) and that $\lim_{n \to \infty} T(n)/n = 1/2$.

17. Let the sequence $\{D_n\}_{n=2}^{\infty}$ be defined by the relation

$$D_n = \ln n - 2 \sum_{k=2}^{n} \frac{1}{2k - 1}.$$

Prove that the sequence is an increasing sequence, and find the value of $\lim_{n \to \infty} D_n$.

18. A sequence $\{x_n\}_{n=1}^{\infty}$ is defined by

$$x_n = \begin{cases} \mathrm{Tan}^{-1} 2 & n = 1 \\ x_{n-1} + \mathrm{Tan}^{-1} \frac{8n}{n^4 - 2n^2 + 5} & n > 1. \end{cases}$$

Find the value of $\lim_{n \to \infty} x_n$.

19. Define the sequence $\{D_n\}_{n=1}^{\infty}$ by the statement that D_n is the greatest common divisor of the four entries in $\mathbf{M}^n - \mathbf{I}$, where $\mathbf{M}^n$ implies matrix multiplication and

$$\mathbf{M} = \begin{pmatrix} 3 & 2 \\ 4 & 3 \end{pmatrix}, \quad \mathbf{I} = \begin{pmatrix} 1 & 0 \\ 0 & 1 \end{pmatrix}.$$

For example, $\mathbf{M}^2 - \mathbf{I} = \begin{pmatrix} 16 & 12 \\ 24 & 16 \end{pmatrix}$ and $D_2 = \mathrm{GCD}(12, 16, 24) = 4$. Prove that $\lim_{n \to \infty} D_n = \infty$.

20. A sequence of rectangles is constructed as follows: Rectangle 1 is a square of area 1. Then we adjoin, alternately alongside of or on top of the previous rectangle, rectangles of unit area so as to create new rectangles of area 2, 3, 4, 5, and so on.

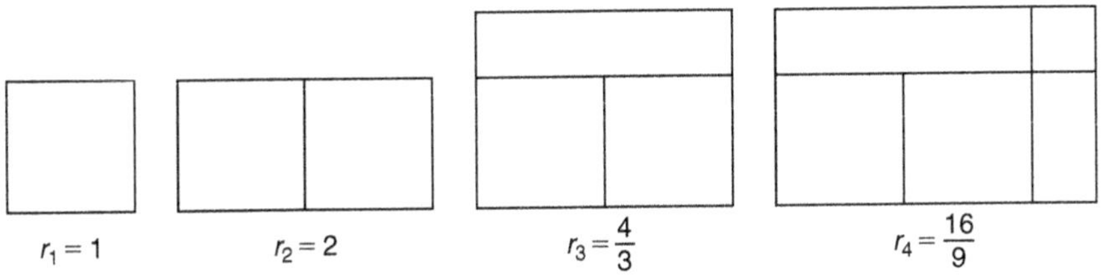

Each term of the sequence $\{r_n\}_{n=1}^{\infty}$ is the ratio of the length to the height of rectangle n. The first four terms of this sequence are as follows:
$r_1 = 1 \qquad r_2 = 2 \qquad r_3 = 4/3 \qquad r_4 = 16/9.$
Find $\lim_{n\to\infty} r_n$.

21. The sequence $\{x_n\}_{n=2}^{\infty}$ is defined by the statement that x_n is the value of the determinant of the $(n-1) \times (n-1)$ matrix $\mathbf{M}_n$ shown below. Now consider the allied sequence $\{x_n/n!\}_{n=2}^{\infty}$. Is it bounded?

$$\mathbf{M}_n = \begin{pmatrix} 3 & 1 & 1 & 1 & \dots & 1 \\ 1 & 4 & 1 & 1 & \dots & 1 \\ 1 & 1 & 5 & 1 & \dots & 1 \\ 1 & 1 & 1 & 6 & \dots & 1 \\ \vdots & \vdots & \vdots & \vdots & & \vdots \\ 1 & 1 & 1 & 1 & \dots & 1 \end{pmatrix}$$

22. A sequence $\{a_n\}_{n=1}^{\infty}$ is defined as follows:

$$a_n = \begin{cases} \sqrt{2} & n = 1 \\ \sqrt{2 + a_{n-1}} & n > 1. \end{cases}$$

Determine $\lim_{n\to\infty} 4^n(2 - a_n)$.

23. Let $\{P_n(x)\}_{n=1}^{\infty}$ be the sequence of polynomials defined by

$$P_n(x) = \begin{cases} x - 1 & n = 1 \\ x^2 - x - 1 & n = 2 \\ xP_{n-1}(x) - P_{n-2}(x) & n > 2. \end{cases}$$

Prove that for each n the roots of the equation $P_n(x) = 0$ are the n real numbers $2\cos[(2k-1)\pi/(2n+1)]$, $k = 1, 2, \dots, n$.

24. For the sequence $\{r_n\}_{n=0}^{\infty}$, defined by

$$r_n = \begin{cases} 3 & n = 0 \\ r_{n-1}^2 - 2 & n > 0, \end{cases}$$

evaluate the limit

$$\lim_{N\to\infty}\left[\prod_{n=0}^{N-1} r_n\right]^{1/2^N}.$$

25. A sequence $\{\beta_n\}_{n=1}^{\infty}$ is defined by the requirement

$$\left[\frac{(n+1)^2}{n(n+2)}\right]^{\beta_n} = \left[\frac{n(n+2)}{(n+1)^2}\right]^{n+1}\left(\frac{n+1}{n}\right).$$

The allied sequence $\{a_n\}_{n=1}^{\infty}$ is defined by $a_n = (1+1/n)^{n+c}$, where $0 < c < 1$. Establish the following:

(a) For each n, the value of β_n is the value of c required for $a_n = a_{n+1}$.

(b) The sequence $\{\beta_n\}_{n=1}^{\infty}$ is an increasing sequence.

(c) $\lim_{n\to\infty} \beta_n = 1/2$.

26. Let the sequence $\{z_n\}_{n=0}^{\infty}$ be defined by

$$z_n = \begin{cases} 1 & n = 0 \\ (1 + z_0^2 + z_1^2 + \cdots + z_{n-1}^2)/n & n > 0. \end{cases}$$

The table below gives the first 10 members of the sequence. Is z_n always a natural number?

n	z_n	n	z_n
0	1	5	28
1	2	6	154
2	3	7	3520
3	5	8	1551880
4	10	9	267593772160

C.2 INFINITE SERIES

1. Let $\{F(n)\}_{n=0}^{\infty}$ be the famous Fibonacci sequence:

$$F(n) = \begin{cases} 0 & n = 0 \\ 1 & n = 1 \\ F(n-1) + F(n-2) & n > 1. \end{cases}$$

Evaluate the sum of the series

$$\sum_{k=0}^{\infty} \frac{1}{F(2^k)}.$$

2. The following infinite series was designated by D.H. Lehmer in 1985 as "interesting":

$$1+\frac{1}{4}+\frac{3}{32}+\frac{5}{128}+\cdots+\frac{\binom{2n}{n}}{8^n}+\cdots.$$

Evaluate the sum of this series.

3. Let the function L be defined by the statement that $L(n)$ is the number of large digits in the positive integer n, where the large digits are 5, 6, 7, 8, 9. For example, $L(156172)=3$. Prove that the following series has the indicated sum:

$$\sum_{k=0}^{\infty}\frac{L(2^k)}{2^k}=\frac{2}{9}.$$

4. Establish the following summation result:

$$2\sqrt{3}\sum_{n=0}^{\infty}\frac{(-1)^n}{3^n(2n+1)}=\pi.$$

5. Prove that the following series converges, and then find its sum.

$$\sum_{n=1}^{\infty}\mathrm{Tan}^{-1}(2/n^2).$$

6. The triangular numbers, $\{T_n\}_{n=1}^{\infty}$, have the form $T_n=n(n+1)/2$. Find the sum of the infinite series

$$\frac{3}{T_1}+\frac{1}{T_2}-\frac{1}{T_3}-\frac{1}{T_4}+\cdots,$$

where all the succeeding numerators are 1's, the denominators are the T_n's, and the signs change after every two terms $(+,+,-,-,+,+,$ etc.).

7. The infinite series expansion of $\sqrt{2x^2-3x^3}$, $|x|<2/3$, is written down. Express in closed form the coefficient of x^{1997}.

8. The Fibonacci polynomials, $\{U_n(x)\}_{n=1}^{\infty}$, are defined by

$$U_n(x)=\begin{cases}1 & n=1\\ x & n=2\\ xU_{n-1}(x)+U_{n-2}(x) & n>2.\end{cases}$$

Find a generating function for the $U_n(x)$'s, that is, obtain a function $f(x,y)$ such that formally (no consideration of convergence), we have

$$f(x,y)=\sum_{n=1}^{\infty}U_n(x)y^n.$$

9. Prove that

$$\sum_{k=1}^{\infty}\left(1+\frac{1}{2}+\frac{1}{3}+\cdots+\frac{1}{k+1}\right)\frac{1}{k(k+1)}=2.$$

10. The following is a nice generalization of the alternating harmonic series for ln 2:

$$\ln k=\sum_{n=1}^{\infty}\left[\left(\sum_{j=1}^{k-1}\frac{1}{kn-(k-j)}\right)-\frac{k-1}{kn}\right],\qquad k=2,3,4,\ldots.$$

Establish it.

11. Establish that for any positive integer N and any fixed $x>0$

$$\sum_{n=1}^{N}\frac{1}{(1+nx)[1+(n+1)x]}=\frac{N}{(1+x)[1+(N+1)x]}.$$

Then make use of this to prove that

$$\frac{1}{1^2\cdot 2}+\frac{1}{2^2\cdot 3}+\frac{1}{3^2\cdot 4}+\frac{1}{4^2\cdot 5}+\cdots=\frac{\pi^2-6}{6}.$$

12. Does the following rearrangement of the alternating harmonic series converge?

$$1-\frac{1}{2}-\frac{1}{4}-\frac{1}{6}-\frac{1}{8}+\frac{1}{3}-\frac{1}{10}-\frac{1}{12}-\frac{1}{14}-\frac{1}{16}+\frac{1}{5}-\cdots.$$

If it does, then to what does it converge?

13. A rather beautiful sum is obtained for the series

$$\sum_{n=0}^{\infty}\frac{1}{(4n+1)(4n+3)}.$$

What is it?

14. Again, if the sequence $\{F(n)\}_{n=1}^{\infty}$ is defined as in Problem 1, evaluate the sum of the alternating series

$$\sum_{n=1}^{\infty}\frac{(-1)^{n-1}}{F(n)F(n+2)}.$$

15. Show that the series $\sum_{m=1}^{\infty}\left[(2m+1)\ln\left(1+\frac{1}{m}\right)-2\right]$ converges to a positive number less than 1/6.

16. For each integer $n \geq 1$, let $D(n)$ be the number of digits in the base-10 expansion of 2^n. Prove that

$$\sum_{n=1}^{\infty}\frac{D(n)}{2^n} > \frac{1169}{1023}.$$

17. Let $\sum_{n=1}^{\infty} a_n$ be a convergent series of nonnegative numbers whose sum is A, and let s_n denote the nth partial sum. Show that the related series $\sum_{n=1}^{\infty} na_n$ converges iff $\sum_{n=1}^{\infty}(A-s_n)$ converges.

18. Prove that the two series

$$\sum_{n=1}^{\infty}\frac{\sin(n)}{n} \quad \text{and} \quad \sum_{n=1}^{\infty}\left[\frac{\sin(n)}{n}\right]^2$$

amazingly converge to the same number, namely, $(\pi-1)/2$.

19. Define $I(n)$, for each positive integer n, to be the integer closest to $\sqrt[3]{n}$. For example, $I(10)=2$ and $I(100)=5$. Evaluate the sum $\sum_{n=1}^{\infty}[I(n)]^{-4}$.

20. Evaluate: $\lim\limits_{x\to 1^-}\sum_{n=0}^{\infty} x^n \ln\left[\frac{1+x^{n+1}}{1+x^n}\right]$.

21. The sum of the series $\sum_{n=4}^{\infty}\left[\sum_{k=2}^{n-2}\binom{n}{k}^{-1}\right]$ is a rational number. What number is it?

22. Let a satisfy $0<a<1$ and β satisfy $0<\beta<\pi/2$. Obtain the sum of the series

$$\sum_{k=0}^{\infty} a^k(\sin\theta + k\beta)$$

as a function of θ and the parameters a, β. When $\theta=\pi/6$ and $\beta=\pi/3$, which values of a in $(0,1)$ will yield a sum of 1 for the series?

23. Seemingly unimportant changes in the terms of a series can have a dramatic effect on the nature of the sum. Thus, $\sum_{n=2}^{\infty}\frac{1}{n^2}$ has a transcendental sum, but if n^2+3n-4 replaces n^2, the new series has a rational sum. Find it.

24. The unique real solution of $x-\cos x=0$ has been referred to as the Dottie number, d. If this number is expressed as a power series in π, $d=\sum_{k=1}^{\infty} c_k\pi^k$, then determine the first three nonzero terms.

25. Evaluate $y=\lim\limits_{n\to\infty}\prod\left(1+\frac{k}{n}\right)^{\sqrt{n/k^3}}$.

C.3 INTEGRALS

1. For each natural number n let $I_n = \int_1^\infty \frac{dx}{1+x^{n+1}}$. Prove that

$$\frac{\ln 2}{n} < I_n < \frac{\ln 2}{n} + \frac{1}{4n^2}.$$

2. Let n be a natural number. Show that the value of the integral

$$\int_0^{\pi/2} \frac{\sin^n x}{\sin^n x + \cos^n x} dx$$

is independent of n.

3. In the figure shown, the line $y = y_0$ is drawn so that area A = area B. Determine y_0.

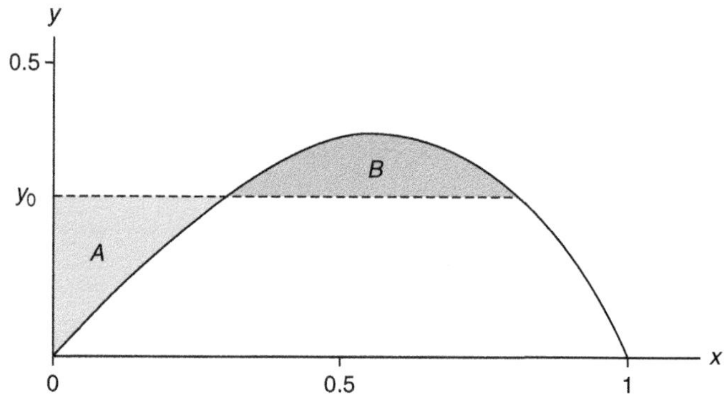

4. Suppose that the function $F : [1, \infty) \to (e, \infty]$ is an increasing function and that $\int_1^\infty \frac{dx}{F(x)}$ diverges to ∞. Prove that the allied integral

$$\int_1^\infty \frac{dx}{x \ln[F(x)]}$$

also diverges to ∞.

5. Determine the value of $\int_0^{\pi/2} \frac{dx}{1+[\tan x]^{\sqrt{2}}}$.

6. Let $\{m_1, m_2, m_3\}$ be any set of three natural numbers. What is the maximum possible value for

$$\int_0^{2\pi} \cos(m_1 x) \cos(m_2 x) \cos(m_3 x) dx?$$

7. Show that the following integral has a value that is the product of a rational number and a power of π:

$$\int_0^{\pi} \frac{x^3}{e^x + 1} dx.$$

8. Let $c \in \mathbf{R}$ be arbitrary but fixed. Determine the value of

$$\int_0^{\infty} \frac{dx}{(1 + x^2)(1 + x^c)}.$$

9. Show that if n is any natural number, then the integral $\int_0^3 \frac{1+x^n}{(1+x)^{n+2}} dx$ has the value $\frac{1}{n+1}\left[1 + \frac{3^{n+1}-1}{4^{n+1}}\right]$.

10. Let the symbol $\{x\}$ denote $x - \lfloor x \rfloor$, that is, the decimal part of x. Prove that the integral

$$I = \int_0^1 \left\{\frac{1}{x}\right\} \ln x \, dx$$

has the value $\gamma + \gamma_1 - 1$, where $\gamma = \lim_{n\to\infty} \left[\sum_{k=1}^{n} \frac{1}{k} - \ln n\right]$ is Euler's constant and $\gamma_1 = \lim_{n\to\infty} \left[\sum_{k=1}^{n} \frac{\ln k}{k} - \frac{(\ln n)^2}{2}\right]$ is the first generalized Euler's constant.

11. Prove that $\int_0^1 \frac{\ln(x+1)}{x^2+1} dx = \frac{\pi}{8} \ln 2.$

Index of Special Symbols

Symbol	Meaning	Section No.
$\oplus$	Addition (of, in): improper integrals; infinite series; a ring; sequences; a vector space	6.17; 3.1; Exercise 5.7; 2.3; 1.5
$\approx$	Approximately equal to	2.2
$A(\mathfrak{D})$	Area of set $\mathfrak{D} \subset \mathbf{R}^2$	6.8
$\mathbf{B}(a; r)$; $\mathbf{B}_d(a; r)$	Ball of radius r and centered at a in a metric space $<\mathbf{M}, d>$	2.1
$\mathbf{B}_n(\mathbf{p}_0; r)$	Ball in $\mathbf{R}^n$, $n > 1$	1.5
$\mathbf{B}_n(a; \delta)\backslash\{a\}$	Deleted n-ball	1.9
$\{B_n\}_{n=0}^{\infty}$	Sequence of Bernoulli numbers	3.6
$\{B_n(x)\}_{n=0}^{\infty}$	Sequence of Bernoulli polynomials	Exercise 3.48
$\binom{n}{j}$	Binomial coefficient; $\binom{n}{j} = \frac{n!}{(n-j)!\,j!}$	Exercise 2.6
$\text{Bd}(\mathbf{S})$	Boundary of a set $\mathbf{S}$	1.8
$\{\mathfrak{B}_n\}_{n=1}^{\infty}$	Sequence of closed, bounded boxes in $\mathbf{R}^m$	2.4
$\mathbf{D} \times \mathbf{S}$	Cartesian product of sets $\mathbf{D}$, $\mathbf{S}$	1.6
$\lceil x \rceil$	Ceiling function of $x \in \mathbf{R}$	2.5
$\overline{\mathbf{S}}$	Closure of a set S; $\mathbf{S} \cup \text{Bd}(\mathbf{S})$	1.8
$\mathfrak{C}$	Collection of sets; parameterized curve	4.5; 5.4
C	Set of all complex numbers	1.1
$g[f]$	Composition of functions g, f	Exercise 1.42
$\wedge$	Conjunction of two propositions	1.1
Corollary X.Y	Corollary number	2.4
$(D)\int_a^b f$	Darboux integral over the interval $[a, b]$	6.4
D	Differentiation operator	6.1
$\mathfrak{D}$	Closed, bounded set in $\mathbf{R}^2$, on which a function $f(\mathbf{p})$ is to be integrated	6.8
D^{-1}	Antidifferentiation operator	6.1

Symbol	Meaning	Section No.
$\{D_n(x)\}_{n=1}^{\infty}$	Sequence of Debye functions	Exercise 6.44
$\cong$	Is defined by	3.1
$D_1 f, D_2 f$	Partial derivatives of $f(x_1, x_2)$ with respect to x_1, x_2, respectively	4.7; 5.10
$\mid D_{2,1}f, D_{1,2}f$	Mixed second-order partial derivatives of the function f	Exercise 5.77
$\vee$	Disjunction of two propositions	1.1
$d_n(\mathbf{x}, \mathbf{y})$	Metric distance in $\mathbf{R}^n$ between vectors $\mathbf{x}$, $\mathbf{y}$	1.5
$\mathbf{D}(f)$	Domain of a function f	1.6
$\leftrightarrow$	Double implication	1.1
$F(x, m)$, $E(x, m)$, $\Pi(x, n, m)$	Elliptic integral of the first kind, the second kind, and the third kind, respectively	Exercise 6.15
$\emptyset$	Empty set	1.1
$\text{erf}(x)$	Error function; $\frac{2}{\sqrt{\pi}}\int_0^x e^{-t^2}dt$	Exercise 6.45
γ	Euler's constant	Exercises 2.11, 3.32, 5.56
$\phi(n)$	Euler phi-function of $n \in \mathbf{N}$	1.7
$\exists$	Existential quantifier	4.2
$\exp(x)$	$\operatorname{Lim}_{k\to\infty}\left(1+\frac{x}{k}\right)^k = \lim_{k\to\infty}\left(1-\frac{x}{k}\right)^{-k}$; e^x	5.7
$\text{Ext}(\mathbf{S})$	Exterior of a set $\mathbf{S}$	1.8
$\mathbf{F}$	Closed set	4.6
$\lfloor x \rfloor$	Floor function of $x \in R$	3.3
$C(x)$, $S(x)$	Fresnel integrals	6.5
$f : \mathbf{D} \to \mathbf{S}$	Function from set $\mathbf{D}$ into set $\mathbf{S}$	1.6
$f'(a)$	Derivative of $f(x)$ at $x = a$	5.1
$f''(a)$, $f'''(a)$, $\cdots$, $f^{(n)}(a)$	Second, third, and higher-order derivatives of $f(x)$ at $x = a$	5.5
$\{F_k\}_{k=1}^{\infty}$	Fibonacci sequence	2.1
f^{-1}	Inverse function of f	Exercise 1.43
$f^{-1}(\mathbf{H})$	Inverse image of a set $\mathbf{H}$ under a function f	Exercise 1.39
$f\mid\mathbf{S}$	Restriction of a function with respect to its domain ($\mathbf{S} \subset \mathbf{D}(f)$)	1.6
$f(\mathbf{S})$	Direct image of a set $\mathbf{S}$ under the function f	1.6
$\Gamma(\alpha)$	Gamma function; $\int_0^{\infty} t^{\alpha-1}e^{-t}dt$	Exercise 6.55
$\binom{c}{n}$	Generalized binomial coefficient	Exercise 2.10
$\text{grad}\, f$; $\nabla f(\mathbf{p})$	Gradient of f at $\mathbf{p}$	5.10
$x > y$ $(y < x)$	x is greater than y; $[x + (-y)] \in \mathbf{P}$	1.2

Symbol	Meaning	Section No.
$<\mathbf{G},{}^{*}>$	Group	Exercise 4.19
$\{H_k(x)\}_{k=1}^{\infty}$	Sequence of Hermite polynomials	2.1
$I_{\mathfrak{D}}$	Characteristic function for a set $\mathfrak{D}$	6.8
$\rightarrow$	Implication	1.1
$I(f)$	Improper integral	6.7
$\int_a^{\infty} f,\ \int_{-\infty}^{b} f,\ \int_{-\infty}^{\infty} f$	Improper integrals of the first kind	6.7
$\inf \mathbf{S}$	Infimum of the set **S**	1.3
$\sum_{n=1}^{\infty} a_n$	Infinite series; sum of a series	3.1
$\mathbf{x} * \mathbf{y}$	Inner product of vectors **x**, **y**	1.5
$\mathrm{Int}(\mathbf{S})$	Interior of a set **S**	1.8
(a, b); $[a, b]$; $(a, b]$; $[a, b)$	Intervals: open, closed, neither (half-open)	1.5
$\{\mathbf{I}_n\}_{n=1}^{\infty}$	Sequence of intervals	2.4
$\{L_n(x)\}_{n=0}^{\infty}$	Sequence of Laguerre polynomials	Exercise 5.36
$L[f(t)]$	Laplace transform of $f(t)$; $\int_a^{\infty} K(s,t)f(t)\,dt,\ K(s,t) \cong e^{-st},\ a = 0$	Exercise 6.52
Lemma X.Y.Z.	Lemma number	2.3
$\lim\limits_{\mathbf{x}\to\mathbf{a}} f = L$	The limit of the function f as **x** approaches **a** is L	1.9
$\lim\limits_{x\to a^+} f$ $(\lim\limits_{x\to a^-} f)$	Right-hand (left-hand) limit of f as x approaches a from above (below)	Exercises 1.57, 1.58
$\lim\limits_{n\to\infty} \inf x_n$	Limit inferior of the sequence $\{x_n\}_{n=1}^{\infty}$; $\inf \mathbf{E},\ \mathbf{E} = \left\{x \in \mathbf{Re} : \lim\limits_{k\to\infty} x_{n_k} = x\right\}$	2.6
$\lim\limits_{n\to\infty} \sup x_n$	Limit superior of the sequence $\{x_n\}_{n=1}^{\infty}$; $\sup \mathbf{E},\ \mathbf{E} = \left\{x \in \mathbf{Re} : \lim\limits_{k\to\infty} x_{n_k} = x\right\}$	2.6
$\underline{\int_a^b} f$	Lower Darboux integral; $\sup\limits_{P} L(P, f)$	6.4
$L(P, f)$	Lower Darboux sum; $\sum\limits_{k=1}^{n} m_k \,\Delta\, a_k$	6.4
$<\mathbf{M}, d>$	Metric space (general)	2.1
$\otimes$	Multiplication (in, of): a ring; infinite series (Cauchy); sequences in $\mathbf{R}^1$	Exercise 5.7; 3.5; Exercise 2.43
$\bullet$	Scalar multiplication (in, of): an improper integral; an infinite series; a sequence; a vector space	6.7; 3.1; 2.3; 1.5
$\mathbf{N}$	Set of all natural numbers	1.1

Symbol	Meaning	Section No.
$\sim p$	Negation of a proposition p	1.1
$\mathfrak{N}$	Norm of a partition	6.1
$\|\|\mathbf{x}\|\|$	Norm of a vector $\mathbf{x}$	1.5
$\mathbf{\Theta}$	Open cover of a set $\mathbf{S}$ in a metric space; $\{\mathbf{O}_\alpha\}$	4.7
$\mathbf{P}$	Set of all real positive numbers	1.2
P	Partition	6.1
$\overline{P}$	Tagged partition	6.1
P^*	Refinement of a partition P	6.4
$(P)\int_a^b f$	Principal value (Cauchy) of $\int_a^b f$	6.7
$\mathbf{Q}$	Set of all rational numbers	1.1
$\mathbf{Q}^+$	Set of all positive, rational numbers	1.7
R	Radius of convergence of a power series	5.8
$\mathbf{R}$	Set of all real numbers	1.1
$\mathbf{R}^1, \mathbf{R}^2, \ldots, \mathbf{R}^n$	Real Euclidean vector spaces	1.5
$R_n(x)$	Remainder at x when f is approximated by its Taylor polynomial out to a term that involves the nth derivative of f	5.6
$<\mathbf{R}^n, d_n>$	Real Euclidean metric space, $n \in \mathbf{N}$	1.5
$\mathbf{R}(f)$	Range of a function f	1.6
$\mathbf{Re}$	Extended real number system	1.9
$\int_a^b f$	Riemann integral over the interval $[a, b]$	6.1
$\iint\limits_{\mathfrak{D}} f$	Riemann double integral of f over $\mathfrak{D}$	6.8
$\sum\limits_{k=1}^{n} f(x_k)\Delta a_k$	Riemann sum	6.1
$S(\mathfrak{D}, \overline{P}, f)$	Riemann sum for f in $\mathbf{R}^2$, given the region $\mathfrak{D}$, the tagged partition $\overline{P}$, and f	6.8
$<\mathbf{S}, \oplus, \otimes>$	Ring	Exercise 5.7
$\{x_k\}_{k=1}^\infty$	Sequence	2.1
$\mathbf{S}\backslash\mathbf{T}$	Sets: complement; elements in set $\mathbf{S}$ but not in set $\mathbf{T}$	1.1
$\mathbf{S} \cap \mathbf{T}$	Sets: intersection	1.1
$\mathbf{S} \subset \mathbf{T}$, $\mathbf{S} \subseteq \mathbf{T}$	Sets: proper subset, subset	1.1
$\mathbf{S} \cup \mathbf{T}$	Sets: union	1.1
$\in$ ($\notin$)	Set membership (nonmembership)	1.1
$\text{sgn}(x)$	Signum function of x	Exercise 4.11
$\sigma(x)$, $\rho(x)$	Step-functions defined on $[a, b]$	6.3
$\sigma(x, y)$	Step-function defined on a closed, bounded rectangle $\mathbf{T}$	Exercise 6.65

Author Index

O

P

R

S

T

U

V

W

Y

Z

Subject Index

A

B

C

D

J

K

L

M

Printed and bound by CPI Group (UK) Ltd, Croydon, CR0 4YY
03/10/2024
01040313-0001